Biomass: Methods for Volume Calculation and Energy Valorization

Biomass: Methods for Volume Calculation and Energy Valorization

Editor: Nina Goldman

MURPHY & MOORE
www.murphy-moorepublishing.com

Murphy & Moore Publishing,
1 Rockefeller Plaza,
New York City, NY 10020, USA

Visit us on the World Wide Web at:
www.murphy-moorepublishing.com

ISBN: 978-1-63987-076-9 (Hardback)

Cataloging-in-Publication Data

Biomass : methods for volume calculation and energy valorization / edited by Nina Goldman.
p. cm.
Includes bibliographical references and index.
ISBN 978-1-63987-076-9
1. Biomass. 2. Biomass energy. 3. Renewable energy sources. I. Goldman, Nina.
TP339 .B56 2021
662.88--dc23

Table of Contents

Preface

Over the recent decade, advancements and applications have progressed exponentially. This has led to the increased interest in this field and projects are being conducted to enhance knowledge. The main objective of this book is to present some of the critical challenges and provide insights into possible solutions. This book will answer the varied questions that arise in the field and also provide an increased scope for furthering studies.

Any plant or animal material used by humans as fuel is referred to as biomass. Wood is an example of biomass which is used in the form of pellets and briquettes. Biomass feedstock are the substances used to generate energy or develop biofuels. Some of its examples are agricultural waste, municipal waste, sewer sludge and manufacturing waste. These are converted into suitable energy sources using various techniques such as thermal conversion and biochemical conversion. Biomass such as sugarcane and corn are used to manufacture bioethanol or first generation biofuel. Second generation biofuels use energy crops as their resource and third generation biofuels are produced using algae. The process of recycling, reusing, or composing waste materials and transferring them into useful products such as chemicals and fuels is known as valorization. This book strives to provide a fair idea about this discipline and to help develop a better understanding of the latest advances within this field. It consists of contributions made by international experts. This book is appropriate for students seeking detailed information in this area as well as for experts.

I hope that this book, with its visionary approach, will be a valuable addition and will promote interest among readers. Each of the authors has provided their extraordinary competence in their specific fields by providing different perspectives as they come from diverse nations and regions. I thank them for their contributions.

Editor

1

Modeling Biomass Substrates for Syngas Generation by using CFD Approaches

Nuno Couto and Valter Silva

Abstract

Recent reports from top universities state that in spite of having great national importance, there are dozens of fields of study that are suffering due to a lack of funding. Perhaps the greatest tool available to assist researchers with this regard is numerical simulation. This tool allows cutting costs, decreasing the necessary design cycle and allows an enormous amount of physical insight on the process itself). Numerical model's ability to correctly predict a complex system was tested in this chapter by drawing from a previously developed computational fluid dynamics model for biomass gasification. Numerical results were compared with both experimental results (pilot scale plant) and available literature. Results from common Portuguese biomass substrates were found to be within a satisfactory margin of error of 20%. Influence of all major operational conditions was then investigated and the model was once again able to predict all the expected trends. All the relevant process products were also analyzed. Finally, the numerical model was coupled with an optimization model. Maximum efficiency value was found at 900°C with a SBR of 1.5 for MSW and 1 for forest residues. Results showed that numerical models could have a preponderant impact on biomass gasification field.

Keywords: numerical simulation, biomass, municipal solid wastes, gasification, optimization

1. Introduction

In the last few years, biomass has become an important source of energy and it most often refers to any organic matter derived from plant-based materials. Biomass includes, among others, forestry and agricultural residues, organic waste, energy crops, sewage sludge, and woody plants [1, 2].

Among all the thermochemical conversion processes, gasification has been emerging as a very promising technology due to being environmental friendly, having high efficiency and flexible enough to use different substrates with a wide range of applications for heat, electricity, chemicals, and transportation purposes [3]. Gasification can be defined as the conversion of a solid waste to synthetic gas or fuel by the partial oxidation of the feedstock under stoichiometric combustion conditions. The synthetic gas is generally called "producer gas" or syngas and contains mainly carbon monoxide and hydrogen. However, some undesired products can be also found such as tar, alkali metals, chloride and sulfide, among others [3]. Gasification presents several advantages over waste combustion, namely and among others [3]: (a) effective response to increasingly environmental restrictive regulations; (b) syngas can be used not only in highly efficient internally fired cycles but also in producing valuable products as chemicals and fuels; and (c) flexible use under different operating conditions and reactors.

However, some drawbacks need further investigation. Besides high ranges of operating and capital costs than those of conventional combustion-based plants, the syngas generated from biomass and especially from municipal solid wastes (MSWs) gasification is unstable due to changes in the feedstock properties. Indeed, the high heterogeneous nature of MSW implies significant variations in syngas yield and quality [4]. The negative impact from the heterogeneous nature of MSW can be reduced by implementing strategies as MSW preprocessing, where undesired MSW components are removed before sending them to the gasifier. Further improvements can also be found by blending MSW with other feedstock with more favorable characteristics [5]. Portugal has a major potential considering biomass resources, only forestry and pruning residues have potential to produce 13,800 GWh, about 13% of the total primary energy demand in Portugal [6]. Also, operating the MSW gasifier under certain operating conditions will allow reliable operation with stable and improved syngas generation.

Because experimental runs conducted on industrial gasification plants or even on pilot scale gasification plants are very expensive, predictable models able to simulate the syngas composition and other responses of interest are required. Accurate predictions by gasification models require the simulation of different kinetic and hydrodynamic phenomena and we should not forget that gasification always involves complex chemistry. Concerning this degree of complexity, and also, the high investment costs involved to perform the experimental studies, the use of tools involving numerical simulation, such as computer fluid dynamics (CFD), exhibit utmost importance [7–9].

There are different designs to build a gasifier but one with special relevance is the fluidized bed configuration [10]. Basically, in this kind of reactor, both fuel and inert bed material act as a fluid. To obtain this behavior a gas is forced to pass through the solid particles of the reactor. The great advantages of operating such a system are its ability to handle with different feedstock due to the easy control of temperature, to deal with fine grained materials and the fact that generates an intense mixing enhancing the transference of mass and heat. Finally, the fluidized bed gasifiers are flexible enough to use additives with the goal to remove pollutants and increase tar conversion [11].

The fluidized bed gasification is essentially a multiphase problem. There are two main strategies that are commonly implemented to handle this kind of gasification: Euler-Lagrange and Euler-Euler approaches [12]. In the Euler-Lagrange strategy, the dispersed phase is characterized by following a wide number of particles through information given from the continuum phase. Both phases exchange mass, momentum, and energy. One major assumption relies on the fact that volume fraction of the dispersed phase is lower than about 10%, even the mass fraction could be higher.

Snider et al. [13] implemented a 3-D model based on the previous approach to simulate the behavior of a coal gasifier. The model also included energy transport equations and homogeneous and heterogeneous chemistry described by reduced chemistry. The model was able to identify zones with stagnant particles and then provide important data for reactor design purposes.

The Euler-Euler approach considers the different phases as interpenetrating continua, meaning that both volume phases depend on each other regarding time and space. Each phase obeys the conservation laws and constitutive relations from empirical data are required to close the equations system. Regarding the Euler-Euler approach, there are three multiphase models: (a) the volume of fluid; (b) the mixture model and; (c) the Eulerian model, which is indicated to model the discussed gasification system in this chapter.

Xue et al. [14] followed the Eulerian-Eulerian approach to predict the syngas generation from wood gasification using air as a fluidization agent. The developed model was able to provide detailed information on several processes, such as char elutriation, species profiles along the reactor and gas composition at reactor outlet.

This work aims at presenting an advanced modeling strategy within the framework of the CFD ANSYS Fluent program combined with suitable experimental data input and user defined the code to predict a gasification process under several operating conditions and robust enough to be applied both to biomass and MSW. Model development is detailed highlighting the set of complex choices that have to be taken and corresponding implications and consequences. Numerical results for both biomass and MSW were validated against experimental results, and then, the effect of using different operating conditions is deeply discussed as well as the main syngas quality indices. The combined use of the developed numerical model and the design of experiments (DoE) procedure is also discussed for optimization purposes. Finally, a reflection section is included to highlight the efficiency and effectiveness of using numerical methods to describe physical problems.

2. Experimental setup

The numerical model (presented in Section 3) was developed using experimental data collected in the gasification plant from the School of Technology and Management (ESTG) of the Polytechnic Institute of Portalegre (IPP). The plant is based on fluidized bed technology, with a 200 kW reactor that produces an average syngas flow rate close to 100 Nm^3/h. **Figure 1** depicts a diagram of the biomass gasification unit used in the experiments.

The system comprises two biomass silos connected to a worm screw, which forms the feeding system. The gasifier itself is just slightly above 4 m high and half a meter wide with a biomass intake capacity of 100 kg/h. The synthetic gas is cooled by two heat exchangers, with the first

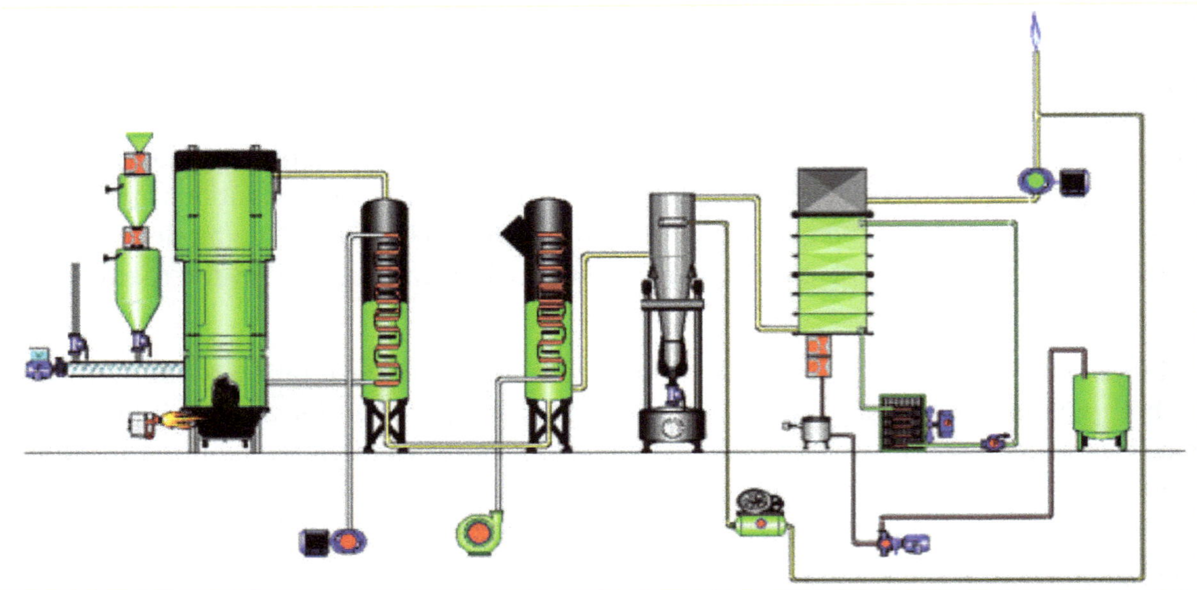

Figure 1. Schematics of the gasification plant.

one being also responsible for preheating the gas. The ash and char are removed to the bag filter, a system with seven filters cleaned with shots taken from synthetic gas through a compressor downstream of the vacuum pump. The purpose of the third heat exchanger is to remove tars from the system by condensation, which are then forwarded to a deposit. The last component of the system is a vacuum pump, which guarantees that the entire system control.

Once the predetermined temperature and biomass input flow are secured, the system stabilizes for about 2 h and then the test starts. The test lasts 2 h each, including syngas collections (usually taking two different samples) and recording temperatures, flows, ash, and tars. The schematics as well as an extensive description of the gasification plant can be found elsewhere [15, 16].

The described system was first used to analyze the gasification process of Portuguese biomass substrates and their potential for fossil fuel replacement [15, 17]. Recently, to counter the improper disposal of municipal waste in landfills, the focus shifted from Portuguese biomass gasification to Portuguese municipal solid waste (referred from now on as PMSW for simplicity) gasification.

The characterization and analysis of Portuguese MSW were carried out using data from the Oporto metropolitan area obtained from LIPOR, entity responsible for the management, treatment, and recovery of solid waste municipal produced in the city. From the pretreatment of MSW conducted by LIPOR, usually via shredding and dehydration, a refuse-derived fuel (RDF) containing only cellulosic and plastics is obtained [18] (chemical composition of the MSW is presented in **Table 1**).

However, since at that moment, the reactor could not handle LIPOR's wastes, the model had to be validated using data collected from the literature. To properly assess the potential of PMSW,

Category	% Weight	Chemical formula
Cellulosic material	85.42	*
Polyethylene	10.99	$(C_2H_4)_n$
Polyethylene terephthalate	2.02	$(C_{10}H_8O)_n$
Polypropylene	0.81	$(C_3H_6)_n$
Polystyrene	0.76	$(C_8H_8)_n$

Table 1. Chemical composition of the MSW in Oporto in 2014.

Substrate properties	Forest residues	PMSW
Elementary analysis (dry ash free)		
N (%)	2.4 ± 0.3	1.4 ± 0.2
C (%)	43 ± 4.1	48 ± 4.4
H (%)	5 ± 0.6	6.3 ± 1.4
O (%)	49.6 ± 5.2	43.6 ± 3.6
Humidity (%)	11.3 ± 1.7	17.6 ± 2.3
Density (Kg/m^3)	650 ± 70	247 ± 23
Lower heating value (MJ/Kg biomass)	21.2 ± 1.8	14.4 ± 1.1
Mean particle size (mm)	5 ± 2	20 ± 10
Proximal analysis (%)		
Ash	0.2 ± 0.1	14.9 ± 1.2
Volatile matter	79.8 ± 3.1	76.62 ± 2.9
Fixed carbon	20 ± 2.5	8.46 ± 1.5

Table 2. Ultimate and proximate analyses of forest residues and PMSW.

a previously studied Portuguese biomass substrate will be used as benchmark. Forest residues [17] (in pellet form) were selected for this purpose, since it revealed relevant energetic as well as economic benefits.

Prior to the actual gasification process, biomass analysis was carried out in the Laboratory of Chemistry of the High School of Technology and Management located in Portalegre, Portugal, since biomass characteristics can provide valuable information on how the gasification process will occur. This kind of analysis also provides crucial data to treat the implemented numerical model. The instruments used in the performed analysis are: thermal gravimetric analysis—data for proximal analysis; elemental analysis—determination of biomass composition with respect to the percentage of C, H, N, and O; humidity—sample moisture content assessment; calorific value—appraisal of energy contained in biomass. Data regarding proximate and ultimate analysis for the referred substrates are presented in **Table 2**.

It should be noted that regarding the experimental gasification runs, as well as the analyses for the studied substrates, every run was performed twice in order to avoid measurements. When deviation was higher than 5%, extra runs were performed to ensure reproducibility below 5%, which is typical for this kind of system.

3. Mathematical model

Experimental studies conducted in pilot scale or industrial reactors like the one presented in the previous section are fairly absent from the available literature. The reason is due to the difficulty in regulating operating parameters but primarily due to the high cost of a gasification plant, which can reach tens of millions of Euros depending on the generated power [19].

Mathematical models, with the ability to theoretically simulate any physical condition, allow studying the gasification process without resorting to major investments and/or the need for long-waiting periods (with all the bureaucratic and logistical problems associated). However, due to the extreme complexity of the gasification process, largely due to the chemical and physical interactions that occur throughout, the ability of numerical models to correctly predict experimental data collected from pilot scale or industrial reactors is usually very limited. In fact, the lack of reliable models to describe the gasification process in the open literature was the main motivation for our team to develop a new CFD model able to predict the syngas generation from biomass gasification in a pilot scale gasifier [20].

3.1. Gas-solid interaction

When modeling bubbling fluidized-bed reactors, like the one described, the two-phase flow theory of fluidization is usually applied for the description of the process hydrodynamics. Because of this, correctly modeling the interaction between gas and solid phases is crucial since they exchange heat by convection, mass over the heterogeneous chemical reactions, and momentum due to the drag between gas and solid phase. The main equations governing both phases are depicted in **Table 3**.

3.1.1. Granular Eulerian model

According to Goldschmidt et al. [21], two-phase flows can be modeled using two different approaches: the Lagrangian-Eulerian and the Eulerian-Eulerian models. With considerable similarities, the fundamental difference between them lies in the way the particles are treated. The former describes the solid phase at the particle level while the latter treats the particles as a continuum. In industrial applications, typically composed of millions of particles, following individual particles become excessively time consuming and for this reason the Lagrangian approach tends to be less used [22].

The Eulerian approach not only requires lower computational resources and calculation times, but also allows a detailed analysis of the disperse phase flow field, which is convenient for engineering design applications. For this reason, the granular Eulerian model was applied to our model.

Gas phase	Solid phase
Energy:	
$\frac{\partial(\alpha_g\rho_g h_g)}{\partial t}+\nabla\cdot(\alpha_g\rho_g\vec{u}_g h_g)=-\alpha_g\frac{\partial(p_g)}{\partial t}$ $+\overline{\overline{\tau}}_g:\nabla(\vec{u}_g)-\nabla\vec{q}_g+S_g+\sum_{g=1}^{n}(\vec{Q}_{sg}+\dot{m}_{sg}h_{sg})$	$\frac{\partial(\alpha_s\rho_s h_s)}{\partial t}+\nabla\cdot(\alpha_s\rho_s\vec{u}_s h_s)=-\alpha_s\frac{\partial(p_s)}{\partial t}$ $+\overline{\overline{\tau}}_s:\nabla(\vec{u}_s)-\nabla\vec{q}_s+S_{ps}+\sum_{s=1}^{n}(\vec{Q}_{sg}+\dot{m}_{sg}h_{sg})$
Mass:	
$\frac{\partial(\alpha_g\rho_g)}{\partial t}+\nabla\cdot(\alpha_g\rho_g\vec{u}_g)=-M_C\sum\gamma_C R_C$	$\frac{\partial(\alpha_s\rho_s)}{\partial t}+\nabla\cdot(\alpha_s\rho_s\vec{u}_s)=M_C\sum\gamma_C R_C$
Momentum:	
$\frac{\partial(\alpha_g\rho_g\vec{u}_g)}{\partial t}+\nabla\cdot(\alpha_g\rho_g\vec{u}_g\vec{u}_g)=-\alpha_g\nabla p_g+\alpha\rho_g g$ $+\beta(u_g-u_g)+\nabla\cdot\alpha_g\overline{\overline{\tau}}_g+S_{sg}U_S$	$\frac{\partial(\alpha_s\rho_s\vec{u}_s)}{\partial t}+\nabla\cdot(\alpha_s\rho_s\vec{u}_s\vec{u}_s)=-\alpha_s\nabla p_s+\alpha\rho_s g$ $+\beta(u_s-u_s)+\nabla\cdot\alpha_s\overline{\overline{\tau}}_s+S_{sg}U_S$

Table 3. Governing equations for gas and solid phases.

The granular Eulerian model is described by the following conservation equation for granular temperature:

$$\frac{3}{2}\left[\left(\frac{\partial(\rho_s\alpha_s\Theta_s)}{\partial t}\right)+\nabla\cdot(\rho_s\alpha_s\vec{v}_s\Theta_s)\right]=(-p_s\bar{I}+\overline{\tau}_s):\nabla(\vec{v}_s)+\nabla\cdot\left(k_{\Theta_a}\nabla(\Theta_s)\right)-\gamma_{\Theta a}+\phi_{ls} \quad (1)$$

This expression is obtained from the kinetic theory of gases. The term $(-p_s\bar{I}+\overline{\tau}_s):\nabla(\vec{v}_s)$ describes the generation of energy by the solid stress tensor, ϕ_{ls} stands for the energy exchange between the fluid and solid phase, γ_{Θ_a} stands for the collisional dissipation of energy and $k_{\Theta_a}\nabla(\Theta_s)$ stands for the diffusion energy, in which k_{Θ_a} is the diffusion coefficient.

3.2. Turbulent model

In turbulent flows, like the one being studied, transported quantities like momentum, energy, and species concentration tend to fluctuate. Modeling fluctuations can be to computationally expensive, which is why instantaneous governing equations are usually replaced with their time-averaged, ensemble-averaged, or otherwise manipulated to remove the small time scales.

The standard *k*-ε model was used to simulate the turbulent flow due to its suitability for a wide range of wall-bounded and free-shear flows. The model is the simplest turbulence two-equation model in which the solution of two separate transport equation allows the turbulent velocity and length scales, which are to be independently determined. Turbulence kinetic energy (k) and dissipation rate (ε) are, respectively, given by:

$$\frac{\partial}{\partial t}(\rho k) + \frac{\partial}{\partial x_i}(\rho k u_i) = \frac{\partial}{\partial x_j}\left[\left(\mu + \frac{\mu_t}{\sigma_k}\right)\right] G_k + G_b - \rho\varepsilon - Y_M + S_k \tag{2}$$

$$\frac{\partial}{\partial t}(\rho \varepsilon) + \frac{\partial}{\partial x_i}(\rho \varepsilon u_i) = \frac{\partial}{\partial x_j}\left[\left(\mu + \frac{\mu_t}{\sigma_\varepsilon}\right)\right] + C_{1\varepsilon}\frac{\varepsilon}{k}(G_\varepsilon + C_{3\varepsilon}G_b) - C_{2\varepsilon}\rho\frac{\varepsilon^2}{k} + S_\varepsilon \tag{3}$$

To determine the turbulence kinetic energy as well as the dissipation rate, the following constants were assumed: $G_k = 1.0$ and $G_\varepsilon = 1.3$ stand for the turbulent Prandtl numbers for k and ε, respectively, $C_{1\varepsilon} = 1.44$, $C_{2\varepsilon} = 1.92$, and $C_{3\varepsilon} = 0$ are default constants commonly used in fluent and S_k and S_ε are user-defined source terms.

3.3. Chemical reaction model

In this study, two different chemical reaction models were used in the simulation: the finite-rate/Eddy-dissipation model was used to describe homogeneous reactions while the kinetic/diffusion surface reaction model was employed for heterogeneous ones. The main distinction between these two models is associated with how the carbon species are treated. The homogeneous gas reaction assumes the carbon species gasified straightaway, and that the carbon is treated as a gas, while heterogeneous particle-gas reaction treats carbon as solid particles and they go through finite-rate reaction via a typical reaction at particle surface. **Table 4** presents the main reactions as well as the corresponding reaction rates for these two models.

Above reactions can slightly change if different gasifying agents other than air are considered. Previously published models using steam [23] and carbon dioxide [24] address those exact changes.

Reactions	Reaction rate
Homogeneous reactions:	
$CO + 0.5O_2 \rightarrow CO_2$	$r_1 = 1.0\times10^{15}\exp\left(\frac{-16,000}{T}\right)C_{CO}C_{O_2}^{0.5}$
$CO + H_2O \rightarrow CO_2 + H_2$	$r_2 = 5.159\times10^{15}\exp\left(\frac{-3430}{T}\right)T^{-1.5}C_{O_2}C_{H_2}^{1.5}$
$CO + 3H_2 \leftrightarrow CH_4 + H_2O$	$r_3 = 3.552\times10^{14}\exp\left(\frac{-15,700}{T}\right)T^{-1}C_{O_2}C_{CH_4}$
$H_2 + 0.5O_2 \rightarrow H_2O$	$r_4 = 2780\exp\left(\frac{-1510}{T}\right)\left[C_{CO}C_{H_2O} - \frac{C_{CO_2}C_{H_2}}{0.0265\exp\left(\frac{3968}{T}\right)}\right]$
$CH_4 + 2O_2 \rightarrow CO_2 + 2H_2O$	$r_5 = 3.0\times10^{5}\exp\left(\frac{-15,042}{T}\right)C_{H_2O}C_{CH_4}$
Heterogeneous reactions:	
$C + 0.5O_2 \rightarrow CO$	$r_6 = 596T_P\exp\left(\frac{-1800}{T}\right)$
$C + CO_2 \rightarrow 2CO$	$r_7 = 2082.7\exp\left(\frac{-18,036}{T}\right)$
$C + H_2O \rightarrow CO + H_2$	$r_8 = 63.3\exp\left(\frac{-14,051}{T}\right)$

Table 4. Chemical reaction model.

3.4. Model expansion for MSW

In the above subsection, we purposely did not address the mathematical treatment for the devolatilization phenomenon. The reason why being that the devolatilization section was only properly included when the model was expanded to handle the heterogeneity of MSW. Until that point, a single rate model, developed by Badzioch and Hawksley [25], was assumed which computed reliable devolatilization rates in a simple way.

To cope with the heterogeneity of said substrate a pyrolysis model with secondary tar generation was implemented. MSW is mainly composed of cellulosic and plastic components, and while cellulosic material can be divided in cellulose, hemicellulose, and lignin [26], plastics comprise polyethylene, polystyrene, and polypropylene, among others. To distinguish the several components that comprise the MSW, the pyrolysis reactions of cellulosic and plastic groups are considered individually and following an Arrhenius kinetic expression, as shown in **Table 5**.

3.5. Numerical procedure

Fluent, a finite volume method-based CFD solver, was employed in this work to solve the stated problem. Regarding the geometry modeling there are some simplifications that one can make in order to make the computation less expensive. Since the described reactor type is cylindrical, one can use a 2D axisymmetric problem setup.

Mesh was built using GAMBIT software and a total number of 83,000 quadrilateral cells of uniform grid spacing were generated. The mesh density in a finite element model is an important topic because of its relationship to accuracy and cost. In this particular case, the chosen cell size was about 12 times larger than the average particle size which was shown to be able to effectively capture the hydrodynamics in fluidized bed gasifier [27].

In such a complex model, it is sometimes difficult to define a good initial condition. For this reason, the process was first simulated considering only flow and nonreacting heat transfer (also known as "cold flow") and after reaching conversion reactive multiphase flow was added.

Substrate and air inlet were defined as "velocity inlets". Each velocity-inlet surface was identified by mass fractions, temperature, and a velocity magnitude. The flow direction was kept normal to

Reactions	**Reaction rate**
Cellulose $\rightarrow \alpha_1$volatiles $+ \alpha_2 TAR + \alpha_3$char	$r_9 = A_i \exp\left(\frac{-E_i}{T_s}\right)(1-a_i)^n$
Hemicellulose $\rightarrow \alpha_4$volatiles $+ \alpha_5 TAR + \alpha_6$char	$r_{10} = A_i \exp\left(\frac{-E_i}{T_s}\right)(1-a_i)^n$
Lignin $\rightarrow \alpha_7$volatiles $+ \alpha_8 TAR + \alpha_9$char	$r_{11} = A_i \exp\left(\frac{-E_i}{T_s}\right)(1-a_i)^n$
Plastics $\rightarrow \alpha_{10}$volatiles $+ \alpha_{11} TAR + \alpha_{12}$char	$r_{12} = \left[\sum_{i=1}^{n} A_i \exp\left(\frac{-E_i}{RT}\right)\right]\rho_v$
Primary$TAR \rightarrow$ volatiles + SecondaryTAR	$r_{13} = 9.55\times10^4 \exp\left(\frac{-1.12\times10^4}{T_g}\right)\rho_{TAR1}$

Table 5. Devolatilization model.

the surface. Turbulence of the inlet surfaces was detailed by turbulence intensity and hydraulic diameter. Outlet was set as the pressure outlet, specified by a gauge pressure value of zero.

For the pressure-velocity coupling, the widely used SIMPLE algorithm was enabled. For all simulations presented in this paper, a first-order upwind scheme was used for all equations. The standard scheme was used for interpolation methods of pressure. This means that the solution approximation in each finite volume was presumed to be linear, leading to less computational expenses. In order to properly justify using a first-order scheme, it was necessary to make sure that the grid used in this work had adequate resolution to accurately capture the physics occurring within the domain. In other words, the results needed to be independent of the grid resolution. Nonetheless, for better accuracy, this was later changed to the second-order upwind scheme.

Similarly, the underrelaxation factors were initially set at 0.5 (except for turbulent viscosity, pressure, and body forces, which were kept the same as default), and were then gradually conveyed to their default values with convergence. Convergence criteria were set that normalized residuals for all equations must fall under 10^{-6}.

4. Results and discussion

4.1. Model validation for forest residues

The main goal of researchers when it comes to numerical simulation is to increase the system complexity and accuracy while minimizing, if not complete eliminating, physical testing. There are two key systematic processes for confirming numerical results: the verification and the validation processes. The former tries to answer "did I solve the model right?" while the latter asks "did I solve the right model?" and this is where one checks against experimental data.

In the gasification process, this step becomes crucial since one is dealing with an extremely complex multiphase model where gas and solid phases exchange heat, momentum, and mass. To make matters worse, the hydrodynamic phenomena on a laboratory scale fluidized bed are not the same as on large scales [28]. To overcome this problem, the numerical results were compared with gasification tests performed in a pilot scale gasification plant, installed in the Industrial Park of Portalegre, Portugal. A wide range of operational conditions were tested for several Portuguese biomass substrates. **Table 6** shows three of these tested operating conditions for one particular substrate used to validate the numerical model and respective results.

The presented results show that the developed numerical model has the ability to predict the obtained synthetic gas composition within a satisfactory margin of error of 20%, commonly found in similar studies [29]. The highest deviation was observed for CH_4, which was expected since smaller fractions tend to produce higher relative errors. Furthermore, all light hydrocarbons and tar can lump into CH_4, which can explain the disagreement sometimes found in Ref. [15]. Very similar margins of error were found for different biomass substrates analyzed in the same pilot scale.

Nevertheless, some differences can be observed due to some simplifying assumptions followed by our model:

Experimental conditions		Forest residues		
Run		1	2	3
Temperature (°C)		815	815	790
Admission biomass (Kg/h)		63	74	63
Air flow rate (Nm^3/h)		94	98	98
Syngas fraction (dry and inert basis)				
H_2	Experimental	8.2	8.4	7.6
	Numerical	7.5	7.7	6.8
CO	Experimental	18.6	18	17.9
	Numerical	20.9	20.6	20.1
CH_4	Experimental	4.6	4.4	4.4
	Numerical	3.9	3.7	3.7
CO_2	Experimental	16.7	17.1	17.1
	Numerical	15.9	16.5	16.2

Table 6. Operating conditions for validation proposes and respective results.

- A more detailed devolatilization approach should be attempted.
- In addition, tar decomposition and chemical reactions for ash and some light hydrocarbons were not included.
- Ideal gas principles apply for the gases.
- Syngas is only formed by H_2, CO, CO_2, and C_nH_m and it is at chemical equilibrium.
- Heat losses from the components are neglected.

Some of these simplifications were corrected when the model was expended to deal with the heterogeneity of MSW as will be shown in the next subsection.

4.2. Model validation for PMSW

As stated in the mathematical model, to cope with the heterogeneity of MSW, the devolatilization model had to be restructured. Ideally, one would like to validate the new upgraded model with the experimental setup used earlier. However, due to unfortunate logistical and bureaucratic setbacks this was not possible.

To work around the problem, it was decided to validate the model using data collected from the literature [30]. **Table 7** shows the operating conditions used to validate the numerical model.

Figure 2 shows the comparison of the composition of obtained gas, estimated by the model, with that measured in the experiments.

Comparison between **Figure 2** and **Table 6** shows that the advancement in the model allowed a more complex system to perform in a similar manner and in some cases to predict syngas

Run	1	2	3	4	5	6	7	8	9
Temperature (°C)	493	705	602	507	687	593	691	593	507
MSW admission (Kg/h)	2.3	3	3	3	4	4	6	6	6
ER	0.5	0.4	0.4	0.4	0.3	0.3	0.2	0.2	0.2
Preheated air (°C)	290	352	296	281	352	307	352	308	279

Table 7. Operating conditions for the experimental gasification runs [30].

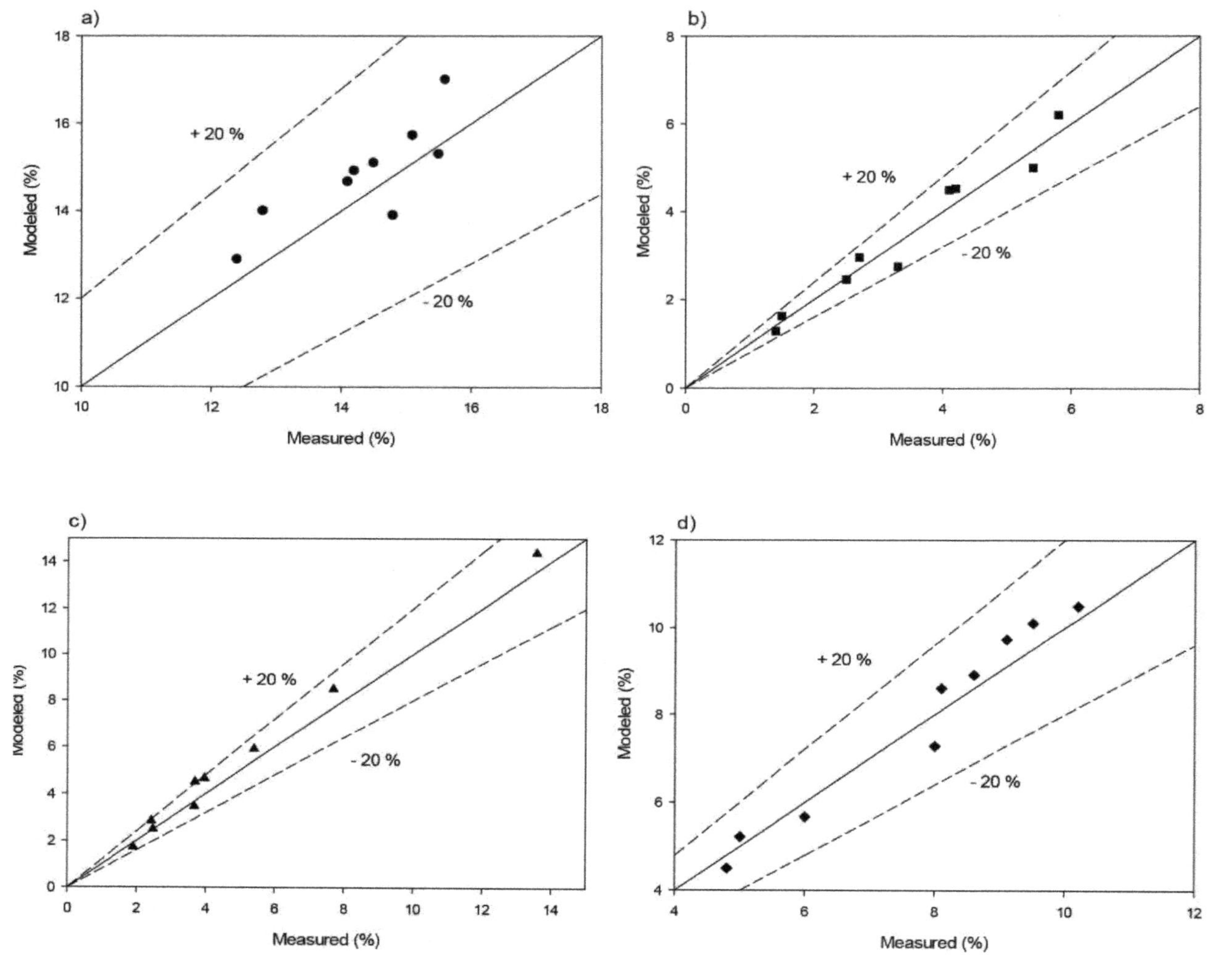

Figure 2. Comparison of the modeled and measured composition for (a) CO_2, (b) H_2, (c) C_nH_m, and (d) CO.

composition slightly better. This was due to a more realistic devolatilization model and the inclusion of light hydrocarbons.

4.3. Assessment of operational conditions

When scrutinizing a system (whatever it may be) it is imperative to devote a significant effort determining the influence of the main parameters on the system's output. In fact, the aim of the present subsection is to analyze the influence of several gasifier operating conditions including equivalence ratio (ER), steam-to-biomass ratio (SBR), carbon dioxide-to-biomass (CO_2/B) ratio, and reactor temperature on the obtained gas composition.

4.3.1. Equivalent ratio

Perhaps the most common, and the most analyzed, parameter regarding the gasification process is the air flow rate. Air is the cheapest gasifying agent but it also produces the poorest gas because it is highly diluted in N_2 [31]. To solve this problem, pure oxygen can be used to obtain a much higher quality gas (since no nitrogen will be present or at least a much lower fraction). However, implementation is only possible through high investment causing uncertainty when it comes to large facilities [32].

Furthermore, the O_2content in an air mixture can sometimes be misleading due to nitrogen dilution and the effects of small variations on substrate admission in the syngas composition going unnoticed, hence our decision to study ER. ER is one of the most significant parameters, which have effect on the gasification process including syngas composition. In accordance with previous gasification studies [33], the equivalence ratio can be defined as:

$$\mathrm{ER} = \frac{\text{oxygen mass/dry MSW mass}}{\text{stoichiometric oxygen/MSW ratio}} \tag{4}$$

The ratio was maintained between 0.15 and 0.35 since all of the experiments conducted to validate the model must be in this range and also because the ER values most suitable for gasification found in the current literature range between 0.2 and 0.4 [34].

The model predictions for the described reactor about the influence of ER on syngas molar fraction are shown in **Figure 3**.

A quick analysis shows that increasing ER suppresses the formation of combustible gases (H_2, CO, and C_nH_m) while promoting the formation of CO_2 contents. Since increasing the ER leads to more inert gas to enter the reactor, the obtained gas will be more diluted in nitrogen resulting in a poorer gas. A decrease in H_2 and CO can also be explained by a decrease in the residence time, considering that as air flow rate increases, it is no longer sufficient for CO and H_2 formation reactions to occur. Although to a smaller degree, ER negatively affects the C_nH_m content by enhancing steam reactions at higher temperatures leading to methane decomposition [35]. Finally, carbon dioxide fraction is expected to increase since combustion reactions (that consume CO and H_2 to produce CO_2) will be promoted. Results are consistent with the current literature [36, 37].

Despite presenting similar trends, studied substrates exhibit significant differences in the relative syngas molar fraction. One may explain this difference by recurring to the fuel's chemical composition. As demonstrated by Silva et al. [15], higher biomass calorific values result in higher calorific syngas production. This relationship between the biomass calorific content and the syngas lower heating values (LHV) can be explained considering, first, that the biomass calorific value is related to the amount of carbon (C) and hydrogen (H) present in the biomass, and second, a larger amount of these two elements allows production of larger quantities of hydrogen and carbon monoxide, the major contributors for the calorific value of

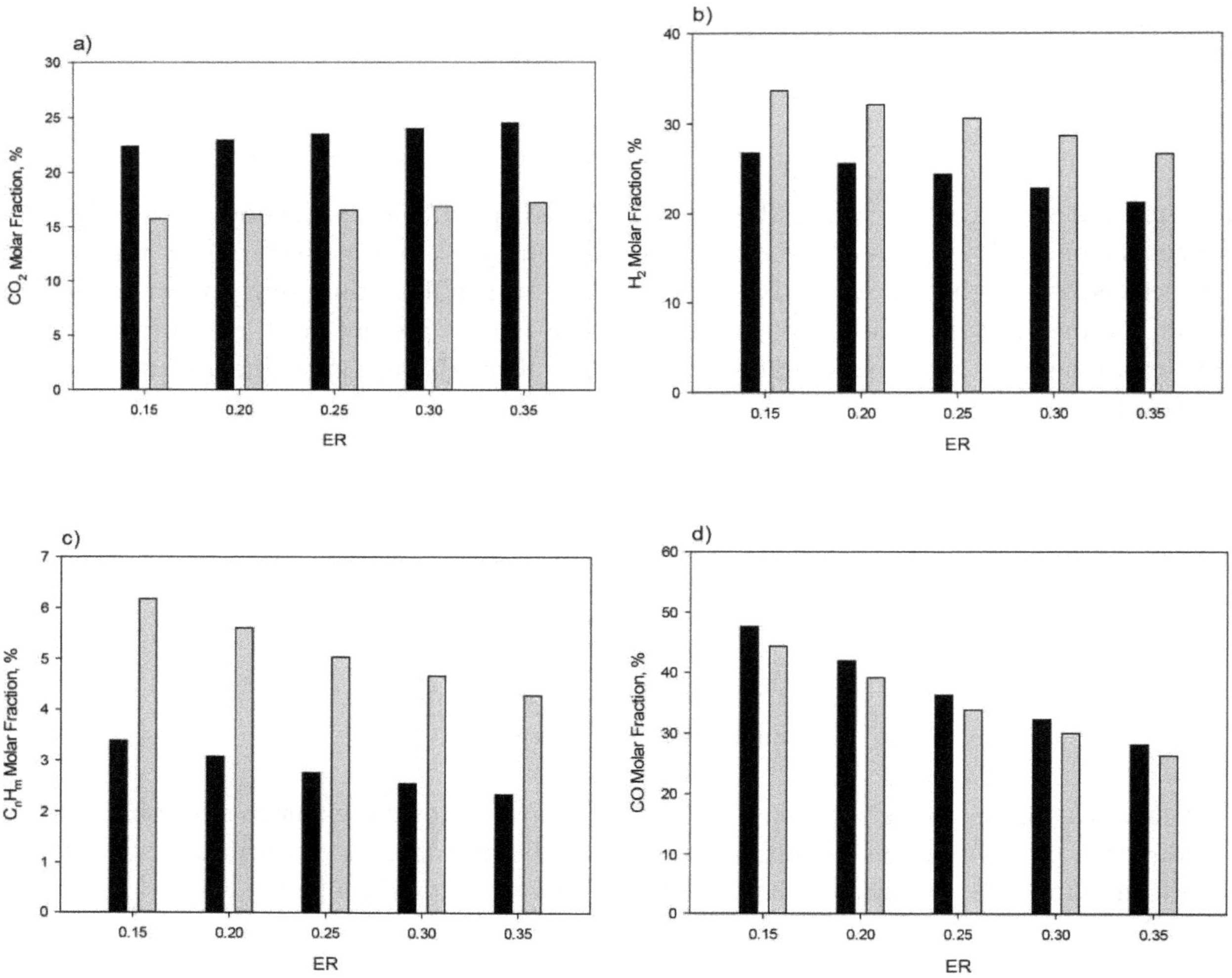

Figure 3. Comparison between PMSW (black columns) and forest residues (gray columns) as a function of ER for (a) CO_2, (b) H_2, (c) C_nH_m, and (d) CO (operating conditions: fuel feed rate = 25 kg/h; gasification temperature = 750°C).

the syngas. Nevertheless, since there are other biomass properties that can greatly influence the gasification process one must not take these conclusions as absolutes.

4.3.2. Steam-to-biomass ratio

Among oxidizing agents, steam gasification has received special attention since it produces a fuel gas with medium lower heating values of 12–18 MJ/Nm^3 [38], which is considerably higher than those from air gasification, while being less costly than oxygen gasification.

Steam-to-biomass ratio (SBR) is used in order to emphasize the effects of small variations on biomass admission, which often go unnoticed [39]. The SBR can be defined as the steam mass flow rate divided by the fuel mass flow rate (dry basis):

$$\mathrm{SBR} = \frac{\text{Steam mass flow rate}}{\text{Biomass substrate mass flow rate}} \tag{5}$$

The SBR was varied over a range of values from 0 to 2 by holding the other variables constant. The range was selected based on previous findings from our research team using the same facilities (**Figure 4**).

An increase in SBR leads to an increase in H_2 and CO_2 and a decrease in CO and C_nH_m. SBR will mostly favor char and tar steam reforming as well as the water-gas shift reaction, which in turn will lead to an increase in CO_2 and H_2 content at the expense of CO and C_nH_m. In fact, according to Hernández et al. [40], for steam gasification, the water-gas shift reaction will dominate over the Boudouard one and CO will be consumed to produce CO_2 and H_2. These results are consistent with the current literature [41].

An increase in the CH_4 content relates to the decrease in oxidation of volatile matter, which is not balanced out by the consumption of CH_4 in the reforming reactions. These reactions have lower rates than oxidation ones but are most favored by low temperatures. However, at higher steam levels, the steam reforming can in fact shift CH_4 consumption will also be affected.

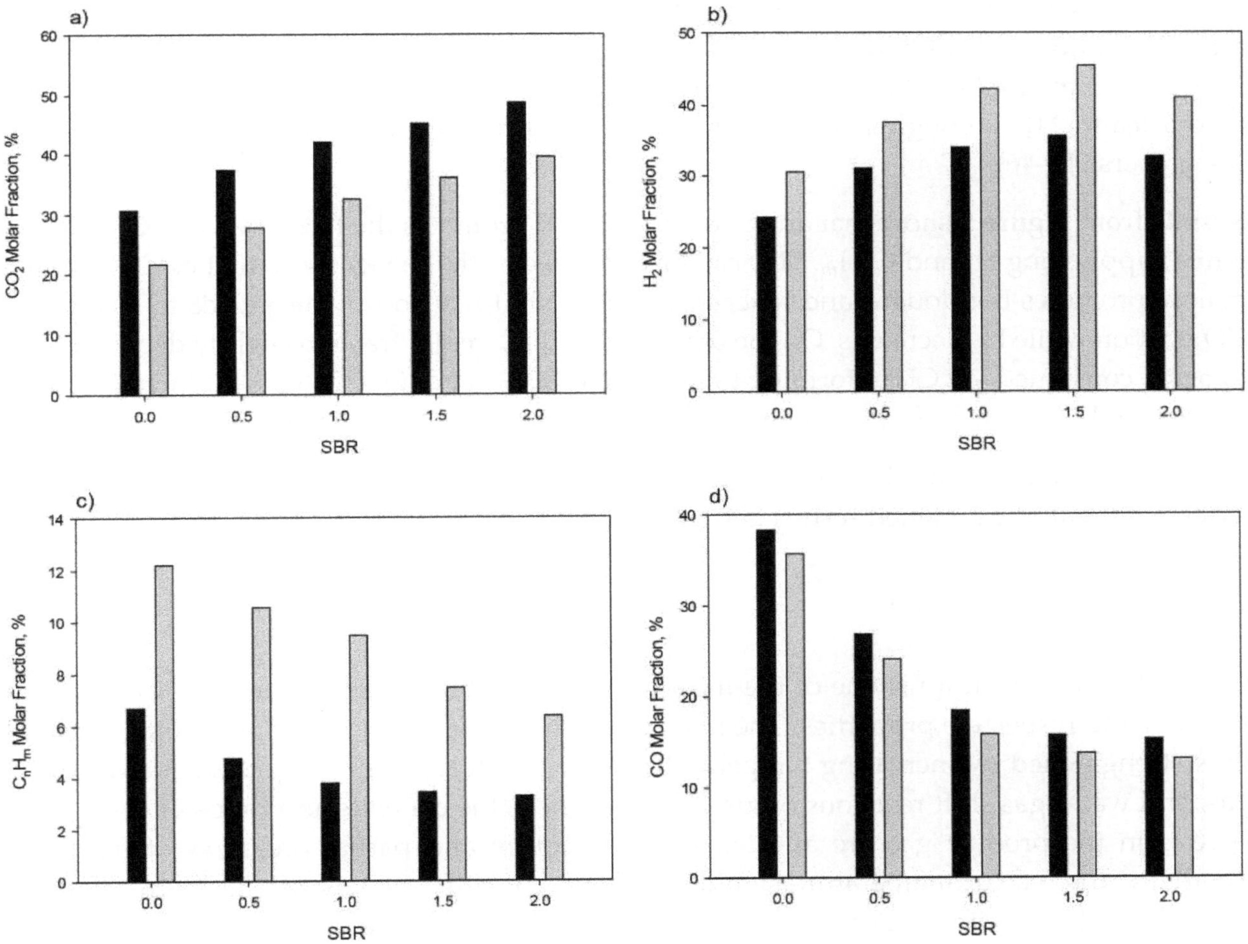

Figure 4. Comparison between PMSW (black columns) and forest residues (gray columns) as a function of SBR for (a) CO_2, (b) H_2, (c) C_nH_m, and (d) CO (operating conditions: fuel feed rate = 25 kg/h; gasification temperature = 750°C).

Excessive steam intake will lead to a significant decrease in gasification temperature, which in turn will have a negative effect on endothermic reactions, impairing product generation, which explains the decrease in H_2 after SBR=1.5, and producing insufficient heat to promote steam reforming and primary water-gas reactions. Furthermore, excessive steam could shift the steam reforming and water-gas reactions backward, consuming CO and H_2 to produce CO_2 and H_2O [42].

4.3.3. Carbon dioxide-to-biomass ratio

Even though steam gasification presents several advantages over air gasification, there are still some associated setbacks, such as the consumption of H_2O, which is an increasingly scarce resource.

Although poorly studied, carbon dioxide as a gasifier agent has been showing promising results, for starters, it consumes an unwanted end product (CO_2) of various industrial processes [43]. Furthermore, it also enhances both char gasification and pyrolysis and has the ability to act as a catalyst to enhance the thermal cracking of volatiles leading to tar mitigation.

The effect of carbon dioxide as a gasifying agent was studied using carbon dioxide-to-biomass ratio. The ratio was varied over a range of values from 0 to 1 by maintaining the other variables constant. To the best of our knowledge, regarding MSW, this ratio has only been investigated by our team [24], although some work has been published using biomass substrates by other researchers [44–46].

Results from **Figure 5** show that an increase in CO_2/B promotes the formation of CO and CO_2 while suppressing H_2 and C_nH_m. This can be explained with the fact that a higher CO_2 content mainly promotes Boudouard and reverse water-gas shift reactions, which leads to increase in CO fraction while H_2 decreases. On the other hand, C_nH_m molar fraction slightly decreases due to being consumed via CH_4reforming to produce CO and H_2 [47]. CO_2 content increases since a considerable fraction of the gasifying agent leaves the reactor unreacted.

Despite no results in the literature were found for MSW, those available for biomass are in agreement with the obtained results [44, 45, 48].

4.3.4. Gasification temperature

Gasification temperature is one of the most influential factors affecting the product gas composition and respective properties. The main reactions of the gasification are endothermic and thus strengthened by increasing temperature. Since the steam reforming, Boudouard, water-gas, and water-gas shift reactions occur simultaneously, the contents and ratios of considered species in the product gas are affected by temperature and partial pressures of reactants. Therefore, the reactor temperature significantly influences the syngas compositions. **Figure 6** show the influence of reactor temperature on final syngas composition for both substrates.

According to the Le Chatelier's principle, an increase in gasification temperature will favor products in endothermic reactions. Furthermore, promotion of endothermic reactions will lead to an increase in CO and H_2 content formation while decreasing CO_2 and C_nH_m [49].

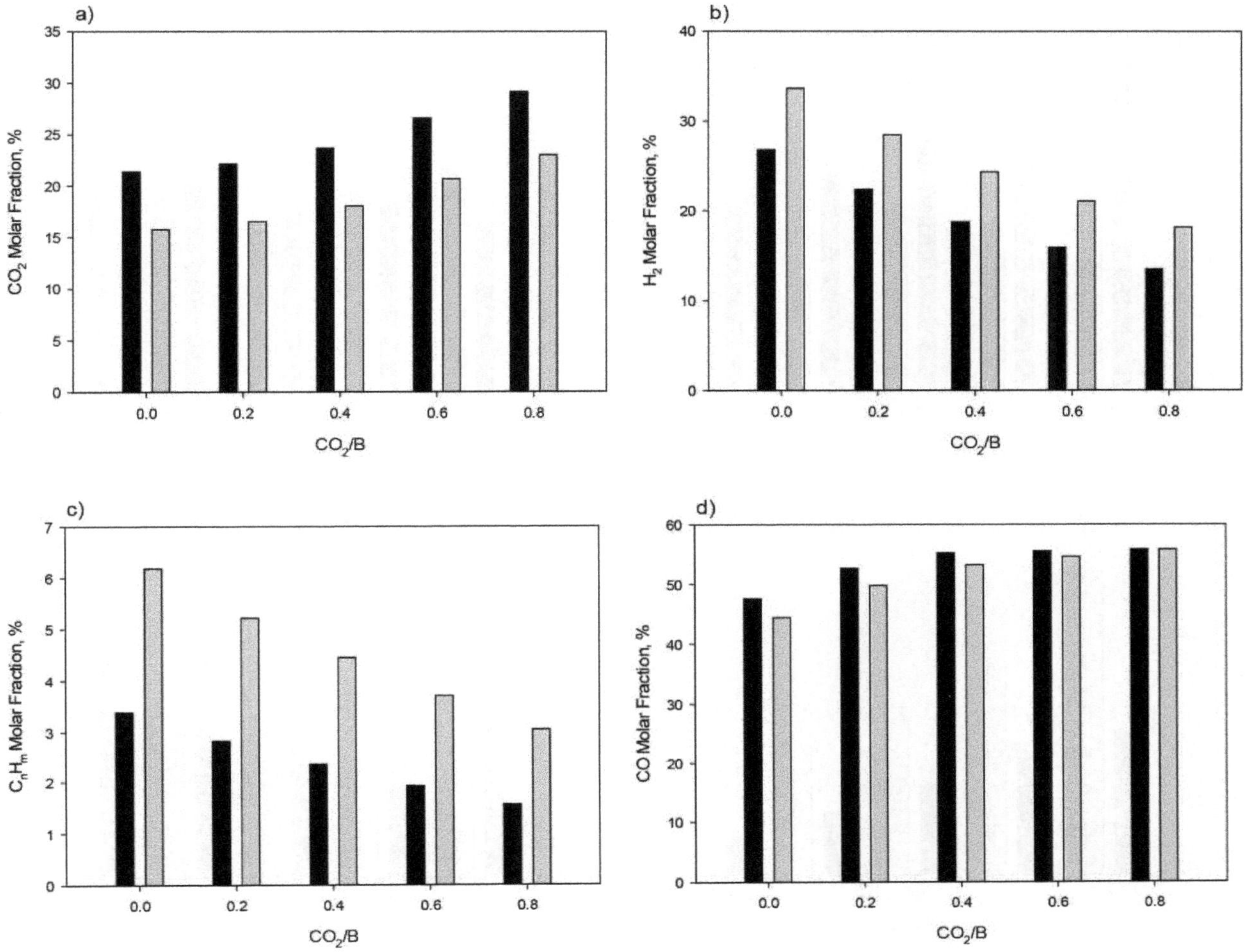

Figure 5. Comparison between PMSW (black columns) and forest residues (gray columns) as a function of CO_2/B for (a) CO_2, (b) H_2, (c) C_nH_m, and (d) CO (operating conditions: fuel feed rate = 25 kg/h; gasification temperature = 750°C).

Careful analysis shows that above 750°C hydrogen production becomes less pronounced while carbon monoxide further increases. In fact, in this range, the standard Gibbs free energy of the Boudouard reaction (responsible for CO production) becomes less than that of the water-gas reaction (main responsible for hydrogen production at lower temperatures), meaning that the former dominates over the latter as temperature increases. This is in agreement with the current literature [50].

Comparison between substrates was intentionally not discussed in the two previous subsections. The reason behind it is simply that all operating conditions share the same trends when it comes to different substrates, the underlining key is the biomass properties themselves.

4.4. Syngas quality indices

Syngas quality indices such as CH_4/H_2 and H_2/CO not only give a good indication of process efficiency, but also give its most suitable application. For instances, syngas with high CH_4/H_2 ratios tends to be used in domestic purposes while high H_2/CO ratios tend to be preferred in the chemical industry [51].

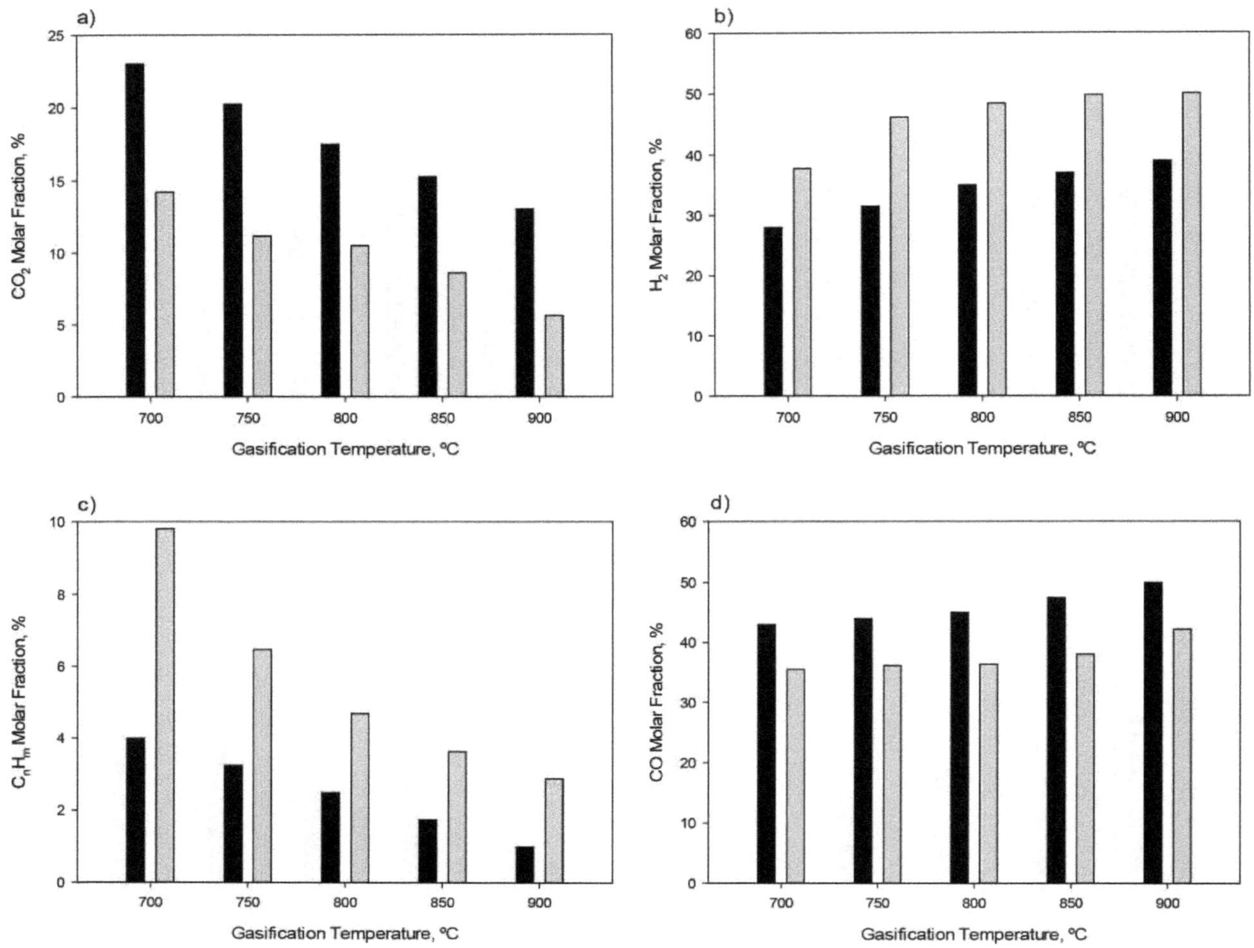

Figure 6. Comparison between PMSW (black columns) and forest residues (gray columns) as a function of gasification temperature for (a) CO_2, (b) H_2, (c) C_nH_m, and (d) CO. Dry and N_2-free basis (operating conditions: ER – 0.25; SBR – 1; CDMR – 0.4; MSW admission – 25 kg/h).

Besides syngas quality indices, there are other gasification products that can help determine process quality, namely carbon conversion (CC), cold gas efficiency (CGE), and tar content.

4.4.1. Methane-to-hydrogen ratio

Figure 7 displays the effect of gasification temperature on the syngas CH_4/H_2 ratio using the operating conditions from **Figure 6**.

Figure shows that the CH_4/H_2 ratio suffers a steep decrease with an increase in temperature. As can be seen from **Figure 6** (and corresponding explanation), an increase in gasification temperature will promote the formation of H_2. Meanwhile, due to the strengthening of the endothermic steam-methane reactions a decrease in CH_4 fraction is also expected.

At higher temperatures, usually above 800°C, one may notice that ratio decrease becomes less pronounced. This can be explained with the fact that, simply because within the studied

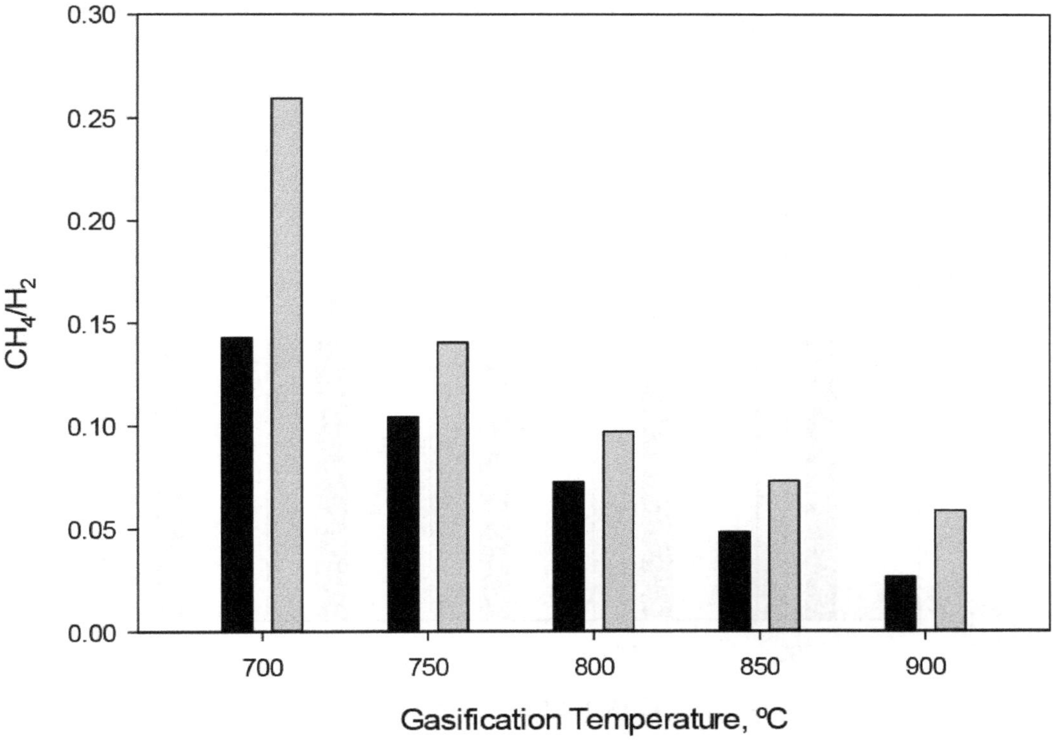

Figure 7. Syngas CH_4/H_2 ratio as a function of the gasification temperature for PMSW (black columns) and forest residues (gray columns) (dry and N_2-free basis).

temperature range, a decrease in CH_4 is somewhat constant while at higher temperature hydrogen production will also be less pronounced [33].

Forest residues presented a higher CH_4/H_2 ratio, especially for lower temperatures. Pinpointing exactly why this is can be extremely challenging due to a number of biomass properties that influence the gasification process. One can argue that since forest residues have a higher carbon and hydrogen content (according to our data, if we consider the proximate analysis as received from PMSW, C content is just 33.66 and H 4.42%) which will lead to more combustible gases. Furthermore, an increase in the CH_4 content is related to higher levels of volatile matter, which also might explain why forest residues present the higher ratio.

4.4.2. Hydrogen-to-carbon monoxide ratio

As estimated by Butterman and Castaldi [43], the H_2/CO ratio has an obvious impact on the ideal application for a given substrate. Higher H_2/CO ratios permit for the operation of solid oxide fuel cells (SOFC) [52] while medium H_2/CO ratios are suitable for FT synthesis of liquid fuels. Mid-to-lower ratios are mainly appropriate for catalyst-based FT synthesis while very low ratios are particularly appropriate for the production of a specific biomass-derived liquid chemical [53].

Figure 8 displays the effect of gasification temperature on the syngas H_2/CO ratio using the operating conditions from **Figure 6**.

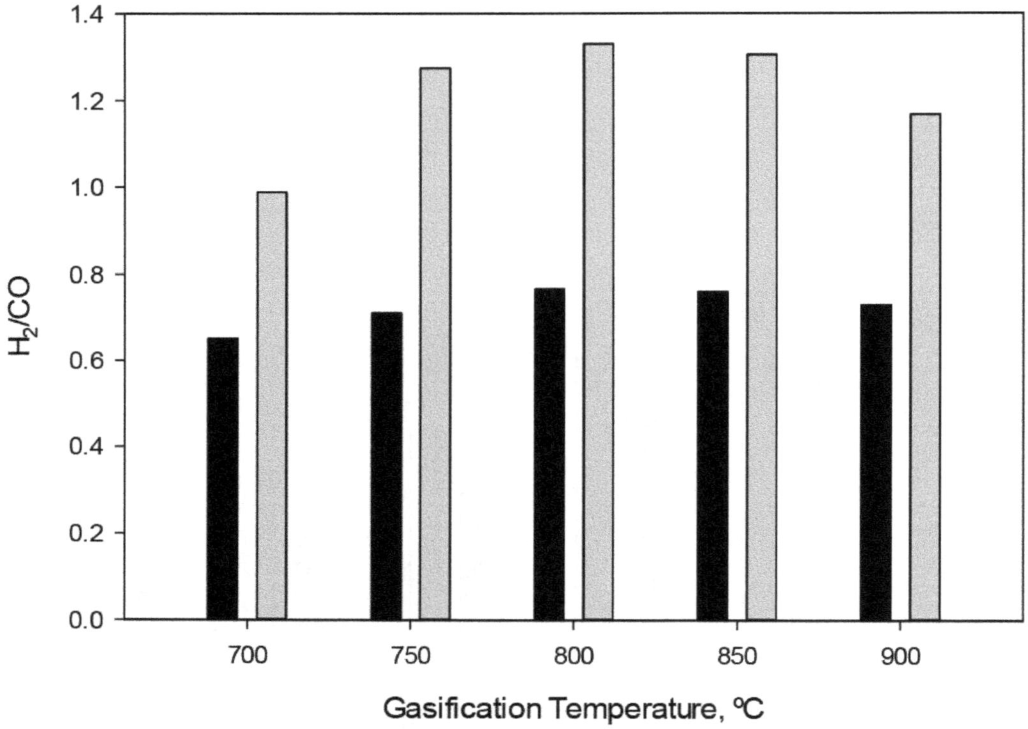

Figure 8. Syngas H_2/CO ratio as a function of the gasification temperature for PMSW (black columns) and forest residues (gray columns) (dry and N_2-free basis).

Conversely to CH_4/H_2, the H_2/CO ratio presents two distinct trends: from 700 to 800°C, gasification temperature has a positive influence on the H_2/CO ratio whereas an increase beyond 800°C shows the opposite trend. This can be explained with the fact that at low temperatures H_2 production is enhanced by primary water-gas reaction as well as steam-methane reforming reactions while at higher temperatures primary water-gas as well as Boudouard reaction will favor CO production.

Perhaps most importantly, in the range between 750 and 800°C, the standard Gibbs free energy of the Boudouard reaction (responsible for CO production) becomes less than that of the water-gas reaction (main responsible for hydrogen production at lower temperatures), indicating that the former dominates over the latter as temperature increases, leading to a decrease in the H_2/CO ratio, which is in agreement with the literature [54].

To make it useful for the chemical industry to synthesize products such as methanol and virgin naphtha, H_2/CO ratios higher than 1.70 should be presented, which neither of these two substrates were able to achieve, although forest residues were able to reach close to 1.4. The H_2/CO ratio would be increased by an injection of water [51].

4.4.3. Carbon conversion

Carbon conversion can be defined by the fraction of carbon from the substrate converted into carbon in syngas composition, and can give an good indication of the amount of unconverted materials, thus providing a measure of chemical efficiency of the process. It can be expressed as:

$$\text{Carbon Conversion} = \frac{12 \times M}{X_c \times m} \tag{6}$$

where M stands for the total molar flow rate of carbon in syngas composition, X_c stands for the carbon fraction in the MSW, and m stands for the MSW flow rate into the gasifier.

Figure 9 displays the effect of gasification temperature on carbon conversion using the operating conditions from **Figure 6**.

Figure shows that gasification temperature has a positive effect on carbon conversion. Higher temperatures will favor tar reforming leading to an increase in gas yield and carbon conversion [55]. Furthermore, an increase in temperature enhances steam reforming reactions, which in turn promote carbon conversion [41].

Although operational conditions were kept constant to ensure uniform residence time, substrates with different size particles lead to different residence times [47, 52]. In addition, the increasing residence time promotes gasification and carbon conversion reactions, leading to a higher gas yield [53]. Since forest residues have a smaller particle it leads to higher carbon conversion values.

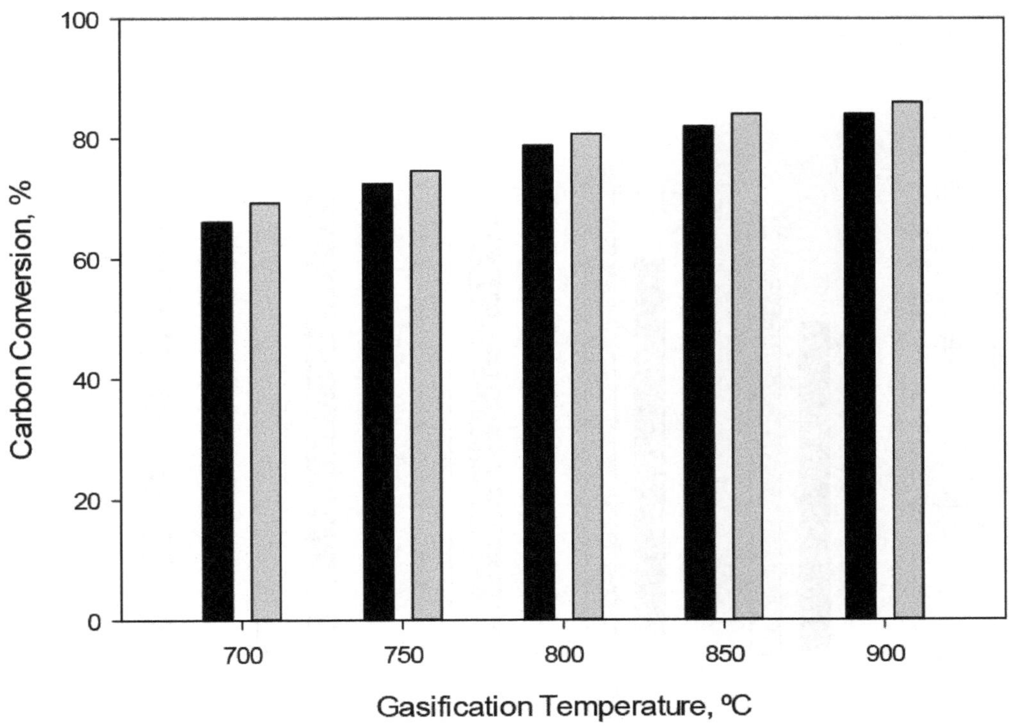

Figure 9. Carbon conversion as a function of the gasification temperature for PMSW (black columns) and forest residues (gray columns) (dry and N_2-free basis).

4.4.4. Cold gas efficiency

CGE can be defined as the percentage of the heating value of MSW converted into the heating value of the product gas. It can be computed as follows:

$$CGE = \frac{\text{Gas yield} \times \text{HHV of product gas}}{\text{HHV of fuel} + \text{Heat addition}} \tag{7}$$

Figure 10 displays the effect of gasification temperature on CGE using the operating conditions from **Figure 6**.

Similarly to CC, gasification temperature also has a positive influence on CGE. This is expected since the main gasification reactions are endothermic and thus strengthened by increasing temperature as well as gas yield, the two main factors responsible for CGE increase. Results are in agreement with the current literature [56, 57].

As expected, the calculated CGE for MSW is significantly lower due to a combination of low gas yield and poor syngas LHV. Efficiency values for both forest residues and MSW are within range from what is commonly found in the current literature [58].

4.4.5. Tar content

One of the major problems to deal with during biomass gasification is tar formation. At reduced temperature, tar condenses, blocking and fouling process equipment (such as engines

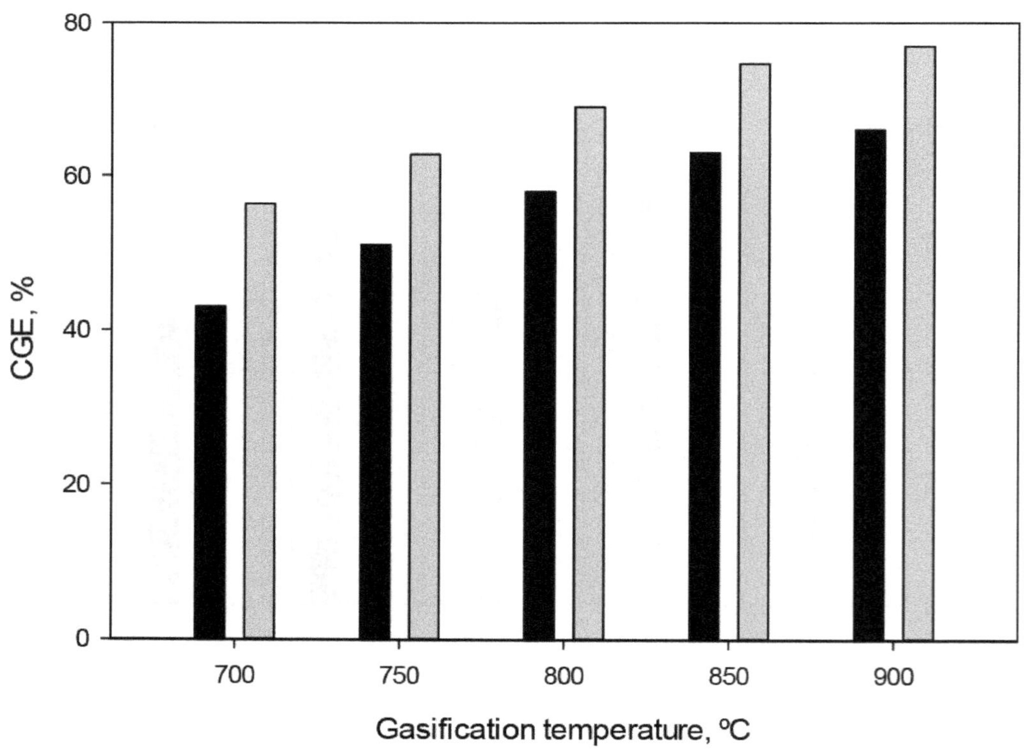

Figure 10. CGE as a function of the gasification temperature for PMSW (black columns) and forest residues (gray columns) (dry and N_2-free basis).

and turbines). According to Devi et al. [59], tar should be prevented or eliminated in the gasifier by manipulating factors such as operating conditions, addition of active bed materials, and possible reactor design modifications.

Figure 11 displays the effect of gasification temperature on the tar content using the operating conditions from **Figure 6**.

Figure shows that gasification temperature strongly influences the tar content. This can be explained with the fact that a high temperature favors destruction and reforming of tar, leading to a decrease in tar content [60].

As previously mentioned, a higher volatile content leads to an increase in residence time that, in turn, favors gasification reactions [61]. According to Aljbour and Kawamoto [62], an increase in residence time can lead to a reduction in the tar content. Since forest residues have a slightly higher volatile content, it comes with no surprise that it also consist a slightly lower tar content.

4.5. Process optimization

As shown all throughout the paper, gasification parameters have a strong influence on the overall process quality. Trying to have a better understanding of the underpinning mechanisms requires performing several gasification runs studying different operating conditions and parameters. Without a systematic approach such as the design of experiments this can become extremely time-consuming as well as expensive. Combining optimization methodologies such as the DoE with numerical models avoids expensive and time-consuming

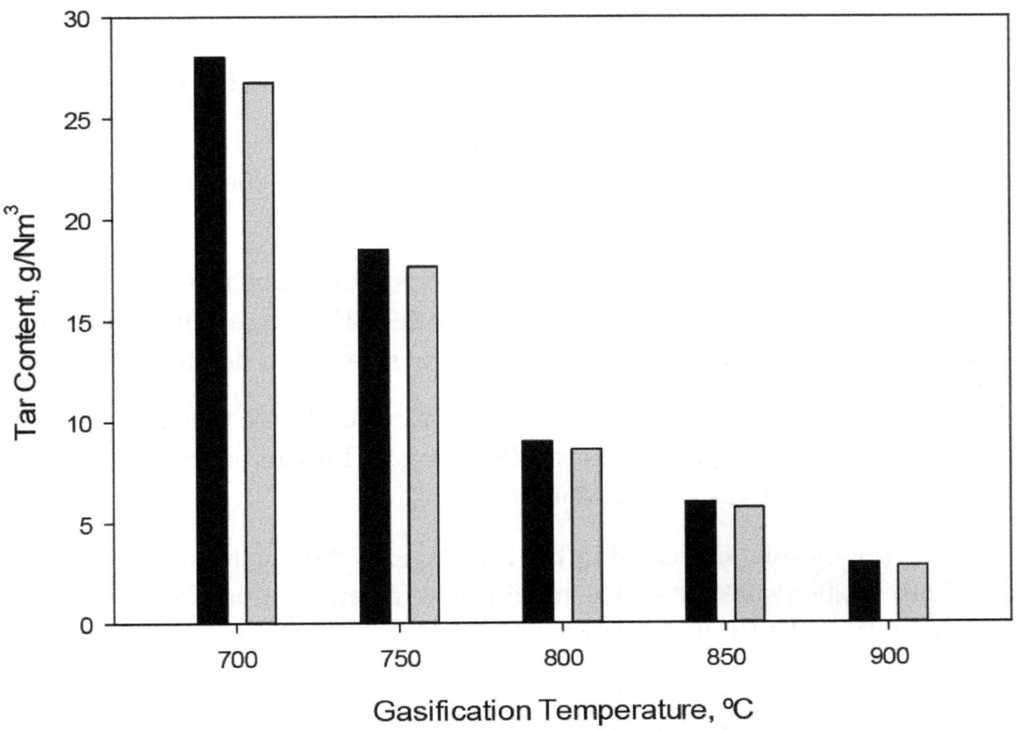

Figure 11. Tar content as a function of the gasification temperature for PMSW (black columns) and forest residues (gray columns) (dry and N_2-free basis).

experiments while obtaining the optimal operational condition set according to the desired response [63].

To test this hypothesis, an exergy efficiency optimization model was built using data obtained from the numerical model. The optimization model aimed to establish where the maximum exergy efficiency is according to the substrate and operating conditions chosen.

Particularly, in order to promote a more hydrogen-rich gas, the optimization model focused on an exergy flow rate of the produced hydrogen instead of syngas.

The empirical model was built minimizing the sum of the residues square to give the parameters of a second-order response model.

$$Y = B_0 + \sum_{i=1}^{3} B_i \times X_i + \sum_{i=1}^{3} B_{i,i} \times X_i^2 + \sum_{j=2}^{3} \sum_{i<j}^{3} B_{i,j} \times X_i \times X_j \quad (8)$$

where Y is the response, the X_i terms are the main factors ($-1 \leq X_i \leq 1$), temperature (1), steam-to-biomass ratio (2), and biomass type (3) and the B_i terms are the equation coefficients related to the main factors. The B_0 term is the interception coefficient, the $B_{i,i}$ terms are the quadratic effects (give the curvature to the response surface), and the $B_{i,j}$ terms symbolize the cross-interactions between factors. The present design does not consider the use of replicates because the results are obtained by computer simulations. In this case, a test on lack of fit and analysis concerning pure error are not provided [64]. Despite these circumstances much of the standard statistical analyses remain relevant, including measurements of model-fit such as PRESS (predicted residual sum of squares).

ANOVA analysis with high values for PRESS and "R-squared predicted" (not shown) also reinforced that the exergy efficiency response is well described by the empirical model in the design space.

Figures 12 and **13** show the hydrogen exergy efficiency as a function of the temperature and SBR for MSW and forest residues, respectively.

Similarly to what was shown in previous subsections, both substrates present very similar trends. In general, hydrogen efficiency increases with SBR since adding steam increases its chemical energy and exergy content. However, adding steam also demands additional exergy.

Regarding influence of gasification temperature one can clearly see that it has a positive effect on hydrogen efficiency all throughout the studied range. This can be explained by an increase in gas yield and enthalpy of gas component [56, 65].

The interpolating polynomial indicated in Eq. (8) provides the maximum values for hydrogen efficiency for both studied substrates. The maximum efficiency value was found at 900°C with a SBR of 1.5 for MSW and 1 for forest residues, with respective values of 50.6 and 50.2%.

Surprisingly, forest residues and MSW presented virtually the same maximum hydrogen efficiency. This can be explained with forest residues having the higher hydrogen molar composition and the higher gas yield, while MSW presenting the lower substrate exergy value,

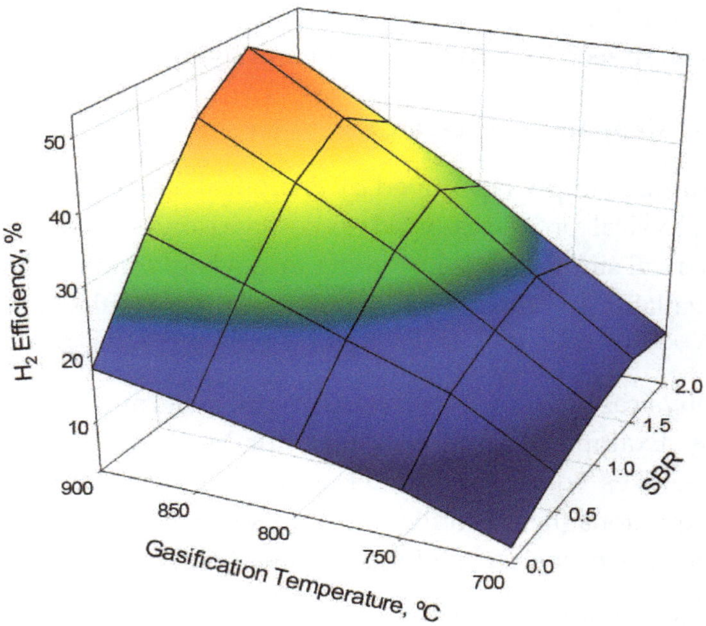

Figure 12. Hydrogen exergy efficiency as a function of the temperature and SBR for MSW.

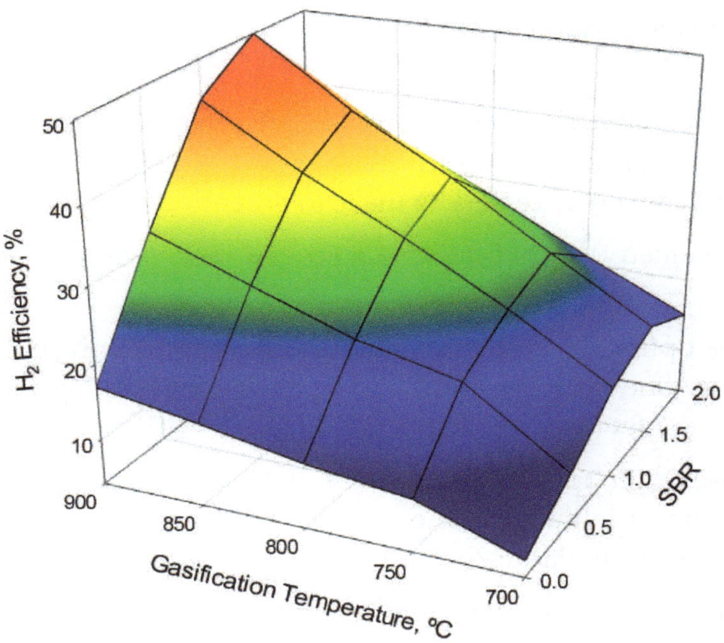

Figure 13. Hydrogen exergy efficiency as a function of temperature and SBR for forest residues.

leading to the latter having a slightly higher maximum hydrogen exergy value. Perhaps more important is the fact that forest residues have a much higher C_nH_m fraction (which is much more calorific than H_2 or CO), leading to a substantially higher CGE. When considering hydrogen efficiency, we are only considering hydrogen fraction and gas yield, so higher hydrocarbons do not influence the results. Available literature supports the conclusions made in this study [56, 65].

Even though results from MSW were not on par with those from forest residues in regard to CGE, exciting results were seen in both tar content and hydrogen efficiencies.

4.6. Role of numerical simulation in research assistance

Most real-life problems are simply not solved by analytical methods, especially if they involve interaction between physical processes or when comparison is made with real-experimental data. The use of numerical simulation is becoming increasingly imperative with the increase in computing power available by allowing researchers to get results not achievable by other means.

Most simulators of physical phenomena nowadays are using numerical methods rather than closed-form solutions. Examples would be solvers for Maxwell's equations (electromagnetic phenomena), Navier-Stokes equations (fluid mechanics), Schrödinger's equation (quantum mechanics), Fourier equations (heat conduction), and so on. Even "simple" simulators like the SPICE circuit simulator use numerical methods for nearly all but the simplest resistor circuits.

Numerical solutions open new possibilities for analysis but solutions can sometimes be difficult to interpret. Furthermore, simulators such as fluent are not 100% accurate or certain to give the user a physically real answer, one still has to know the basic physics to even setup most such simulators and one needs to know something about numerical analysis to know how and why such methods typically fail to simulate if you want identify or debug simulation failures.

The key determining factor is the level of complexity the researcher is willing to take (in other words what computational cost is the researcher willing to take). In the presented work, there were several choices that were necessary to make in order to ensure work feasibility:

- What kind of geometry simplifications could one make without jeopardizing physical model?
- What would the optimal grid density be? A very courser mesh could lead to very poor results, but a very fine mesh could mean dozens or even hundreds of hours extra just to improve results by a few percentage points.
- Is it better to use faster approaches like PDF flamelet models even though their fundamental assumption of infinitely fast chemistry renders them inapplicable for modeling low-Btu, low-exit velocity flaring?
- Using a radiation model can definitely improve the simulation results but requires a considerable amount of additional computational time, is it worth?
- When using a chemistry set, some authors opt to use detailed mechanisms (sometimes containing hundreds of reactions and dozens of species), while others prefer to use reduced or even simplified versions.

These are just a small fraction of all choices necessary to make when dealing with such a complex system as the one described in Section 3. Depending on system complexity, even with all the necessary adjustments to ensure faster convergence, some simulations will require optimal planning and patience. Some known examples by ANSYS:

- NETL fluidization challenge problem, able to simulate 16 s of flow time per day on 12 processors with a complete particle size distribution (PSD);
- Flow of solids in a riser, approximately 1 s of flow time simulated in a day in a single processor.

However, all this effort is rewarded when researchers are able to apply their results in creating credible model descriptions of new systems, which can be used in new designs or to improve the parameters on working systems and decision making for its modernization.

Nevertheless, perhaps the greatest benefit of using numerical models is the impact that these can bring not only for industrial purposes but also in important areas such as financial, social, and environmental areas.

5. Conclusions

Ideally, researchers would always want to test their hypothesis using the latest equipment in the most advanced laboratories with a large team of experts available to assist them. However, only an insignificant fraction has access to such conditions, and even those who do still have to go through tedious bureaucratic and logistical setbacks. Numerical simulation methods are used to study the behavior of systems whose mathematical models are too complex to provide analytical solutions in a time and cost-efficient way.

In this chapter, one of the most challenging energetic systems to predict was analyzed, the gasification process. In fact, gasification is considered to be one of the most difficult processes to model due to the chemical and physical interactions that occur throughout. A previously developed numerical model was used and its results validated using data collected from the literature and from a semiindustrial gasifier. Forest residues (one of the most abundant substrates in Portugal) and municipal solid wastes (one of the greatest challenges facing modern society) were analyzed. The model able not only to satisfactory the experimental results but also to correctly predict the trends of all studied parameters.

Finally, the numerical model was coupled with an optimization model designed to predict the optimal operation conditions for obtaining a more hydrogen-rich gas. This further states the ability of numerical simulations not only to assist analyzing the key trends of a given process but also how they can be used to predict the optimal operating point for a given optimal response.

Nomenclature

A, B	calibration constants
A_i	preexponential factor
$C_{1\varepsilon}, C_{2\varepsilon}, C_{3\varepsilon}$	constants
C_p	specific heat capacity
D_0	diffusion rate coefficient

E_i	activation energy
G_k	generation of turbulence kinetic energy due to the mean velocity gradients
G_b	generation of turbulence kinetic energy due to buoyancy
h_q	specific enthalpy of phase
h_{pq}	heat transfer coefficient between the fluid phase and the solid phase
k	thermal conductivity
Nu	Nusselt number
$\dot{m}$	biomass flow entering into the gasifier
M	total mole flow of carbon in the syngas components
M_i	molecular weight of each the species
M_c	molecular weight
$M_{w,i}$	molecular weight of i component
p	gas pressure
Pr	Prandtl number
p_s	particle phase pressure due to particle collisions
$\vec{Q}_{pq}$	heat transfer between p^{th} and q^{th} phases
$\vec{q}_q$	heat flux
q^{th}	specific enthalpy
R	universal gas constant
R_i	net generation rate of specie i due to homogeneous reaction
Re	Reynolds number
R_c	reaction rate
S_i	source term of the species i production from the solid heterogeneous reaction
S_k	user-defined source terms
S_q	source term due to chemical reactions
$S\varepsilon$	user-defined source terms
T	temperature
t_s	particle phase stress tensor
U	mean velocity
v	instantaneous velocity
X_C	carbon fraction in the biomass (obtained from the ultimate analysis)
Y	mass fraction
Y_M	contribution of the fluctuating dilatation in compressible turbulence to the overall dissipation rate

Other symbols

α	volume fraction
β	gas-solid interphase drag coefficient
γ_c	stoichiometric coefficient
γ_{Θ_a}	collisional dissipation of energy
ε	dissipation rate
ρ	density

φ_{ls}	energy exchange between gas and solid phases
k_{Θ_a}	diffusion coefficient
$k_{\Theta_a}\nabla(\Theta_s)$	diffusion energy
$(-p_s\bar{I}+\bar{\tau}_s):\nabla(\vec{v}_s)$	generation of energy by the solid stress tensor.
τ	tensor stress
μ	viscosity

Subscripts

g gas phase
s solid phase
i component

Acknowledgements

We would like to express our gratitude to the Portuguese Foundation for Science and Technology (FCT) for the support to the grant SFRH/BD/86068/2012 and the project PTDC/EMS-ENE/6553/2014 as well as IF/01772/2014. We also would like to thank Dr. Paulo Brito for providing experimental data needed to complete this paper.

Author details

Nuno Couto and Valter Silva*

*Address all correspondence to: vsilva@inegi.up.pt

INEGI-FEUP, Faculty of Engineering, University of Porto, Porto, Portugal

References

[1] V. Silva, E. Monteiro, A. Rouboa, An analysis on the opportunities, technology and potential of biomass residues for energy production in Portugal, in: A. Méndez-Vilas (Ed.) Materials and Processes for Energy, Formatex Research Center, Spain 2013, pp. 190–201.

[2] W.C. Tuckenger, Renewable energy technologies, in: U.W. UNDP (Ed.) World Energy Assessment of the United Nations, USA 2000, pp. 219–272.

[3] U. Arena, Process and technological aspects of municipal solid waste gasification. A review, Waste Management, 32 (2012) 625–639.

[4] G. del Alamo, A. Hart, A. Grimshaw, P. Lundstrøm, Characterization of syngas produced from MSW gasification at commercial-scale ENERGOS Plants, Waste Management, 32 (2012) 1835–1842.

[5] M. Lapuerta, J.J. Hernández, A. Pazo, J. López, Gasification and co-gasification of biomass wastes: Effect of the biomass origin and the gasifier operating conditions, Fuel Processing Technology, 89 (2008) 828–837.

[6] S. Ferreira, N.A. Moreira, E. Monteiro, Bioenergy overview for Portugal, Biomass and Bioenergy, 33 (2009) 1567–1576.

[7] D. Baruah, D.C. Baruah, Modeling of biomass gasification: A review, Renewable and Sustainable Energy Reviews, 39 (2014) 806–815.

[8] F. Tabet, I. Gökalp, Review on CFD based models for co-firing coal and biomass, Renewable and Sustainable Energy Reviews, 51 (2015) 1101–1114.

[9] S. Kraft, M. Kuba, F. Kirnbauer, K. Bosch, H. Hofbauer, Optimization of a 50 MW bubbling fluidized bed biomass combustion chamber by means of computational particle fluid dynamics, Biomass and Bioenergy, 89 (2016) 31–39.

[10] M. Siedlecki, W. De Jong, A.H.M. Verkooijen, Fluidized bed gasification as a mature and reliable technology for the production of bio-syngas and applied in the production of liquid transportation fuels: A review, Energies, 4 (2011) 389.

[11] G. Mirmoshtaghi, H. Li, E. Thorin, E. Dahlquist, Evaluation of different biomass gasification modeling approaches for fluidized bed gasifiers, Biomass and Bioenergy, 91 (2016) 69–82.

[12] R.I. Singh, A. Brink, M. Hupa, CFD modeling to study fluidized bed combustion and gasification, Applied Thermal Engineering, 52 (2013) 585–614.

[13] D.M. Snider, S.M. Clark, P.J. O'Rourke, Eulerian–Lagrangian method for three-dimensional thermal reacting flow with application to coal gasifiers, Chemical Engineering Science, 66 (2011) 1285–1295.

[14] Q. Xue, R.O. Fox, Multi-fluid CFD modeling of biomass gasification in polydisperse fluidized-bed gasifiers, Powder Technology, 254 (2014) 187–198.

[15] V. Silva, E. Monteiro, N. Couto, P. Brito, A. Rouboa, Analysis of syngas quality from Portuguese biomasses: An experimental and numerical study, Energy & Fuels, 28 (2014) 5766–5777.

[16] N. Couto, V. Silva, E. Monteiro, S. Teixeira, R. Chacartegui, K. Bouziane, P.S.D. Brito, A. Rouboa, Numerical and experimental analysis of municipal solid wastes gasification process, Applied Thermal Engineering, 78 (2015) 185–195.

[17] N. Couto, V. Silva, E. Monteiro, P. Brito, A. Rouboa, Using an Eulerian-granular 2-D multiphase CFD model to simulate oxygen air enriched gasification of agroindustrial residues, Renewable Energy, 77 (2015) 174–181.

[18] S. Teixeira, E. Monteiro, V. Silva, A. Rouboa, Prospective application of municipal solid wastes for energy production in Portugal, Energy Policy, 71 (2014) 159–168.

[19] Z.A.B.Z. Alauddin, P. Lahijani, M. Mohammadi, A.R. Mohamed, Gasification of lignocellulosic biomass in fluidized beds for renewable energy development: A review, Renewable and Sustainable Energy Reviews, 14 (2010) 2852–2862.

[20] N. Couto, V. Silva, E. Monteiro, P.S.D. Brito, A. Rouboa, Experimental and numerical analysis of coffee husks biomass gasification in a fluidized bed reactor, Energy Procedia, 36 (2013) 591–595.

[21] M.J.V. Goldschmidt, J.A.M. Kuipers, W.P.M. van Swaaij, Hydrodynamic modelling of dense gas-fluidised beds using the kinetic theory of granular flow: Effect of coefficient of restitution on bed dynamics, Chemical Engineering Science, 56 (2001) 571–578.

[22] P. Cornejo, O. Farías, Mathematical modeling of coal gasification in a fluidized bed reactor using a Eulerian granular description. International Journal of Chemical Reactor Engineering 9 (1) 2011.

[23] N. Couto, V. Silva, A. Rouboa, Assessment on steam gasification of municipal solid waste against biomass substrates, Energy Conversion Management 124 (15) 2016.

[24] N. Couto, V. Silva, A. Rouboa, Municipal solid waste gasification in semi-industrial conditions using air-CO2 mixtures, Energy, 104 (2016) 42–52.

[25] S. Badzioch, P.G.W. Hawksley, Kinetics of thermal decomposition of pulverized coal particles, Industrial & Engineering Chemistry Process Design and Development, 9 (1970) 521–530.

[26] O. Onel, A.M. Niziolek, M.M.F. Hasan, C.A. Floudas, Municipal solid waste to liquid transportation fuels – Part I: Mathematical modeling of a municipal solid waste gasifier, Computers & Chemical Engineering, 71 (2014) 636–647.

[27] S.J. Gelderbloom, D. Gidaspow, R.W. Lyczkowski, CFD simulations of bubbling/collapsing fluidized beds for three Geldart Groups, AIChE Journal, 49 (2003) 844–858.

[28] N. Couto, V.B. Silva, C. Bispo, A. Rouboa, From laboratorial to pilot fluidized bed reactors: Analysis of the scale-up phenomenon, Energy Conversion and Management, 119 (2016) 177–186.

[29] V. Silva, A. Rouboa, Combining a 2-D multiphase CFD model with a response surface methodology to optimize the gasification of Portuguese biomasses, Energy Conversion and Management, 99 (2015) 28–40.

[30] G. Xiao, B.-S. Jin, Z.-P. Zhong, Y. Chi, M.-j. Ni, K.-F. Cen, R. Xiao, Y.-J. Huang, H. Huang, Experimental study on MSW gasification and melting technology, Journal of Environmental Sciences, 19 (2007) 1398–1403.

[31] M. Campoy, A. Gómez-Barea, F.B. Vidal, P. Ollero, Air–steam gasification of biomass in a fluidised bed: Process optimisation by enriched air, Fuel Processing Technology, 90 (2009) 677–685.

[32] K. Maniatis, Progress in biomass gasification: An overview, in: Progress in Thermochemical Biomass Conversion, Blackwell Science Ltd., United Kingdom 2008, pp. 1–31.

[33] N.D. Couto, V.B. Silva, E. Monteiro, A. Rouboa, Assessment of municipal solid wastes gasification in a semi-industrial gasifier using syngas quality indices, part 1, Energy, 93 (2015) 864–873.

[34] L. Wang, C.L. Weller, D.D. Jones, M.A. Hanna, Contemporary issues in thermal gasification of biomass and its application to electricity and fuel production, Biomass and Bioenergy, 32 (2008) 573–581.

[35] Z. Wang, T. He, J. Qin, J. Wu, J. Li, Z. Zi, G. Liu, J. Wu, L. Sun, Gasification of biomass with oxygen-enriched air in a pilot scale two-stage gasifier, Fuel, 150 (2015) 386–393.

[36] S.S. Thanapal, K. Annamalai, J.M. Sweeten, G. Gordillo, Fixed bed gasification of dairy biomass with enriched air mixture, Applied Energy, 97 (2012) 525–531.

[37] N. Couto, E. Monteiro, V. Silva, A. Rouboa, Hydrogen-rich gas from gasification of Portuguese municipal solid wastes, International Journal of Hydrogen Energy, 41 (2016) 10619–10630.

[38] P. McKendry, Energy production from biomass (part 3): Gasification technologies, Bioresource Technology, 83 (2002) 55–63.

[39] Q. Miao, J. Zhu, S. Barghi, C. Wu, X. Yin, Z. Zhou, Modeling biomass gasification in circulating fluidized beds: Model sensitivity analysis, International Journal of Energy and Power Engineering, 2 (2013) 57.

[40] J.J. Hernández, G. Aranda, J. Barba, J.M. Mendoza, Effect of steam content in the air–steam flow on biomass entrained flow gasification, Fuel Processing Technology, 99 (2012) 43–55.

[41] J. Wang, G. Cheng, Y. You, B. Xiao, S. Liu, P. He, D. Guo, X. Guo, G. Zhang, Hydrogen-rich gas production by steam gasification of municipal solid waste (MSW) using NiO supported on modified dolomite, International Journal of Hydrogen Energy, 37 (2012) 6503–6510.

[42] C. Franco, F. Pinto, I. Gulyurtlu, I. Cabrita, The study of reactions influencing the biomass steam gasification process, Fuel, 82 (2003) 835–842.

[43] H.C. Butterman, M.J. Castaldi, CO2 as a carbon neutral fuel source via enhanced biomass gasification, Environmental Science and Technology, 43 (2009) 9030–9037.

[44] T. Renganathan, M.V. Yadav, S. Pushpavanam, R.K. Voolapalli, Y.S. Cho, CO_2 utilization for gasification of carbonaceous feedstocks: A thermodynamic analysis, Chemical Engineering Science, 83 (2012) 159–170.

[45] L. Garcia, M.L. Salvador, J. Arauzo, R. Bilbao, CO_2 as a gasifying agent for gas production from pine sawdust at low temperatures using a Ni/Al coprecipitated catalyst, Fuel Processing Technology, 69 (2001) 157–174.

[46] A.M. Parvez, I.M. Mujtaba, T. Wu, Energy, exergy and environmental analyses of conventional, steam and CO_2-enhanced rice straw gasification, Energy, 94 (2016) 579–588.

[47] O. Corigliano, P. Fragiacomo, Technical analysis of hydrogen-rich stream generation through CO_2 reforming of biogas by using numerical modeling, Fuel, 158 (2015) 538–548.

[48] Y. Cheng, Z. Thow, C.-H. Wang, Biomass gasification with CO_2 in a fluidized bed, Powder Technology, 296 (2016) 87–101.

[49] C. Loha, P.K. Chatterjee, H. Chattopadhyay, Performance of fluidized bed steam gasification of biomass – Modeling and experiment, Energy Conversion and Management, 52 (2011) 1583–1588.

[50] T. Song, J. Wu, L. Shen, J. Xiao, Experimental investigation on hydrogen production from biomass gasification in interconnected fluidized beds, Biomass and Bioenergy, 36 (2012) 258–267.

[51] P. De Filippis, C. Borgianni, M. Paolucci, F. Pochetti, Prediction of syngas quality for two-stage gasification of selected waste feedstocks, Waste Management, 24 (2004) 633–639.

[52] M. Marco, M. Giovanni, P. Lorenzo, R. Rosario, A quasi-3D computer model of a planar SO fuel cell stack, in: 3rd International Energy Conversion Engineering Conference, American Institute of Aeronautics and Astronautics, San Francisco, CA 2005.

[53] X. Peng, B. Toseland, A. Wang, G. Parris, Progress in development of LPDME process: Kinetics and catalysts, in: Coal Liquefaction & Solid Fuels Contractors Review Conference, Pittsburgh, Pennsylvania, 1997.

[54] J. Xiao, L.H. Shen, X. Deng, Z.M. Wang, X.L. Zhong, Study on characteristics of pressurized biomass gasification, Zhongguo Dianji Gongcheng Xuebao/Proceedings of the Chinese Society of Electrical Engineering, 29 (2009) 103–108.

[55] S. Luo, B. Xiao, Z. Hu, S. Liu, X. Guo, M. He, Hydrogen-rich gas from catalytic steam gasification of biomass in a fixed bed reactor: Influence of temperature and steam on gasification performance, International Journal of Hydrogen Energy, 34 (2009) 2191–2194.

[56] Y. Zhang, B. Li, H. Li, B. Zhang, Exergy analysis of biomass utilization via steam gasification and partial oxidation, Thermochimica Acta, 538 (2012) 21–28.

[57] Y. Wu, W. Yang, W. Blasiak, Energy and exergy analysis of high temperature agent gasification of biomass, Energies, 7 (2014) 2107.

[58] K.J. Ptasinski, M.J. Prins, A. Pierik, Exergetic evaluation of biomass gasification, Energy, 32 (2007) 568–574.

[59] L. Devi, K.J. Ptasinski, F.J.J.G. Janssen, A review of the primary measures for tar elimination in biomass gasification processes, Biomass and Bioenergy, 24 (2003) 125–140.

[60] X. Zhang, S. Deng, J. Wu, W. Jiang, A sustainability analysis of a municipal sewage treatment ecosystem based on energy, Ecological Engineering, 36 (2010) 685–696.

[61] F. Pinto, R.N. André, C. Carolino, M. Miranda, P. Abelha, D. Direito, N. Perdikaris, I. Boukis, Gasification improvement of a poor quality solid recovered fuel (SRF). Effect of using natural minerals and biomass wastes blends, part B, Fuel, 117(2014) 1034–1044.

[62] S.H. Aljbour, K. Kawamoto, Bench-scale gasification of cedar wood – Part II: Effect of operational conditions on contaminant release, Chemosphere, 90 (2013) 1501–1507.

[63] V. Silva, A. Rouboa, Optimizing the gasification operating conditions of forest residues by coupling a two-stage equilibrium model with a response surface methodology, Fuel Processing Technology, 122 (2014) 163–169.

[64] M.J. Anderson, P.J. Whitcomb, RSM Simplified: Optimizing Processes Using Response Surface Methods for Design of Experiments, Productivity Press, New York, 2005.

[65] A. Abuadala, I. Dincer, Efficiency evaluation of dry hydrogen production from biomass gasification, Thermochimica Acta, 507–508 (2010) 127–134.

2

Fatal Accidents during Marine Transport of Wood Pellets Due to Off-Gassing: Experiences from Denmark

Frank Huess Hedlund and

Øssur Jarleivson Hilduberg

Abstract

The atmosphere in unventilated wood pellet storage confinements, such as the cargo hold of marine vessels transporting pellets in solid bulk, can be severely oxygen deficient and contain deadly concentrations of harmful gasses, of which the most feared is the poisonous and odour-less carbon monoxide. The hazard has been known for over a decade and has been responsible for many accidents. We examine three fatal accidents on marine vessels in or near Danish waters and argue that they share strikingly similar aetiologies, if not repetitive patterns. It is generally recognized that accidents should be thoroughly investigated and lessons learned shared widely in order to minimize the number of times the same lessons have to be learned. The three Danish cases suggest that this learning process is deeply troubled for the solid biomass segment. The International Maritime Organization IMO/SOLAS has recently revised its guidance on entering enclosed spaces aboard ships in response to the ongoing problem of confined space incidents. We argue that the interpretation of the concept of an "enclosed space" is of utmost importance because accidents take place in rooms that are not considered enclosed by the crew.

Keywords: wood pellets, off-gassing, confined space, accident investigation

1. Introduction

This paper pays homage to what may be man's oldest accident prevention strategy: Learning from own—and better, other people's—past accidents and misfortune. In its modern form, the key characteristics of this strategy are that past mishaps shall be recorded and analysed to extract lessons learned which in turn shall be disseminated through broad feedback loops in order to prevent future similar, and not just identical, accidents.

Paraphrasing Petroski [1], the profession of accident prevention professionals is Janus-faced, looking always both backwards and forwards. The history of past accidents offers a repository of knowledge gained from painful experience. If heeded, the past thus provides caveats and lessons for future safe operations. If shunned, it will still haunt the future, always lurking in the shadows everyday work.

Having paid the price of an accident, we should use the opportunity to learn from it. The benefits of such learning are obvious—to avoid repetition, and to share the lessons learned in order to minimize the number of times the same lessons have to be learned. What is not so obvious, however, is how to make this seemingly simple and straightforward idea work in practice [2]. This challenge is reflected in common aphorisms, such as Santanyana's: Those who cannot remember the past are condemned to repeat it.

This paper examines three fatal accidents on marine vessels transporting sustainable, environmentally friendly and carbon-neutral wood pellets in bulk, in or near Danish waters, and argue that they share strikingly similar aetiologies, if not repetitive patterns—offering support to Sanatnyana's dismal adage. The paper examines some of the barriers that impede the capacity to learn from accidents on marine vessels in this particular segment.

2. Wood pellets in bulk

2.1. Solid biomass fuel in Danish thermal power plants

To reduce global carbon dioxide emissions, Denmark has converted many coal-fired thermal power plants to burn green and sustainable solid biomass. Solid biomass refers to non-fossilized carbon-based solid materials derived from plants, typically wood, but also straw, and certain energy crops grown primarily for the purpose of fuel utilization. Because Denmark's own woodlands are limited and covered by environmental protection policies, the domestic production of solid biomass is insignificant. Hence, essentially all biomass is imported, mostly from loggings in the Baltic countries and Russia, but also from North America and Africa [3].

Thermal power plants consume very large amounts of fuel; a single block can easily burn half a million metric ton per year. In order for the logistics chain to handle such quantities in an efficient manner, the irregularly sized and bulky biomass is processed into uniformly sized pellets, typically about 8 mm in diameter and 20–25-mm length.

In the production process, the wood is dried, milled to a fine powder and compacted in a pellet press that often employ die ring press and compression roller technology. The pellets are shipped in bulk in marine vessels to their final destination.

2.2. Wood pellet fire and explosion risks

The fire and explosion hazard profile of biomass pellets is much different from that of coal [3]. Pellets inevitably generate fine dust when handled in the logistics chain, e.g. in screw conveyors, band conveyors, cup elevators and when dumped into storage areas or hoppers. If made airborne and ignited, the wood dust is capable of producing severe dust explosions.

In addition, this dust is prone to ignition, e.g. from overheated electric motors or conveyor bearings, or from mechanical friction heat between conveyer belts and accumulated pellets, fines and/or dust. Small pieces of smouldering material are difficult to detect and they may migrate in the band conveyor systems and start smouldering fires in the storage areas which may develop into massive storage fires. Smouldering fires are furthermore notorious for their capacity to generate flammable pyrolysis gasses that travel and accumulate, leading to flash fires or explosion hazards elsewhere in the system [4].

2.3. Wood pellet chemical off-gassing risks

The milling process, the compaction and extreme shear forces in the pellet press, cause plant cell walls to rupture. This exposes cell cytoplasm and chemically reactive compounds such as resins, oils and fatty acids to air making them available to chemical oxidation and other reactions. As a result, freshly produced wood pellets are probably best thought of as a chemically active material.

The chemically active pellets may emit a range of gasses such as carbon monoxide, carbon dioxide, methane and volatile organic compounds (VOCs) that can accumulate in storage confinements and reach dangerous levels [5]. The release of the highly poisonous and odourless carbon monoxide is of particular concern. In addition, the chemical reactions may deplete the atmosphere in storage confinements of the oxygen required to sustain life. The topic has been investigated in several recent publications, e.g. [6, 7] but the exact nature of the chemical processes at play do still not appear to be well-understood.

Chemical oxidation may also explain why wood pellets can self-ignite. Heat loss is largely a surface-based phenomenon and because of the low surface-to-volume ratio of a large pile, any process that generates heat will slowly increase the temperature inside the pile. Pockets may form where the temperature of the contents can rise to the temperature necessary to produce spontaneous ignition. This produces an oxygen deficient smouldering fire deep inside the pile and the associated problems with flammable pyrolysis gasses that travel and accumulate, leading to flash fires or explosion hazards elsewhere in the system.

It is true that some biomass feedstocks species rich in resins are desirable in the pellet production process. The mechanical stability of pellets is an important property as this limits the amount of dust generated when they are handled in the logistics chain. Pellet stability is enhanced if the mill feed composition includes some biomass species rich in resins, e.g. pine wood, which can act as a natural binder system. Naturally occurring oils and resins can have other beneficial properties during production, for instance acting as lubricants in the pellet press. It will probably be difficult to reduce the problem with chemical reactivity of wood pellets through substitution of resin-rich species in the feedstock unless, of course, non-bio but safer additives are tolerated by pellet consumers.

What is clear though is that the massive increase in wood pellets consumption entails novel risks—wood pellets have a much different hazard profile than fossil fuel cargoes.

3. Case 1—Fatal accident on AMIRANTE, 2009

3.1. AMIRANTE

The ship AMIRANTE (IMO 7425334) was a general cargo ship enabling it to carry a variety of bulk and mixed cargos. It was built in 1978/1982 and had a gross tonnage of 4083 and a length of 82 m. The vessel was classed by DNV.

At the time of the accident, the vessel had a crew of eight, seven from the Russian Federation and one from Belarus. The flag State was St. Vincent & The Grenadines. AMIRANTE was mainly in a coastal trade in northern Europe and the Mediterranean.

3.2. Sources

Information on this accident is based on records from the Danish police [8] obtained through the Danish equivalent of a Freedom of Information Act unless otherwise noted.

3.3. The accident

On 15 July 2009 at 0300 (all times are local time), AMIRANTE left the port of Riga, Latvia, after having loaded a cargo of 2600 t of wood pellets in bulk. The port of destination was Copenhagen, Denmark, were the cargo was to be unloaded at the Amagerværket power plant owned by Wattenfall, which had recently converted one of its coal fired blocks from coal to solid biomass. The shipper was Granul Invest in Tallinn, Estonia and the operator was Alga Chartering in Copenhagen.

On the afternoon 15 July, during the passage of the Baltic Sea, four of the vessel's crew, three able seamen and the motorman, worked on the deck readying new mooring ropes on the forward part of the ship.

Shortly after 1800, the master was alerted by the ship's cook that one of the able seamen and the motorman had not shown up for dinner and their whereabouts were unknown. A search was initiated and at 1835, a crew member spotted the missing crew members lying lifeless at the bottom of the stairwell to the forepeak compartment. The crew member did not attempt to reach them fearing for his own safety.

The forepeak is a wedge-shaped stowage room at the bow of the ship, about 5 m deep, and a width ranging from 5 to 1.5 m. The room is reached from the deck through a stairwell climbing a narrow 18-m long fixed vertical steel ladder. The forepeak was used for stowing, e.g. paint containers, floor brushes, buckets, surplus wooden planks and mooring ropes.

The master rushed to the forepeak and climbed down the steel ladder. Halfway down he began to experience laboured breathing. Upon reaching the forepeak floor his breathing had become so laboured that he realized the atmosphere might be dangerous. The motorman was lying about 1 m from the ladder and the able seaman was lying just at the foot of the ladder, as if he had attempted to climb it, still wearing work gloves. Froth could be seen from the corner of the mouth of both men. The master briefly tried to awaken them and then hurried up the ladder to the deck. The vessel was about 16 km north of Bornholm, a Danish island in the Baltic Sea about midway between Sweden and Poland. At 1845, the master radioed the Danish authorities requesting immediate medical assistance.

The crew worked frantically to extract the two men from the forepeak compartment. They switched the room's mechanical ventilation on and crew members climbed down the ladder to tie a rope around the chest of the victims. Hoisting the victims through the narrow stairwell proved difficult. When a doctor arrived by helicopter from the Danish Admiral Fleet, at 1945, one victim has been extracted and there was a work frenzy to hoist the other. The doctor could soon declare both men dead as they presented with livor mortis and beginning rigor mortis. The doctor estimated that death had occurred 2–3 h earlier.

3.4. The investigation

The vessel was redirected to the port of Rønne at Bornholm by Danish Admiral Fleet, where it arrived at 2300. The police was waiting at the quay and began a standard crime scene procedure, securing evidence and interviewing the crew. The forepeak was inspected by the fire and rescue services done full breathing protection at about 0100. At the time of the inspection, the compartment's mechanical ventilation had been switched off for about 40 min upon request. Oxygen levels were normal, CO levels were 100 ppm. The inspection team noticed a pair of glasses which had been placed on some stowed construction timber and an unopened packet of cigarettes lying on the floor.

A door in the forepeak compartment allowed access to the front cargo hold. The door had three closing hinges but only two were engaged. It was speculated that gasses might travel from the cargo hold to the forepeak through a crevice in the door opening where the third hinge was not engaged (**Figure 1**).

The deceased presented no signs of violence or trauma and there was no other evidence to suggest criminal actions, e.g. if they should have been pushed while standing in the door opening on the deck and then fallen the 18 m down the forepeak stairwell. There was briefly some speculation if red stains on the floor of the forepeak were bloodstains but they turned out to be red paint. Follow-up interviews with the master and other crew members were carried out in the afternoon to resolve some inconsistencies in reported times and observations, this time with the assistance of a certified interpreter. Only the master spoke some English, and he had acted as interpreter during the first interviews.

A sample of the wood pellets was sent for chemical analysis which found 150 ppm CO in the headspace of the specimen jar. The actual concentration is likely much higher, however, as 150 ppm simply appears to be the upper range of the measuring instrument. An autopsy confirmed death by CO poisoning, COHb blood saturation levels were measured at 52% (the motorman) and 60% (the able seaman). Tests for cyanide (blood) and alcohol (lacrimal fluid) were negative.

Figure 1. The door in the forepeak compartment, which led the front cargo hold. The door had three closing hinges but only two were engaged. Gasses travelled from the cargo hold to the forepeak through a crevice where the third hinge was not engaged. (AMIRANTE). Source: Police photo file #13 [8].

The crew was at pains to explain why the two seamen would go down the forepeak. They had no business down there whatsoever. It was suggested that they might want to check if there was sufficient room for stowage of the old mooring ropes.

The two deceased were experienced seamen. The motorman was born in 1981 in Russia, the able seaman in 1971 in Belarus. Records showed that both seamen had duly complied with regulations concerning rest and work time. Both seamen had received the proper safety instruction for working in enclosed rooms in 2008 (July 8 and April 20, respectively) and they were not newcomers to the ship. Police records state that they should have known that full breathing protection was required before entering a compartment such as the forepeak. In the absence of evidence of the contrary, the police ruled the two deaths an unfortunate workplace accident and the crime scene investigation was closed.

3.5. Reflections

The direct cause of the accident is probably simple and understandable. Working new and bulky mooring ropes, often weighing several 100 kg, is a laborious and tedious task. According to the weather data from Rønne airport, the afternoon weather conditions were mostly cloudy, temperature 17–19°C, light winds about 3 m/s and occasional rain showers and light rains. While such ambient weather conditions definitely are tolerable to seasoned seamen, it is easy to imagine that the nearby forepeak might offer a convenient refuge for a clandestine cigarette break, in particular during a rain shower.

Important questions aimed at understanding *why* the accident happened were never raised, simply because they were out scope of a judicial police investigation. Were the crew aware of hazardous properties of the cargo? Was the forepeak considered to be a compartment with properties similar to those of enclosed cargo spaces? The starkest circumstantial evidence may be that the two seamen chose to enter the forepeak without switching on the mechanical ventilation. In all likelihood, they were clueless about the danger—otherwise, they would not have entered. Were the rest of the crew and in particular the master aware of the hazard? Although it is pure conjecture, the answer is probably negative. In the police interrogation reports, none of the crew mentioned the ability of wood pellets to generate carbon monoxide. The closest mention was the master stating that he suspected something was "wrong" with this particular cargo, because it had a "funny smell" unlike any of the 15 earlier occasions where AMIRANTE had carried wood pellets.

Was information about the hazardous nature of the cargo available to the crew? The question too was never raised and although the answer is uncertain it is probably negative. The manifest of cargo simply stated: "Wood Pellets in bulk—Cargo Status C—European Community Cargo" and the "Dangerous" field on the manifest form was left blank.

The criminal investigation was satisfied that the two seamen had received the proper safety instruction for working in enclosed rooms and concluded that they should have known that full breathing protection was required before entering a compartment such as the forepeak. The question was never raised, however, if the seamen, or the rest of crew for that matter, would consider the forepeak "an enclosed room". The police interrogation reports state that the practice on board the vessel was to lock open the door on the deck that provided access to the stairwell to the forepeak. The door was only closed during bad weather. The reason for this practice is unexplained. But in the absence of specific precautionary instructions or warnings, the crew may well have assumed that a room with a door locked open is not "enclosed".

3.6. Information sharing and learning

The local news media at Bornholm briefly covered the event, reporting that a coaster with two dead sailors had arrived at the port of Rønne. At the time of media reporting the cause had not yet been determined, and the media speculated that CO poisoning or oxygen deficiency in an enclosed room might be responsible for a workplace accident. The vessel's cargo was correctly identified as wood pellets, but there is nothing in the media coverage to suggest that there might be a causal link between the type of cargo and the presence of toxic gasses. For an accident prevention professional, the media coverage is worthless.

Because the police opened the case as a crime scene investigation, the case was technically registered as a criminal case. Criminal cases are kept confidential in order to protect the privacy of the individuals concerned. To date, very little information, if any, relevant to accident prevention professionals is available in open sources. The source material for this paper could only be obtained through persistent requests and the Danish equivalent of a Freedom of Information Act.

It is true that there is sporadic and passing mention of the AMIRANTE accident in the wood pellet literature, e.g. [9, 10] but descriptions are hazy and basic facts are corrupted. This is also true for the papers produced by authors in Denmark. For example [11] states that the seamen entered the cargo hold, which they did not, and [12] states that they died from asphyxia, which is also incorrect.

Because proper sharing of correct information about an accident is a basic precondition for learning, we conclude that learning processes were derailed at a very early stage. Therefore, it may be expected that repeat accidents will take place. As we will argue shortly, this is precisely what happened. Lamentably, the broader accident history reveals that the AMIRANTE accident was no first, but only one in a string of previous similar accidents with similar aetiologies.

4. Case 2—Fatal accident at "LADY IRINA", 2014

4.1. LADY IRINA

The ship LADY IRINA (IMO 9137038) was a general cargo ship with capacity to carry containers enabling it to carry a variety of bulk and general cargo. LADY IRINA was built in 1997 with a gross tonnage of 3323 and a length of 88 m. The vessel was classed by Lloyd's Register. At the time of the accident, the vessel had a crew of seven. The flag state was The Netherlands.

4.2. Sources

Information on this accident is based on a report from the Dutch Safety Board [13] unless otherwise noted.

4.3. The accident

On 5 July 2014, LADY IRINA departed from the Port of Arkhangelsk, Russia, with a cargo of wood pellets in bulk which was to be unloaded in Kolding, Denmark. Arkhangelsk is LADY

IRINA's permanent loading port for this type of cargo, and it has a dozen discharge ports in Europe, of which Kolding is one.

On the evening of 13 July 2014, the crew prepared for LADY IRINA's arrival in the Port of Kolding. At about 1900, the chief engineer met with the chief officer and decided to drain the anchor chain locker compartment before LADY IRINA expected arrival at about 2400 h. The anchor chain locker is located in the forward part of the ship. Access to the anchor chain locker is from the deck, entering the forecastle compartment, opening a door and climbing a vertical steel ladder in a stairwell to the bow thruster room, located directly beneath the forecastle room.

The forecastle compartment holds maintenance and safety equipment and other frequently used items such as tools, paint and brushes. The door to the forecastle room is therefore normally open throughout the day and only closed at the end of work, typically at about 1700, to secure the watertight integrity of the ship while the crew were off duty.

On 13 July, sometime after 1900, the chief engineer went to the engine room to start the pump draining the anchor chain locker and then went to the bow thruster room, probably to check if the draining operation was complete. The door to the forecastle had been closed since end of work Saturday, i.e. for at least 26 h.

At 2145, the chief officer went looking for the chief engineer. The door to the forecastle room was open and he found the chief engineer lifeless on the floor of the deck lower, on the floor of the bow thruster room. He examined the chief engineer and found that he was not breathing and had no heartbeat.

The chief officer rushed to the bridge to alert the master, who had the watch, and to summon assistance. The second mate, who happened to be on the bridge was ordered to go to the accident site with a stretcher. The master immediately decided to divert to the nearby port of Fredericia (Denmark), only 4 nautical miles away, and requested emergency assistance to be ready at the quay.

The chief officer went to the crew accommodation to alert the rest of the crew and directed them to the bow thruster room. Probably at about 2215, all crew members except the master were involved in the rescue operation in the bow thruster room. They first worked to resuscitate the chief engineer and then attempted to evacuate him through the stairwell on the stretcher, which proved unsuccessful as he was a heavily built man. The rescue involved six crew members: the first and second mates, the apprentice engineer, the cook and two seamen. They did not use the breathing apparatus that they had taken with them because they thought that the chief engineer had fallen from the ladder and that air quality was not a problem.

Around 2245, the vessel arrived in the Port of Frederica and the apprentice engineer and a seaman left the bow thruster room to assist with berthing the vessel. The bow thruster was required for berthing and had to be operated from the engine room. The second mate also left the room to collect a neck brace because the crew still thought that the chief engineer had fallen from the ladder, and an oxygen kit because the chief engineer was not breathing.

Upon returning to the forecastle room, a few minutes later, the second mate saw the chief officer unconscious, a crew member trying to carry him away and another crew member walking around as if drunk. The second mate climbed down ladder to the bow thruster room, placed a breathing apparatus over the mouth of the chief officer and urged the two other crew members to leave the room. He then left the bow thruster room and forecastle himself because he did not feel well. Before leaving the forecastle, he opened the medical oxygen kit and lowered it down the bow thruster room with a rope, in order to generate extra oxygen there.

The vessel berthed at 2250 and the local fire and rescue services evacuated the chief engineer and the chief officer from the bow thruster room. The other two crew members had managed to climb the ladder and get outside by themselves but were much weakened. The chief officer was given oxygen and regained consciousness on deck.

The chief officer and the two seamen were admitted to hospital and were diagnosed with carbon monoxide poisoning. The chief engineer had died from carbon monoxide poisoning. Danish news media [14] report that four, not three, seamen were admitted to hospital and two of them were treated in a hyperbaric oxygen therapy chamber. All crew appear to be of Russian nationality.

The local fire and rescue services ventilated the forecastle and bow thruster room with electrical ventilators for 1½ h and then took CO measurements. CO levels in the forecastle were measured at 80 ppm, bow thruster room at 20 ppm. After having the forecastle and bow thruster rooms closed for 36 h, the measurements were 690 and 555 ppm, respectively. The LADY IRINA is equipped with a fixed CO_2 system, which can inject carbon dioxide in the cargo hold for firefighting purposes. CO concentrations in the neighbouring carbon dioxide storage room, which had not been ventilated, were in excess of 2000 ppm, the upper limit of the measuring device.

4.4. The investigation

The investigation by the Dutch Safety Board [13] concluded that carbon monoxide generated by the wood pellets in the unventilated cargo hold could potentially have reached the bow area in several ways, e.g. via piping or ventilations ducts:

The CO_2 room, forecastle and bow thruster room are adjacent to each other and gasses might travel from one room to another through a number of connecting conduits. The shipping company contracted the classification society to investigate how carbon monoxide could have travelled from the cargo hold to the bow area. The investigation identified a minor gap in a section of the rim of an inspection cover on a ventilation duct that runs from the cargo hold through the forecastle as the most likely route (**Figure 2**).

The atmosphere in the bow area had been identified as potentially unsafe and warning signs on the forecastle access door stated: "DANGER. Low Oxygen Level". The bow area had no mechanical ventilation. In principle, according the rulebook, the bow area should probably be considered a confined space. At LADY IRINA, however, the daily work practice had evolved and adapted to the need to enter the room on a routinely and daily basis. The crew simply opened the door for 15–20 min to allow natural ventilation before entering and they would normally leave the door open.

Figure 2. The inspection hatch on a ventilation duct that ran from the cargo hold through the forecastle. A minor gap in a section of the rim (knife blade inserted) allowed gasses to travel from the cargo hold to the bow area. (LADY IRINA). Source: [13].

In response to the fatal accident, the shipping company acquired multi-gas meters and introduced a new work procedure mandating that the person making "First Entry" must carry an operational multi-gas meter. New warning placards with the instruction "ATTENTION—No first entry without use of multi-gas meter" were been put up on access doors.

The investigation report states that: "*The [Dutch Safety] Board is of the opinion that, in the light of this accident, it is good that the shipping company examines the working method that was common in practice. However, the new working method desires focus on definition and crew discipline during use. The working method should also be evaluated and improved, where necessary*" [13].

4.5. Reflections

The most striking feature of this accident is probably that the *entire crew*, except the master on the bridge, climbed down the ladder to the bow thruster room where dangerous levels of carbon monoxide were present and stayed there for more than 30 min, *completely oblivious of the danger*. It is easy to imagine that under just marginally different circumstances, the entire crew would have succumbed to the poisonous gasses down there, which in turn would have hampered the master's ability to berth the vessel and hence delayed the arrival of the rescue services. It can safely be stated that the death toll could have been higher.

The Dutch investigation observes that the relevant International Maritime Solid Bulk Cargoes (IMSBC) Code for wood pellets at the time of the accident stated that: *"all crew members must carry an oxygen and carbon dioxide meter and switch this on when they enter the cargo hold or adjacent confined spaces"*. We observe that there is only superficial and passing mention of carbon monoxide in the Code. We also remark that an atmosphere with deadly levels of CO may present with normal readings for oxygen and CO_2.

Was information about the hazardous nature of the cargo available in a form that made the crew aware of the risk? The investigation does not clearly address this issue. The investigation simply states that the crew *"was informed of the risks of this cargo"* and that they *"knew that the hold should not be entered"*. Furthermore that *"each new crew member had received safety instructions"* which comprised *"confined spaces and ventilation of holds in connection with possible lack of oxygen and other dangerous substances that emerge as a consequence of cargo. Also during the safety committee meetings attention was paid to entering confined spaces"*.

We conclude that the answer probably is negative: information was not available to the crew. In any case, their actions demonstrate that they were clueless of breathing hazards of any kind, be it oxygen depletion, carbon dioxide, carbon monoxide, etc.

For a safety professional working with industrial accident prevention (onshore), it is unusual that rooms routinely entered on a daily basis are, first, identified and labelled as having a potentially unsafe atmosphere, and second, do not have mechanical ventilation. The absence of mechanical ventilation would appear to be a weakness in design, which ought to be eliminated in future designs and retrofitted in existing designs wherever possible. Mechanical ventilation is a *risk prevention* measure, whereas testing before entry is a *risk control* measure, merely to confirm that the preventive measures are successful. It is remarkable that a relatively recently built ship (1997) has no mechanical ventilation where dangerous gasses potentially may be present.

4.6. Information sharing and learning?

The accident has been investigated and the findings exhaustively communicated by the Dutch Safety Board [13]. The coverage in Danish new media correctly identifies the generation of CO from wood pellets as the culprit and has no factual errors [14]. A Maritime Merchant Periodical with international coverage [15] carried an article that summarized the findings of Dutch Safety Board's accident report, highlighting that other crew members could have been killed in the rescue attempt. The article is without factual errors. The information provided is sufficiently detailed to permit safety professionals elsewhere to prevent recurrence.

The basic preconditions for learning: broad sharing of relevant and factually correct information; are met. As we will argue later, this is a necessary, but not a sufficient precondition, for learning to take place. The information must also be delivered locally in a manner so its importance is recognized and it must be acted upon.

5. Case 3—Fatal accident at CORINA, 2015

5.1. CORINA

M/V CORINA (IMO 8908545) was a general cargo vessel with capacity to carry containers enabling it to carry a variety of bulk and general cargo, length 115 m, gross tonnage 5796. The vessel was built in 1990 and classed by Germanischer Lloyd. At the time of the accident, the vessel had a crew of eleven, all of Polish nationality. The flag state was Poland.

5.2. Sources

Information on this accident is based on a report from the Polish State Maritime Accident Investigation Commission [16], and records from the Danish police [17] and the Danish Working Environment Authority [18], both obtained through the Freedom of Information Act.

5.3. The accident

Between 11 April and 20 April 2015, CORINA loaded 6744 t of wood pellets in the Port of Arkhangelsk (Russia). On the 28 April, the vessel arrived at Hanstholm, Denmark, to unload 1921 t before proceeding to The Netherlands.

Company policy required the visible presence of a watchman near the access gangway at all times while at port and an able seaman had the watch from 1600 h. The cook expected him for dinner at about 1700 and eventually informed the chief officer that the able seaman had not shown up. The chief officer searched the ship twice and noticed that the door to the bosun store under forecastle deck was open. Inside, he noticed that the hatch cover to a lashing equipment room at a lower deck in the forepeak was open. He went down and found the seamen there, unconscious. He radioed for help and shortly after, at 1754, the master contacted the port in broken English and requested medical assistance to a heart attack victim.

Within about 15 min a bosun, an electrician and another seaman had come to the accident site to assist and taking turns with cardiac pulmonary resuscitation believing that the unconscious seaman, born in 1947 and overweight at 120 kg, had suffered a heart attack. At about 1810, the port officer arrived at the site with a defibrillator. The port officer soon did not feel well and asked the crew if the air was ok, which was confirmed. After some time one of the crew members, probably the chief officer, attempted to climb the ladder but fell down and received a blow to the head and had to sit down. The first ambulance arrived at 1813 and one of the paramedics soon arrived at the site and began administering oxygen to the able seaman. Six persons were now present in the room. After some time, the paramedic asked the crew if the air was ok, which was confirmed. Shortly after the port officer radioed, the master asking if the air was ok. Only then was the mechanical ventilation switched on. The paramedic ordered everybody out of the room because the air was not ok. Two crew members were unable to climb the ladder and had to be left behind with the unconscious seaman while the other four managed to move up the forepeak and reach the deck, some only with great difficulty.

Exact times are not available, but the on-scene commander arrived at 1845 at which four ambulances and three vehicles from the emergency services were at the site. A doctor arrived by helicopter at 1855. Fire fighters donned full respiratory protection extracted the two unconscious crew members who regained consciousness on the deck. At the entry door to the forepeak, CO levels were measured at 66 ppm. Further down 366 ppm were reported. It should be noted that these measurements were taken after mechanical ventilation had been switched on. The able seaman was difficult to extract through the narrow hatch because of his heavy build. He was declared dead at 1940. An autopsy confirmed CO poisoning with COHb measured at 60%. Four crew members and the port officer were admitted to hospital.

5.4. The investigation

CORINA was general cargo multipurpose vessel adapted to transport of containers. Containers must be lashed, and in order to ease the handling and movement of lashing gear, the vessel was designed with a 2100 mm × 1150 mm opening in the forward bulkhead of cargo hold no. 1. The opening allowed easy access to the neighbouring storage room, where chains, stacking cones and other lashing equipment for containers were kept (**Figure 3**). When carrying bulk cargo, the opening was closed with planks inserted into guide bars. This barrier was not gas tight and gasses from the cargo hold could therefore travel to the forecastle (**Figure 4**).

The crew was at pains to explain why the able seaman would enter the lashing equipment room. He had no business down there whatsoever and had not requested permission or informed anybody of his intention to go there. He was supposed to be on watch near the gangway.

Figure 3. An opening in the front cargo hold allowed easy access to the lashing equipment room. When carrying bulk cargo, the opening was closed with planks of wood inserted into guide bars (CORINA). Source: [16].

Figure 4. The plank barrier seen from the lashing equipment room. Wood pellets in the cargo hold are visible are visible between planks 10 and 11. The barrier is evidently not gastight. (CORINA). Source: Police photo file #23 [8].

In line with the International Maritime Solid Bulk Cargoes (IMSBC) Code, the shipper had declared the hazardous nature the cargo. The wood pellets were declared as a group B cargo and marked as Materials Hazardous only in Bulk (MHB), which covers materials that may possess chemical hazards when carried in bulk. The shipper had provided several safety recommendations, only allowing entry to the cargo hold after 2 h of natural ventilation and measurements of oxygen content and dangerous gasses, carbon monoxide included.

The Polish investigation concluded, however, that crew members likely were unaware of the risks associated with the cargo. Points of entry to hazardous or confined spaces were not marked with warning signs and potential access doors had not been locked to prevent inadvertent entry [16]. To the Danish police, the master explained that he knew that wood pellets could *"suck oxygen out of the air"* but also that he believed that gasses were present in the cargo hold only, although *"he was not a university professor"* [17].

5.5. Reflections

In this recent accident, the hazards associated with wood pellets and preventive measures are clearly stated in the cargo documents. The extent to which this information is fully grasped by the master can be questioned. Seemingly, air quality hazards only became apparent to him after the port officer radioed and asked if the air in the forepeak was ok. Only then did he arrange for the mechanical ventilation to be switched on. Information trickle down from the master to the rest of the crew is obviously doubtful. Again, the actions of the crew demonstrate that they were truly oblivious of the danger. And so were the Danish port officer, the paramedic and probably many more.

Language issues may also have played a role. In particular, the ability of seamen from Eastern Europe with limited proficiency in English to understand hazard codes and guidelines in English, even more so for Russian seamen who may have a preference for information printed in Cyrillic letters.

5.6. Information sharing and learning?

The accident was investigated and the findings exhaustively communicated by the Polish State Maritime Accident Investigation Commission [16]. The coverage in Danish media mainly paid attention to the irony that the would-be rescuers succumbed to gasses and themselves became victims but is otherwise riddled with misleading factual errors. For example, the accident is stated to have taken place in the cargo hold because of oxygen depletion, both facts are plain wrong. The source of the hazard, the wood pellets, is correctly identified, however.

6. Discussion

6.1. Accident history involving off-gassing from wood pellets

Lamentably, strikingly similar serious accidents related to marine transport of solid biomass continue to take place. Publicly available information on these incidents is often scant, however, and as the AMIRANTE case shows, at times incorrect or misleading. The list below (**Table 1**) has been compiled from various sources and does not claim to be exhaustive. But, it is evident that the release of hazardous gasses from wood pellets in bulk qualifies as an emerging risk for which proper control strategies have yet to be developed.

6.2. Repetitive patterns in accident aetiology

The three Danish cases have strikingly similar accident aetiologies:

- the accidents take place in compartments adjacent to the cargo hold
- the compartments are likely covered by IMO's technical definition of an "enclosed room" but the crew does perceive the rooms as being "enclosed"
- dangerous gasses from the cargo hold travel unexpectedly through minor crevices and small gaps unknown to the crew

- the crew appear oblivious of the hazardous properties of the cargo
- would-be rescuers subject themselves to grave danger
- victims in rooms only accessible using a vertical ladder can only be extracted with great difficulties, in particular if the victim is of heavy build

We are of the opinion that similar accident aetiologies apply to many of the other accidents in the list above though scant publicly available information (and time constraints) prevents a rigorous demonstration of this claim.

Year	Vessel name, location, accident	Fatal	Injured	Source
2002	VEAWER ARROW, Rotterdam (NL), entry into cargo hold with wood pellets	1		[10]
2005	EKEN, Gruvön (SE), entry into cargo hold with pulp logs	1		[10]
2005	SAGA FOREST CARRIER, Wilmington (US), working in open cargo hold with green lumber, oxygen depletion	1		[16, 19]
2006	SAGA SPRAY, Helsingborg (SE), entry into cargo hold with wood pellets	1	Several	[10]
2006	NOREN, Skelleftehamn (SE), entry into cargo hold with wood chips	1		[10]
2007	FEMBRIA, Timrå (SE), entry into cargo hold with timber	2		[10]
2009	AMIRANTE, Bornholm (DK), entry intro room adjacent to the cargo hold with wood pellets	2	0	[8]
2010	TPC WELLINGTON, New Zealand, entry into cargo hold with timber	2		[20]
2014	LADY IRINA, Fredericia (DK), entry intro room adjacent to the cargo hold with wood pellets	1	3	[13]
2015	CORINA, Hanstholm (DK), entry intro room adjacent to the cargo hold with wood pellets	1	5	[17, 16]

Table 1. Serious accidents related to marine transport of solid biomass known to the authors.

6.3. The IMO concept of a "confined room" and local sense-making

The Danish accidents illustrate how the explanatory framework for unsafe atmosphere accidents is based on a discourse that certain rooms are considered confined or enclosed. By implication, the hazard is perceived as being a property of the compartment and not a property of the atmosphere itself or the cargo.

This is particularly evident in the recently revised industry recommendations for entering enclosed spaced aboard ships issued by IMO [21]. The resolution's overall definition of an enclosed space is so broad that it can potentially apply to any space on the ship. However, the examples provided of enclosed spaces are characterized by being spaces that are not part of the normal workspace of the seaman, e.g. fuel tanks, ballast tanks, cargo pump rooms (used on e.g. chemical tankers), etc.

In the Danish cases, the accidents took place in spaces that were part of the normal working areas of the ship and were routinely accessed. The doors were kept open for practical purposes or convenience. If such rooms are labelled "enclosed" and warning signs posted, as in the LADY IRINA case, the crew may become desensitized and dulled to what is dangerous, because the warnings signs are inconsistent with daily routine and local sense-making.

6.4. Mental models of accident causation and stop rules

An accident investigation is a tremendously challenging task. The identification of accident causes is often subjective and guided by the tool box of the analyst rather than the nature of the accident. We subscribe to the ideas of Jens Rasmussen that there is a tendency to see what you expect to find [22], and his proposition [23] that this is a simple reflection of the nature of causal analysis and the fact that no objective "stop rule" exists to terminate the causal back tracking in search of a root cause.

In the AMIRANTE case, the stop rule was from what Jens Rasmussen has termed the "prosecutor's perspective"—the investigation was closed when it was not possible to find somebody to punish. In the LADY IRINA case, the stop rule was from the "therapist's perspective"—the search for causes stopped when a cure known to the analyst was identified (a new procedure and larger warning signs with more text).

The CORINA case was a hybrid, the search stopped when violations were identified (no warning signs and insufficient crew safety instructions) which conveniently also pointed to the cure: improve work procedures and safety training. The accident investigation outcomes above align well with the constructivist thinking of Donald Schön, that the moment we name the things to which we will attend we also frame the context in which we will attend to them [24].

We are of the opinion that it is important to be cognizant of such stop rules. In particular, we note that enhanced proceduralization of work may have limited impact on actual workplace safety. The use of administrative risk controls like procedures may from a managerial perspective be perceived as a time- and cost-effective tool to manage risk, but the actual impact on daily safety can be greatly limited when crew members face multiple and sometimes conflicting goals in their everyday work.

6.5. The coaster segment—a challenge to reach

Wood pellets are imported to Denmark on two types of ships: Small tonnage coastal ships (coasters) able to navigate in the shallow and narrow Danish waters calling smaller commercial ports, and larger bulk carriers used for transporting larger quantities to larger ports.

Contrary to coal, wood pellets do not tolerate moisture and cannot be stored in outdoor piles. Wood pellets must be handled under roof or in silos and storage capacity comes at a price. Former coal-fired power plants therefore often face storage capacity limitations. Wood pellets' energy content per unit volume is anywhere from half to one-third of that of coal, which furthermore aggravates storage capacity limitations [3]. The limitations call for frequent arrivals of relatively small cargo sizes, a market well-suited for the small coaster segment.

The coasters are usually general cargo ships, i.e. they are designed to carry many different kinds of cargo, e.g. goods on pallets, machinery and various bulk cargoes such as grain, coal and pellets. The high variability of the type of cargo sets limits to the degree of specialization and cargo domain safety knowledge that can be achieved in the small coaster segment.

In addition, coastal ships have, as most ships in international trade, crews of mixed nationality with varying background in training and experience, often from low-wage countries, and sometimes from Russia accustomed to the Cyrillic alphabet. We argue that cargo documents and safety recommendations in English may not be readily absorbed by this group and the work culture may have an oral rather than a written tradition.

6.6. Media shifting—a sub-optimality

Environmental interventions may be undertaken with insufficient attention paid to the environment of the workers. The concept of "media shifting" refers to a suboptimal situation which occurs if the "resolution" of a problem within the environmental domain gives rise to new, and unforeseen, problems within other domains, specifically the workplace safety domain [25].

Examples of insufficient attention to workplace safety are sometimes apparent in green reporting and environmental statements, if they omit or misrepresent sections covering occupational health and safety. For example, a study examined "Green Statements" of a biomass pellet facility certified to ISO 14001 and OHSAS 18001—a type of reporting which aims to demonstrate commitments towards sustainability performance. Fire station turnout statistics showed that fires at the facility were frequent. In the sustainability reporting, however, fire and explosion risks were only ranked 11 out of 14, long after more benign, if not trivial issues such as bad acoustics, psychosocial issues and indoor environment problems such as drafts, cold spots and radiant heat [3].

In recent years, critical studies that examine the general topic of balancing safety with sustainability have begun to emerge, see, e.g. [12, 26].

Denmark's climate change policy is very ambitious and comprises the conversion of coal-fired thermal power plants to solid biomass fuel in order to reduce carbon dioxide emissions [3]. This chapter points to evidence of media shifting of this climate change intervention that the transport of unprecedented quantities of wood pellets has created new workplace hazards in the marine transportation sector. We argue that the release of hazardous gasses from wood pellets in bulk present as an emerging risk for which proper control strategies have yet to be developed.

The domestic production of wood pellets in Denmark is limited (**Figure 5**), and more than 95% are imported and arrive by ship (**Table 2**). For the period 2001–2015, both years included, we have estimated the domestic production, imports and total consumption 2015 using simple linear inter-/extrapolation, all quantities in SI ton (t). For the same period, we have knowledge of three serious accidents on board marine vessel during the import voyage, which resulted in four fatalities and eight persons hospitalized after having been exposed to grave danger. For the domestic production, we have knowledge of one fatal accident and five persons hospitalized (at least) after having been exposed to grave danger due to fire and explosion, details cannot be disclosed, however, because of confidentiality agreements.

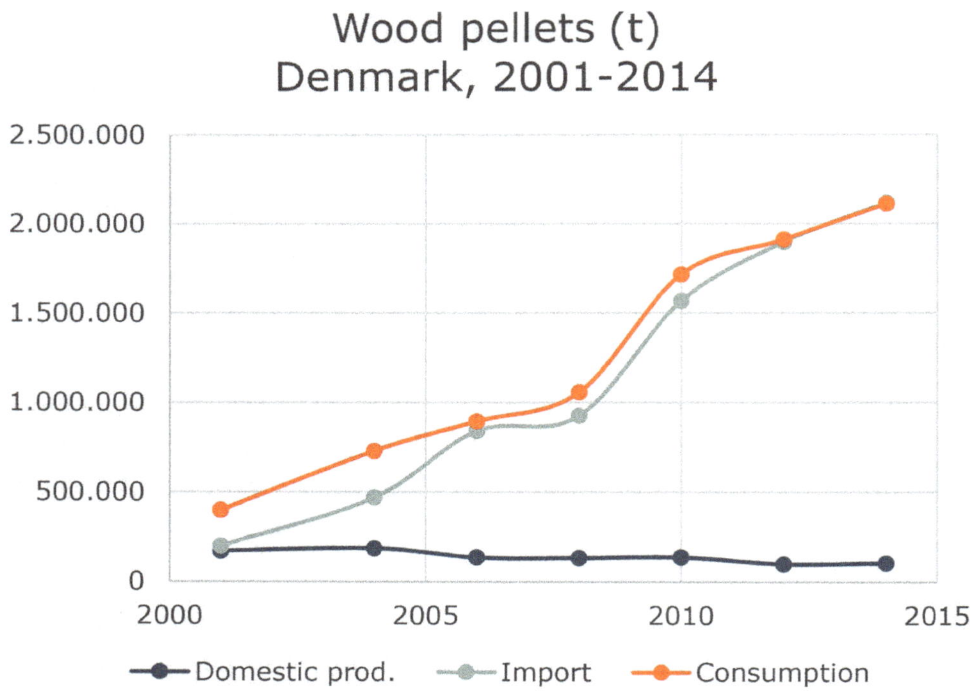

Figure 5. The consumption of wood pellets in Denmark has risen dramatically over the last 15 years.

Year	Domestic	Import	Consumption
2001	173,073	200,871	401,291
2004	187,458	470,588	731,134
2006	137,080	841,132	895,796
2008	134,280	925,401	1,059,519
2010	137,622	1,568,952	1,718,976
2012	99,930	1,898,143	1,914,048
2014	106,000	2,118,750	2,113,908
2001-2015 (est.)	2,100,650	17,452,647	19,191,795

Numbers do not sum due to stockpile changes and (small) exports (Data source: [27]). We estimate the total quantities (t) for 2001–2015.

Table 2. Danish domestic production, imports and consumption (t) of wood pellets for select years 2001–2014.

The numbers allow a crude estimation of the accident rate for imports and domestic production of wood pellets (**Table 3**). The numbers are not comparable because the import figure only comprises accidents (known to us) during the final leg of the marine voyage—excluding for example accidents during biomass harvest, land transport, wood pellet production in the country of origin and loading at the export port.

	Accident rate per Mt wood pellets	
	Fatal	**Grave danger**
Produced domestically	0.48	2.38
During marine voyage (import)	0.23	0.46

Table 3. The accident rate per million t wood pellets for accidents with fatal outcome and accidents where an individual is in grave danger, for wood pellets produced domestically and imports (final leg of marine voyage only).

7. Conclusion

Common sense tells us that having paid the hefty price of a serious accident, we should use every opportunity to learn from it and prevent recurrence. In the case of the fatal accident on AMIRANTE in 2009, such opportunities were wholly missed. The accident was investigated as a criminal case and the legal system is so designed that findings are kept confidential and out of reach of safety professionals and the general public. This approach has been criticized at an earlier occasion as being entirely misplaced in an accident prevention context [4]. It is a systemic deficiency by design—the bureaucratic judiciary system did not malfunction, it worked precisely as intended. The flag state, St. Vincent & The Grenadines, did not investigate. As a direct consequence, there was no learning.

Predictably, the accident recurred, at first not in Denmark, but elsewhere. In fact, it would turn out that similar accidents had happened before the AMIRANTE, but lessons learned had not been effectively shared and the crew at AMIRANTE were ignorant of the dangers of wood pellets in bulk.

After AMIRANTE, evidence slowly began to accumulate that wood pellets in bulk present chemical off-gassing hazards that qualify as an emerging risk. It is difficult not to mention the early work, foresight and leadership of the Canadian Wood Pellet association in shedding light on this novel issue [28].

By 2015, there is a comprehensive body of the literature describing the off-gassing hazards and the IMO has launched initiatives and revised the guidelines for management of entry into confined spaces. Unfortunately, very serious repeat accidents continue to recur. That implies that accidents can no longer be attributed to lack of knowledge but a failure to disseminate and use the knowledge available

The general cargo coaster segment appears difficult to reach. Because of the wide range of cargoes carried, there is little specialization and little accumulation of specific cargo safety knowledge.

We doubt if the recently revised IMO guidelines for management of entry into enclosed spaces will be effective. The guidelines are broad and their implementation will likely result in compartments—that are part of the normal working areas of the crew and routinely accessed—being labelled as potentially confined spaces. The crew may become desensitized and dulled to danger, because the hazard warnings are inconsistent with local sense-making. The accident history supports this prediction—confined space accidents routinely take place in rooms that are not considered confined spaces by the crew.

In addition, coasters generally have crews of mixed nationality, often from low-wage countries, and sometimes from countries accustomed to the Cyrillic alphabet. We are concerned that cargo documents and written safety recommendations in the English language may not be readily absorbed by this group and the work culture may have an oral rather than a written tradition. An entirely new approach must be sought. More research into this issue and into the existing legal and institutional barriers to learning is required.

The situation is special and delicate because the dramatic increase in the consumption of wood pellets arises out of an ambition to do good—to reduce carbon dioxide emissions and prevent climate change and global warming—often at steep costs to society [3]. It is morally problematic, if climate change environmental interventions are pursued so narrow-mindedly that sight is lost of the environment of the workers, and if workplace safety issues are starved of resources.

Utmost care should be taken to avoid media shifting—that the resolution of a problem within one domain, the environmental, creates a new problem in another, the workplace safety domain.

Acknowledgements

This article has been produced as voluntary work· Opinions expressed are those of the authors, not their employers or organizations. We are thankful to COWI for covering publication expenses.

Conflicts of interest

None of the contributing authors have conflicts of interest to declare.

Author details

Frank Huess Hedlund[1,2*] and Øssur Jarleivson Hilduberg[3]

*Address all correspondence to: fhuhe@dtu.dk

1 COWI, Kongens Lyngby, Denmark

2 Technical University of Denmark (DTU/Compute), Kongens Lyngby, Denmark

3 Danish Maritime Accident Investigation Board (DMAIB), Valby, Denmark

References

[1] H. Petroski, *Success through failure: the paradox of design*. Princeton University Press, Princeton, New Jersey, 2006

[2] F. H. Hedlund and H. B. Andersen, "Institutional support of learning from accidents: Some obstacles to getting a useful community-wide database in the EU," in *Society of Risk Analysis, SRA Europe, Annual Conference*, Ljubljana, Slovenia, 2006.

[3] F. H. Hedlund and J. Astad, "Solid biomass climate change interventions examined in a context of inherent safety, media shifting, and emerging risks.," *Human and Ecological Risk Assessment: An International Journal*, vol. 21, no. 5, pp. 1410–27, 2015.

[4] F. H. Hedlund, J. Astad and J. Nichols, "Inherent hazards, poor reporting and limited learning in the solid biomass energy sector: a case study of a wheel loader igniting wood dust, leading to fatal explosion at wood pellet manufacturer," *Biomass and Bioenergy*, vol. 66, pp. 450–459, 2014.

[5] G. Thek, I. Obernberger. *Pellet handbook: the production and thermal utilisation of biomass pellets*. Taylor and Francis, Florence, ISBN: 9781849775328

[6] L. Soto-Garcia, X. Huang, D. Thimmaiah, A. Rossner and P. K. Hopke, "Exposures to carbon monoxide from off-gassing of bulk stored wood pellets," *Energy & Fuels*, vol. 29, no. 1, pp. 218–26, 2014.

[7] F. Meier, I. Sedlmayer, W. Emhofer, E. Wopienka, C. Schmidl, W. Haslinger and H. Hofbauer, "Influence of oxygen availability on off-gassing rates of emissions from stored wood pellets," *Energy & Fuels*, vol. 30, no. 2, pp. 1006–12, 2016.

[8] "Aktindsigt – afsluttet straffesag vedr. arbejdsulykke på skibet Amirante. Journr. 2200–10170-00040-15 Bornholms Politi, Rønne. 2015 (Request for information through the Access to Public Administration Files Act)".

[9] S. Gauthier, H. Grass, M. Lory, T. Krämer, M. Thali and C. Bartsch, "Lethal carbon monoxide poisoning in wood pellet storerooms—two cases and a review of the literature," *Annals of Occupational Hygiene*, vol. 56, no. 7, pp. 755–63, 2012.

[10] U. Svedberg and A. Knutsson, "Hazards and health risk with storage and transport of wood pellets, wood shavings and timber in enclosed spaces. Kunskapsöversikt. Rapport 2011:2 (in Swedish)," Arbets-och miljömedicinska kliniken, Sundsvalls Sjukhus, Sundsvall, Sweden, 2011.

[11] W. Stelte, "Guideline: storage and handling of wood pellets. Resultat Kontrakt (RK) report," Danish technological institute, Energy and climate, Centre for renewable energy and transport, section for biomass, Taastrup, Denmark, 2012.

[12] B. K. Sovacool, R. Andersen, S. Sorensen, K. Sorensen, V. Tienda, A. Vainorius, O. M. Schirach and F. Bjørn-Thygesen, "Balancing safety with sustainability: assessing the risk

of accidents for modern low-carbon energy systems," *Journal of Cleaner Production*, vol. 112, pp. 3952–65, 2016.

[13] Carbon monoxide in bow thruster room," Dutch Safety Board (www.safetyboard.nl), The Hague, The Netherlands, 2015.

[14] "Mystery on ship at Fredericia: One dead and four admitted to hospital by invisible killer," BT, 29 4 2015. [Online]. Available: http://www.bt.dk/danmark/stor-mystik-paa-skib-vedfredericia-en-omkom-og-fire-blev-indlagt-af-usynlig-draeb [Accessed 29 4 2015]

[15] "Officer killed by cargo's fumes". *Telegraph*, vol. 48, no. 12), p. 12, Dec 2015 (Published by Nautilus International, London).

[16] "Very serious marine casualty. M/V Corina. Death of seaman and carbon monoxide poisoning of four other persons during port call in Hanstholm on 28 April 2015. Final report—WIM 13/1.5," Polish State Maritime Accident Investigation Commission, Warsaw, Poland, 2016.

[17] "Aktindsigt vedrørende ulykke på skibet Corina i Hanstholm, journr 4100-10170-00541-16, Midt- og Vestjyllands Politi, Holstebro, 2016. (Request for information through the Access to Public Administration Files Act).".

[18] "Aktindsigt – dødsulykke på skib, Corina, Hanstholm Havn. Sag 20160049016/2. Tilsynscenter Nord, Arbejdstilsynet, København, 2016. (Request for information through the Access to Public Administration Files Act)".

[19] S. Melin, "Research on off-gassing" and self-heating in wood pellets during bulk storage. Report prepared for Ethanol BC," Wood Pellet Association of Canada, 2011.

[20] J. Koppejan, A. Lönnermark, H. Persson, I. Larsson, P. Blomqvist, M. Arshadi, E. Valencia-Reyes, S. Melin, P. Howes, P. Wheeler and D. Baxter, "Health and safety aspects of solid biomass storage, transportation and feeding," *IEA Bioenergy*, 2013.

[21] "IMO Revised Recommendations for Entering Enclosed Spaces Aboard Ships. IMO Resolution A.1050(27).," The International Maritime Organization (IMO), 2012.

[22] J. Rasmussen, A. M. Pejtersen and L. P. Goodstein, *Cognitive Systems Engineering*, New York, USA: Wiley, 1994.

[23] J. Rasmussen, P. Nixon and F. Warner, "Human error and the problem of causality in analysis of accidents," *Philosophical Transactions of the Royal Society, Series B: Biological Sciences.*, vol. 327, no. 1241, pp. 449–62, 1990.

[24] D. A. Schön, The Reflective Practitioner: How Professionals Think in Action, Aldershot, England: Arena, 1995 edition.

[25] N. A. Ashford, "Industrial safety: the neglected issue in industrial ecology," *Journal of Cleaner Production, 5(1), 115–121.*, vol. 5, no. 1, pp. 115–21, 1997.

[26] S. S. Rivera, R. D. C. Olivares, P. A. Baziuk and J. E. N. McLeod, "Assessment of Biofuel Accident Risk: A Preliminary Study," in *Proceedings of the World Congress on Engineering*, London, U.K., July 1–3, 2015.

[27] M. T. Hansen, "The Danish market for wood pellets 2014 (Tasknummer: 115-25037)," FORCE Technology for Energistyrelsen, Kongens Lyngby, Danmark, 2016

[28] S. Melin, U. Svedberg and J. Samuelsson, "Emissions from wood pellets during ocean transportation (EWDOT). Research report.," Wood Pellet Association of Canada (lead), 2008.

3

Microalgal Biomass: A Biorefinery Approach

Luis C. Fernández Linares,

Kevin Á. González Falfán and Citlally Ramírez-López

Abstract

The biorefinery concept has been identified as the most promising way to create a biomass-based industry, which can be defined as the sustainable biomass processing to obtain energy biofuels and high-value products through processes and equipment for biomass. Microalgae can be used as an efficient and economically viable biorefinery feedstock; microalgae could be used in different areas such as human and animal nutrition, nutraceutical and therapeutic products, fertilizers, plastics, isoprenes and biofuels and also in the treatment of wastewaters and CO_2 capture. Microalgae biomass can be used for biofuel production, such as bioelectricity, methane, biohydrogen, bioethanol and biodiesel. In this chapter, an overview of the factors that affect the production of the microalgal biomass yield and value-added by-products production is presented. Likewise, we present the results of the microalgal culture of *Scenedesmus* sp. SCX2 performed in semicontinuous culture (SCC), in 2000 raceway ponds employing Bold's Basal Medium (BBM), under greenhouse conditions. Over the SCC, we monitored biomass concentration, lipid, protein, pigments and sugar production, light irradiance, culture and greenhouse temperature and nitrate concentration in the medium.

Keywords: biomass, microalgae, *Scenedesmus* sp. SCX2, biorefinery, biofuels

1. Introduction

Microalgae have the ability to convert solar energy into chemical forms though photosynthesis, in which CO_2 is captured from the atmosphere and oxygen is produced during the process, furthermore, microalgae offers the potential to produce high-value compounds such as nutraceuticals and compounds for human consumption, and also low-value products as feed, biofertilizers and bioenergy, as well as wastewater treatment, nitrogen fixing and CO_2

mitigation [1, 2]. Microalgae have been taken into consideration as a feedstock for renewable biofuels production, such as bioelectricity, methane produced by anaerobic digestion of the algal biomass, biohydrogen produced under anaerobic conditions, bioethanol (sugar fermentation) and biodiesel derived from microalgal oil [1, 3–5]. Microalgae offer, compared to other renewable sources, higher growth rates and require less water than terrestrial crops [6]; microalgae have a very short harvesting cycle (1–10 days) compared to other land-based feedstock (which are harvested once or twice a year) [7]; algae cultures do not compete with food or arable lands [8]; their phototrophic growth has a favorable environmental impact, since the CO_2 that is released to the atmosphere during hydrocarbons combustion is recycled by microalgae in photosynthetic processes (100 biomass tons require about 183 CO_2 tons) [9]. Microalgae have higher photosynthetic efficiency than crops. Therefore, microalgae biomass cultivation will help in reducing the CO_2 in the atmosphere at a faster rate than land-based crops. Phototrophic metabolism of microalgae is mainly influenced by light irradiance, culture temperature and nutrients supply. However, in large-scale open systems as raceways, it is practically impossible to control diurnal fluctuations of irradiance and temperature [10, 11]. However, raceways are the most commonly employed technology for microalgae biomass production due to their low cost of construction and operation. Microalgae biomass production for obtaining just biofuels is not economically feasible, for which it is necessary to use high-productivity and economic reactors, as well as the use of economic and available nutrients; likewise, the maximum valorization of biomass, including the carbon fixation and wastewater treatment by integrating the concept of industrial ecology. All previous allows the conceptualization of microalgal biorefinery.

This chapter addresses the microalgal biomass production from a biorefinery point of view and describes the improvements of the microalgal biomass production in open systems raceways (RW) of 2000 L capacity operated in semicontinuous culture under greenhouse conditions.

2. Biorefineries

In a conventional refinery, the raw oil undergoes several processes after which the products are used for transportation, electricity and the further production of more complex and valuable chemicals. On the other hand, the biorefineries allow the utilization of raw materials that may be widely available at low cost and sustainable for the production of high-value products that may be considered in terms of their energetic potential. Within a biorefinery there is a wide variety of configurations and generation of several end products. An appropriate definition of biorefinery is as follows: the sustainable processing of the biomass in a wide commercial spectrum of products and energy [12]. Other definition is sustainable biomass processing to obtain energy biofuels and high-value products through processes and equipment for biomass transformation [13]. The biorefinery concept has been identified as the most promising way to create a biomass-based industry.

A biorefinery can utilize all kinds of biomass such as wood (straw and bark), agricultural crops, organic waste (derived from vegetables and animals), industrial and domestic wastes and aquatic biomass (algae and microalgae from sea and freshwater) [14]. The microalgae present a high potential for the biodiesel production compared to the terrestrial biomass [15]. The impact on the utilization of the soil for the microalgal production is significantly lower than the generated by the agricultural biomass based on corn, canola and switchgrass [16]. The versatility of the microalgal metabolism allows the application in different areas such as human and animal nutrition, nutraceutical and therapeutic products, fertilizers, plastics, isoprenes, biofuels and also in the treatment of wastewaters and CO_2 capture [17, 18]. Algae can be used as an efficient and economically viable biorefinery feedstock. An efficient biorefinery using algae can only be constructed through its integration with other industries [19].

2.1. Microalgae as raw material for a biorefinery implementation

Algae are a polyphyletic group of organisms (including organisms that do not have the same origin, but are multiple, independent evolutionary lines) that comprise both, unicellular and multicellular forms and both, prokaryotes and eukaryotes, which are able to capture light energy through pigments (such as chlorophylls, carotenoids, anthocyanins and phycoerythrinsphycobilins) that are necessary to carry out photosynthesis [20, 21]. The classification of algae has five main branches: chromista (brown algae, golden brown algae and diatoms), red algae, dinoflagellates, euglenids and green algae [21]. Microalgae can be eukaryotic (Chlorophyta, Rhodophyta and Bacillariophyta) or prokaryotic (Cyanophyta) [3]. Microalgae are a large group of microscopic algae that are considered primary producers on a global scale, and are involved in all marine and freshwater ecosystems [15, 22]. Microalgae are the largest autotrophic microorganisms of plant life taxa in the world. The biodiversity of microalgae is enormous and they represent an almost untapped resource. It has been estimated that about 20,000–800,000 species exist of which about 40–50,000 species are described [23, 24]. The biomass produces three major biochemical components by *de novo* synthesis consisting of lipids, proteins, carbohydrates, pigments and others.

The proposal to use microalgae for biofuels production is not new and was first suggested in the 1940s–1950s. In the 1970s, the large-scale cultivation of microalgae for producing sustainable liquid fuels was previously investigated and in the 1990s extensive research was carried out, especially at the Solar Energy Research Institute in Golden, Colorado, USA [3, 4, 9, 25].

Specifically, microalgae production offers the potential for the production of high-value compounds, products for human and animal nutrition, bioactive compounds, fluorescent pigments, stable-isotope biochemicals, drug screening, microalgal recombinant proteins, cosmetics, biofertilizers and wastewater treatment, nitrogen fixing, CO_2 mitigation and bioenergy production [2, 9, 21, 25–28]. In this manner, these microscopic organisms have been taken into consideration as a feedstock for the production of renewable biofuels such as bioelectricity, methane produced by anaerobic digestion of the algal biomass, biohydrogen produced under anaerobic conditions, bioethanol (sugar fermentation) and biodiesel derived from microalgal oil [3, 4, 25, 27, 29, 30]. The environmental factors such as illumination, temperature, the amount of nutrients and the salinity affect the photosynthesis and the productivity of the cellular biomass by influencing the pathway and the activity of the cellular metabolism [31, 32].

2.2. Open pond culture systems

Microalgae can be grown in open ponds or in closed systems named photobioreactors. The use of open systems to produce biofuel feedstock is 2.5-times less expensive than using photobioreactors [33]. For that reason, open systems are currently the preferred option for large-scale bioprocessing. There are many types of open ponds used for microalgae cultivation such as raceway, shallow ponds or circular ponds. Raceway pond is the most common open system due to its potential to produce large quantity of microalgae. In a raceway pond system, the cultures are circulated around a race track by using paddle wheels. This will keep the microalgae suspended in the medium and allow the utilization of CO_2 from the atmosphere; CO_2 can also be injected into the medium. The pond is shallow to allow the light penetration into the pond, which in turn maximizes the light exposure of the microalgae for photosynthesis. Open systems make it possible to reduce production costs, this is mainly due to the greater investment required for closed photobioreactors, which increases the depreciation costs (making up as much as 78% of the total microalgae production cost) and also due to the greater power consumption of closed photobioreactors, which likewise increases the operational costs [34]. By using open reactors both the depreciation and the operating costs are reduced. At the same scale and under the same conditions as those considered for tubular photobioreactors, the microalgae production and the reactor costs are reduced almost 4 and 20 fold, respectively, for raceway reactors [35].

The temperature and light irradiance variations are practically impossible to control in outdoor cultures [10]. Such fluctuations can affect microalgal growth at the point to collapse the whole culture [36]. Furthermore, contamination by predators, protozoa, ciliates and other fast growing heterotrophic bacteria have restricted the commercial production of algae in open systems to only those organisms that can grow under specific conditions such as high pH [37], hyper-saline cultures [26] or where suitable irradiance is available. For all these reasons, a crucial factor for the success in culturing open systems is to choose a microalgal strain with the ability to grow in a selective medium, in outdoor environmental conditions with relatively constant productivities [38, 39].

The operational mode of the open systems is important to be considered. The main cultivation modes are in batch, fed-batch, continuous and semicontinuous culture (SCC). The SCC consists of periodic withdrawals of volume from the culture and the removed volume is reinstated with fresh medium. This culture mode is propitious for long-term outdoor cultures, because the growth is maintained in log phase, it avoids the shading effect caused by high cell density [40] and allows the inoculum and biomass availability during each harvesting cycle [41].

SCC cultures have been studied for wastewater removal [42], CO_2 fixation [43, 44] and to enhance biomass productivity in several microalgae species [40, 45, 46], but little work has been published on long-term pilot-scale open systems concerning its effect on the fatty acids methyl esters (FAMES) content, which are the feedstock for microbial biodiesel production.

2.3. Environmental factors and their effect on the microalgal growth

The microalgal growth depends on the availability and efficient utilization of light. A lack of an adequate illumination exerts a negative effect on the productivity, growth and other parameters such as lipids content. The microalgal may be in the respiration phase in the absence of light, in the limiting phase when the illumination is not enough, in a saturation phase if there is a maximum photosynthetic yield or they may be in an inhibition phase due to the suppression of the photosynthetic activity caused by an excess of light (photoinhibition) [47]. High temperatures and concentrations of dissolved O_2, low CO_2 concentrations, as well as a high intensity in illumination and a high pH in the medium, all promote an increase in the photorespiration [48].

The illumination can be supplied to the microalgal culture either by the sun light or by using artificial light such as fluorescent lamps, LEDs, optic fibers, etc. The sources of artificial light may be modified for a specific wavelength emission that leads to metabolic and physiological changes in the microalgae. For a large scale microalgal culture, the utilization of the sun light is recommended in order to operate at a low cost that allows a positive energetic balance. The utilization of artificial sources of light may be necessary for the production of high-value products, but in general, artificial light at large scale should be avoided [49].

2.3.1. *Temperature*

An adequate temperature is necessary for a microalgal photosynthetic activity in which there are no modifications either in the biochemistry or the physiology of the microalgae. The optimum temperature range at which the microalgal growth is between 20 and 25°C for mesophilic microalgal. The thermophilic strains can be cultured at 40°C and the psychrophilic strains can grow at 17°C. The computational modeling designed to estimate the effect of the temperature on the microalgal growth demonstrates that the temperature control at large scale is even more expensive due to the energetic requirement to control the temperature than the energy that eventually could be produced from the microalgal production (e.g. this is the case in the biodiesel production) [38]. The temperature is one of the most difficult factors to be optimized in the open culture systems due to climatic variations and the seasonal changes. This fact may cause a possible decrease in productivity [50].

2.3.2. *Nutrients*

A culture medium for microalgal culturing must contain all of the following: a carbon source, which makes up to 50% of the microalgal biomass. Nitrogen (N), which corresponds to approximately 5–10% dry weight basis of the biomass. Phosphorous (P), which is a component of the DNA, RNA, ATP and the cell membrane. Sulfur (S), which is a component in certain amino acids, vitamins and sulfolipids and is involved in the synthesis of proteins. Potassium (K), which is a cofactor for several enzymes and plays an important role in the osmotic regulation and the synthesis of proteins. Magnesium (Mg), which is found in the core of the chlorophyll molecule andiron (Fe), which is important during the assimilation of nitrogen and is part of the cytochromes. The trace metals, salts and organic components in addition to the other elements mentioned above can be manipulated in order to modify the production of certain desired metabolites [51]. The utilization of wastewater that is rich in N and P as culture medium for microalgae is an economical alternative. Wastewaters can come of sectors such as domestic, commercial-service, industrial and agricultural [52]. One of the advantages that wastewater offers as a culture medium is the possibility of recovery and utilization of inorganic compounds simultaneous to the production of recycling or groundwater recharge water streams [53].

2.3.3. *Salinity*

An excess of sodium ions (Na^+) leads to a saline stress, whereas an osmotic stress may imply the presence or absence of sodium ions. An accumulation of Na^+ and chloride (Cl^-) ions causes an ionic imbalance in the cell. This decreases the capability to absorb minerals such as K, calcium (Ca^{2+}) and manganese (Mn^{2+}) [54]. Under a saline stress, the microalgae present change their metabolism in order to protect themselves from the osmotic damage. The salt concentration that can be tolerated by microalgae, bacteria and cyanobacteria is up to 1.7 M in marine medium [55].

2.4. Selection of the microalgal strain

The selection of the microalgal strain is of upmost importance because each strain has different and specific features. Microalgal strains may be acquired through a specialized collection or they can be isolated from extreme environments. The species isolated from local places present a better adaptation capability and often times can be adapted and grown under adverse and changing conditions and therefore, they tend to survive and grow under stressful conditions even better than the microalgae coming from specialized collections [56]. The biomass of the microalgae can be used in several forms. Biogas can be obtained from the anaerobic digestion of the biomass and the lipids are treated for a transesterification in order to produce biodiesel. The carbohydrates can be used to produce ethanol or butanol, whereas the biomass can be treated thermochemically or by gasification to obtain bio-oil [57]. The decision on the type of microalgae to use depends on its composition and the type of biofuel and other desired valuable products that need to be produced or that are of interest (**Table 1**).

Microalgal species	Mode of cultivation	BR	Biomass productivity (mg L^{-1} d^{-1})	Carbohydrates (%)	Proteins (%)	Lipids (%)	Ref.
Stichococcus sp.	Batch	Erlenmeyer flasks	167.00	40.63	26.25	11.56	[58]
Scenedesmus obliquus	Batch	1-L glass vessel	840.57	46.65	–	22.4	[59]
Scenedesmus obliquus	Semibatch	1-L glass vessel	883.80	52.90	–	–	[60]
Scenedesmus bijugatus	SCC	Vertical tubular	260.00	26.00	–	24	[61]
Chlorella vulgaris	Semibatch	Vertical tubular	268.10	–	52.2	–	[62]

Microalgal species	Mode of cultivation	BR	Biomass productivity (mg L^{-1} d^{-1})	Carbohydrates (%)	Proteins (%)	Lipids (%)	Ref.
Chlorella sp.	Batch	Vertical tubular	154.48	–	–	21.27	[63]
Chlorella ellipsoidea	Batch	Bubble column	31.55	–	41.75	–	[64]
Chlorella variabilis	Batch	Raceway	113.33	–	–	10	[65]
Chaetoceros muelleri	–	Fiberglass cylinders	–	12.7–19.3*	46.9–64.4†	22.1–33.2‡	[66]
Graesiella sp.	SCC	Raceway	36.00	–	–	33.4	[39]

*Winter, 18–19.3; spring, 15.2–15; summer, 12.7–13.6%.
†Winter, 49.3–46.9; spring, 54.2–56.3; summer, 59.2–64.4%.
‡Winter, 30.3–33.2; spring, 29.5–28.2; summer, 24, 22.1%.

Table 1. Microalgal strains for the production of biofuels and high-value products.

In this work, the marine microalgae *Scenedesmus* sp. SCX2 was grown in 2000 L raceway ponds under greenhouse conditions and was used to study the effect of semicontinuous culture mode during summer and autumn seasons on biomass, protein, pigments, sugar and lipid production.

3. Semicontinuous culture of *Scenedesmus* sp. SCX2 in RW under greenhouse conditions

3.1. Materials and methods

The cultivation system of the microalgae *Scenedesmus* sp. SCX2 was performed in semicontinuous culture in 2000 L raceway ponds employing Bold's basal medium (BBM), for 123 days (eight harvesting cycles) during summer and autumn seasons both under greenhouse conditions. Over the SCC, we monitored biomass concentration, lipid, protein, pigments and sugar production, light irradiance, culture temperature, greenhouse temperature and nitrate and ammonium concentrations in the medium as described in the following sections.

3.1.1. *Microalgal strain and growth conditions*

Scenedesmus sp. SCX2 strain was acquired from the Centro de Investigación Científica y de Educación Superior de Ensenada (CICESE). Prior to use it, the strain was grown autotrophically and axenically in BBM medium in a 1 L photobioreactor (PBR) with a working volume of 0.9 L, and bubbled with air at a constant flow rate of 0.5 vvm and then in 20 L PBR. The PBR was illuminated with a light intensity of 180 μmol photons m^{-2} s^{-1} on the surface of the reactor by using cool white fluorescent tubes, with a photoperiod of 12 h of light:12 h of darkness and the temperature was maintained at 24 ± 1°C.

3.1.2. Raceway pond cultures

The open channel raceway reactors dimensions are: two 3.1 m channels, each of 0.72 m wide and connected by 180° bends at both ends to give a total surface area of 6.1 m^2. The raceways were constructed with fiberglass and the culture was circulated and mixed by a two paddle wheel system at a superficial flow velocity of 0.3 ms^{-1}. The operation volume was 1400 L, with a culture medium depth of 0.24 m. The cultures were conducted under greenhouse conditions. The light irradiance and greenhouse temperature were recorded with a data logger T&D RTR-500; the culture temperature was recorded with a data logger T&D RT-200. Additionally, temperature, pH, turbidity, NO_3^-, NH_4^+, conductivity and dissolved oxygen were continuously determined with HYDROLAB DS 5 (Hach) probe. The harvesting cycles were carried out on day 14 or 16, and to do so, half of the culture volume was removed and replaced with the same volume of fresh medium. Water losses due to evaporation were quantified and replaced periodically. The experiments were conducted under greenhouse conditions in Mexico City, Mexico (19.513°N, 99.126°W).

3.2. Analytical methods

3.2.1. Biomass determination

Biomass was determined by absorbance at 600 nm and dry biomass gravimetrically with a thermobalance. The sample was filtered through a glass microfiber membrane (Ahlstrom, 4.7 cm diameter, 0.7 μm pore size) to constant weight.

3.2.2. Determination of nitrate concentration

The nitrate concentration in the culture medium was determined as in Ref. [67] by using 0.5 mL of culture supernatant obtained after centrifugation of a microalgal culture sample at 3823×*g* for 10 min, then this supernatant was placed in a test tube and dried at 100°C for 12 h in an oven Riossa H62. A 0.5 mL portion of diphenyl sulfonic acid was added to the dry sample followed by the addition in four pulses of 2.2 mL of 12N KOH during constant mixing. A mix of 100 μL of supernatant with 900 μL of distilled water was read at 410 nm in a spectrophotometer GENESYS 10S UV-vis (Thermo Scientific, USA). A standard curve with the culture medium was prepared within a range of 0–500 mg L^{-1} of NO_3^- as $NaNO_3$, to obtain the nitrate concentration.

3.2.3. Quantitative lipid determination and FAMEs determination

Cell disruption, lipid extraction, quantitative lipid and lipid composition, determined as fatty acid methyl esters (FAMEs) by gas chromatography after direct lipid transesterification, were realized as in Ref. [68].

3.2.4. Pigment content

Carotenoids and chlorophylls were measured according to the protocols and the equations obtained by Ref. [69]. Culture samples of 1.5 mL were centrifuged for 5 min at 17,000×*g*, after which the supernatant was removed. The cells were suspended in 1.5 mL of methanol and incubated at 45°C in the dark during 30 min. The tubes were centrifuged (17,000×*g*/5 min) and the methanol extract was transferred into plastic cuvettes for measurement at 470, 653, 666 and 750 nm. Carotenoids and chlorophyll A (Chl*a*) and B (Chl*b*) concentrations were estimated according to the following equations (Eqs. 1–4):

$$\text{chl}a\left[\mu\text{g mL}^{-1}\right] = 15.65\left(\text{ABS}_{666}\right) - 7.34\left(\text{ABS}_{653}\right), \tag{1}$$

$$\text{chl}b\left[\mu\text{g mL}^{-1}\right] = 27.05\left(\text{ABS}_{653}\right) - 11.21\left(\text{ABS}_{666}\right), \tag{2}$$

$$\text{Total carotenoids}\left[\mu\text{g mL}^{-1}\right] = \frac{1000\left(\text{ABS}_{470}\right) - 2.86\left(\text{chl}a\right) - 129.2\left(\text{chl}b\right)}{221}, \tag{3}$$

$$\text{Total pigments}\left[\mu\text{g mL}^{-1}\right] = \text{chl}a + \text{chl}b + \text{total carotenoids}. \tag{4}$$

3.2.5. Total sugars

A sample of 5 mg of dried algae was added to 1 mL of 2 N HCl. Hydrolysis was achieved by heating this mixture at 100°C for 120 min. Then total sugars were quantified by the Dubois method [70].

3.2.6. Proteins

A sample of 5 mg of dried algae was added to 1 mL of 1 N NaOH. Hydrolysis was achieved by heating this mixture at 100°C for 120 min. Then proteins were quantified by the Bradford method [71].

All the samples were analyzed in triplicate. The average productivity referenced in [72] was calculated for biomass, lipid, total pigment and sugar and protein content, at the end of every cycle of culture. Productivity is defined as the milligram of product per liter of media per day.

3.3. Results and discussion

The aim of implementing an autotrophic microalgal culture in outdoor conditions was to take advantage of solar energy to convert CO_2 into organic compounds through photosynthesis and then to produce biomass, lipids, proteins, pigments and sugars. The *Scenedesmus* sp. SCX2 culture was performed using BBM medium, under semicontinuous mode, into a raceway reactor of 2000 L. The first culture cycle was considered as an adaptation period. The major growth was observed in the second and third cycles; then the growth was decreasing gradually in subsequent cycles (**Figure 1**). This was certainly due to the weather conditions; in our research group we have observed an increase in biomass with the sequenced cycles [73]. In the second and third cycles, the volumetric biomass productivities of 53 and 5 mg L^{-1} d^{-1} were obtained, respectively, without significant differences between them. These were 15 and 11% higher than volumetric biomass productivity achieved in cycle 1 (**Figure 2**) and up to 50% higher than volumetric biomass productivity obtained in the last two cycles. Ref. [36] reported a volumetric biomass productivities of 30–260 mg L^{-1} d^{-1} with *Scenedesmus* sp. and all productivities obtained in this work are within this range. Areal biomass productivities obtained at the end of the second and third cycles were 12 and 11.5 g m^{-2} d^{-1}, respectively, which were 40% higher than that reported for *Chlorophyta* sp. and similar to the areal biomass productivity obtained with *Chlorella* sp. [74], both microalgae grown in semicontinuous mode in raceways (0.986 m^2), under greenhouse conditions. In addition, higher productivities of this work were half of that obtained by Ref. [35] with *Scenedesmus* sp. cultivated in semicontinuous mode in a raceway of 32 m^2 and CO_2-enriched aeration.

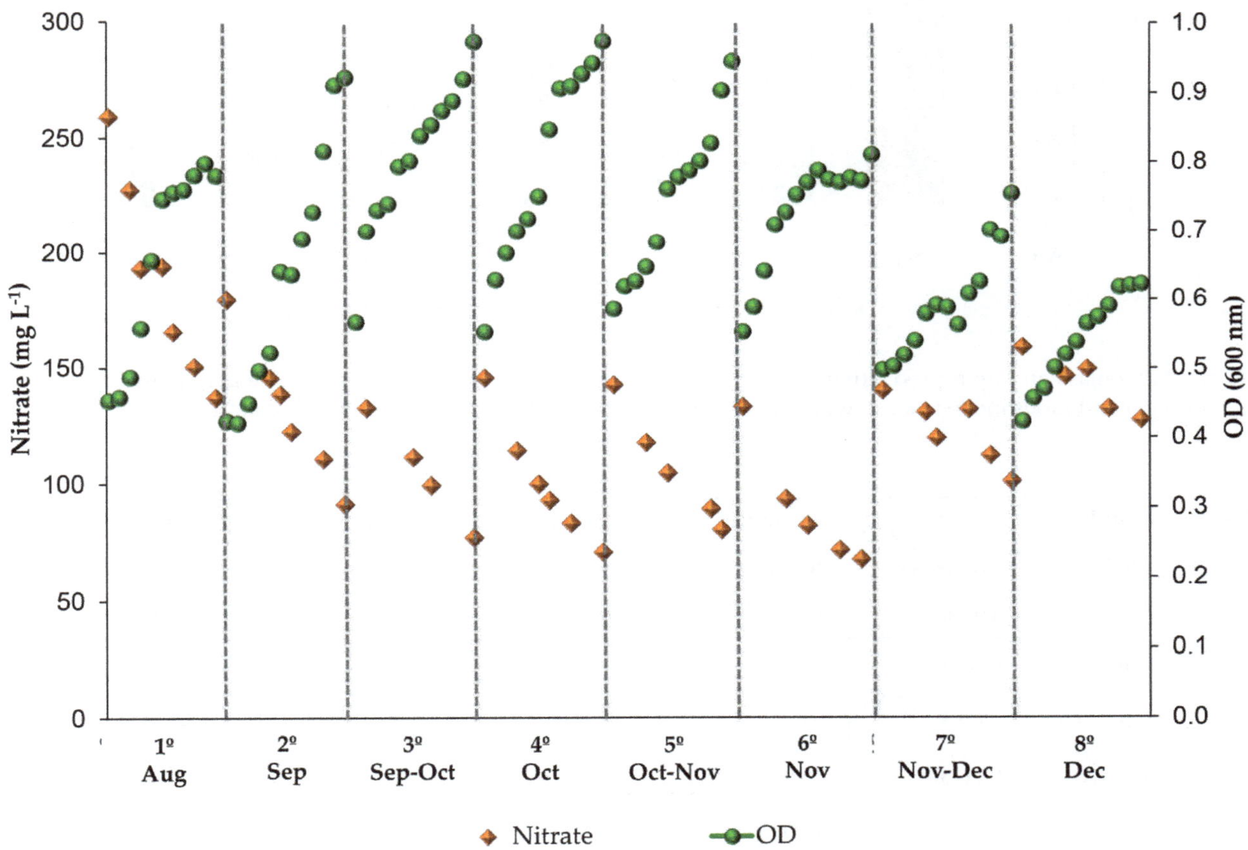

Figure 1. *Scenedesmus* sp. SCX2 growth and nitrate consumption during the eight cycles of culturing under semicontinuous mode in a raceway of 2 m^3 with 1.4 m^3 of medium, under greenhouse conditions.

The first two cycles were developed in the late summer while the other cycles were developed in the autumn, which explains the gradual decrease in both irradiance culture temperatures (**Table 2**); which decreased the biomass productivities (**Figure 2**). Temperatures of the culture of this work were within the range reported by Ref. [75] for the culturing of *Scenedesmus* sp. between 10 and 30°C.

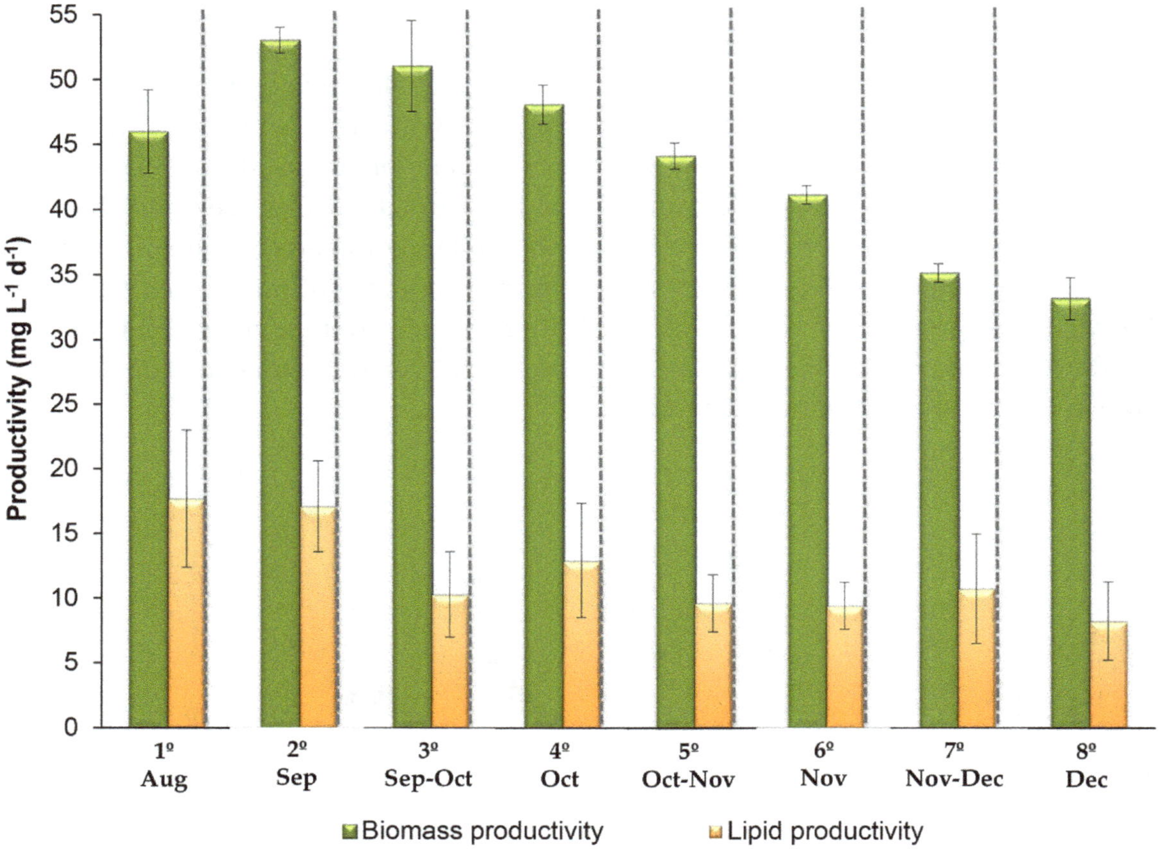

Figure 2. Biomass and lipid productivities from *Scenedesmus* sp. SCX2 cultivated during the eight cycles of culturing under semicontinuous mode in a raceway of 2 m^3 with 1.4 m^3 of medium, under greenhouse conditions.

Cycle	Month (2015)	Maximal irradiance ($\mu E\ m^{-2}\ s^{-1}$)	Average temperature (°C)	Protein productivity ($mg\ L^{-1}\ d^{-1}$)	Sugar productivity ($mg\ L^{-1}\ d^{-1}$)	Pigment productivity ($mg\ L^{-1}\ d^{-1}$)
1	Aug	1251.1	22.0	2.5 ± 0.02	19.2 ± 2.3	1.1 ± 0.07
2	Sep	1279.3	24.7	3.0 ± 0.01	14.8 ± 2.5	1.4 ± 0.11
3	Sep–Oct	1076.3	18.9	2.9 ± 0.01	11.9 ± 1.0	1.3 ± 0.17
4	Oct	1053.5	18.9	2.4 ± 0.02	10.8 ± 0.7	0.8 ± 0.00
5	Oct–Nov	1090.4	18.5	1.1 ± 0.02	12.2 ± 3.5	0.8 ± 0.03
6	Nov	1066.8	18.5	0.8 ± 0.02	9.0 ± 0.9	0.5 ± 0.07
7	Nov–Dec	894.3	17.0	0.8 ± 0.01	11.3 ± 2.0	0.5 ± 0.03
8	Dec	893.2	15.6	0.8 ± 0.01	5.6 ± 0.3	0.4 ± 0.01

Table 2. Irradiances, temperatures and productivities corresponding to the eight cycles of *Scenedesmus* sp. SCX2 culture in the semicontinuous mode.

Lipid production is very important, because lipids are the raw material for biodiesel synthesis by the transesterification reaction. Higher lipid productivities were obtained in the first and second cycles with 17.7 and 17.1 mg L^{-1} d^{-1} (**Figure 2**). The maximum lipid content in the biomass was obtained in the first cycle with 38.4%, which was threefold higher than lipid content obtained by *Chlorella variabilis* grown in a raceway with working volume 450 L [65] and 30% higher than that reported by [76], who cultivated *Scenedesmus* sp. in Erlenmeyer flasks with an operation volume of 800 mL and irradiance of 400 μE m^{-2} s^{-1} at 23°C, using culture media supplemented with nitrogen and phosphate sources, and even vitamins, in different concentrations. In the other culture cycles, the lipid productivities were lower without significant differences among them, with an average lipid productivity of 10.9 mg L^{-1} d^{-1}.

After carbon, nitrogen is the most important nutrient for the microalgae, which is assimilated in the form of NO_3^- or NH_4^+ [77–79]. During the semicontinuous mode, the NO_3^- was always present in the culture medium, and it was not fully assimilated in any cycle (**Figure 1**). Approximately 50% of the initially added NO_3^- for each cycle was consumed in the first six cycles; therefore, none of the cycles reached nitrogen limitation. Nitrate concentration never was less than 68 mg L^{-1} (**Figure 1**). In general, no effect between NO_3^- uptake and lipid productivity was observed, insomuch as the NO_3^- uptake was practically the same in each cycle and the lipid productivity decreased as the semicontinuous culture was progressing. The most significant change was in the last two cycles, when the NO_3^- uptake by *Scenedesmus* sp. was only 44% (**Figure 1**) and the lower productivities of biomass, lipids (**Figure 2**), proteins, total pigments and sugars were obtained (**Table 2**).

The pH of the culture is influenced by several factors such as algal productivity, photosynthesis and ionic composition of the culture medium [79, 80]. The pH is one of the most important factors in the microalgal culture because it determines the solubility and availability of CO_2 and essential nutrients, which has a significant impact on the microalgal metabolism [81]. Generally, a proportional relationship between biomass and pH was observed. In the first three cycles, a pH range of 9.1–11.3 was obtained, which is an advantage insomuch as with high pH values. Only few microorganisms can grow in these pH ranges and a non-contaminated microalgal culture could be guaranteed; however, with pH values above 8.5 the availability of inorganic carbon (CO_2 and HCO_3^-) is limited and the growth of microalgae is affected [82]. In this work, higher pH values (around 11) were achieved in the first two cycles; nevertheless, the pH had no negative effect on microalgal growth. From the third cycle forward, the pH oscillated in a range of 8.7–10.5, which allowed a greater availability of CO_2; however, it was not reflected in the biomass increase.

In regard to the protein content, it was proportional to the biomass content. The higher protein productivities were obtained in the first four cycles; the maximum productivity was achieved in the second cycle with 3.0 mg L^{-1} d^{-1}. From the fifth cycle forward, the protein productivity decreased 61% and in the last two cycles it decreased until 75% from the productivity reached in the second cycle (**Table 2**). It was report that *Scenedesmus dimorphus* has a protein content from 8 to 18% (dry weight basis) [83], and in this work the highest protein content was 5.6%, this was achieved at the end of the second cycle. Ref. [84] obtained six and four times more protein with *Scenedesmus* sp., respectively, that protein content achieved in this work, at the end of the second cycle. *Scenedesmus* sp. was cultured in asbestos tanks, with an operation volume of 150 L and natural lighting, using fertilizer (Nitrofoska Foliar) and residual water fish.

The maximum total sugar productivity (19.2 mg L^{-1} d^{-1}) was obtained in the first cycle (**Table 2**). Total sugar productivities of the other cycles were declining gradually over time. It has been reported that *Scenedesmus* sp. can present a total sugar content of 50% (dry weight basis) [85], similar to what it has been achieved in this work (41.7%). Total pigments presented a similar behavior to the microalgal biomass, insomuch as the first three cycles presented higher productivities. The highest total pigments productivity was achieved in the second cycle (1.4 mg L^{-1} d^{-1}). For the fourth cycle, the total pigment content decreased by 40% and continued to decrease during the subsequent cycles to reach a productivity of 0.4 mg L^{-1} d^{-1} in the eighth cycle (**Table 2**).

The most abundant FAMES were palmitic acid (C16:0), stearic acid (C18:0) and palmitoleic acid (C16:1 Δ9). In several studies C18:0 is usually found in low concentration in microalgae and C16:0, C18:1 and C18:3 are the major fatty acids [86–88].

3.4. Conclusions

Higher productivities of biomass, lipids, proteins, total sugars and pigments were obtained in the first two and three cycles of cultivation, which covered a period of culturing between late summer and early autumn, when the irradiation and temperatures stimulated the *Scenedesmus* sp. SCX2 growth. A *Scenedesmus* sp. SCX2 culture was maintained under semicontinuous mode during 123 days (eight cycles) without NO_3^- limitation, with an average volumetric biomass productivity of 44 mg L^{-1} d^{-1}. The maximum content of protein, sugars, lipids and pigments was 5.6, 41.7, 38.4 and 2.4%, respectively. The lipid productivity decreased after the third cycle by almost 42%. The lowest lipid productivity was obtained in the eighth cycle and decreased by 53% compared with the highest lipid productivity. The total sugar productivity varied throughout of the experiment and decreased by 70% in the eighth cycle, compared with the highest productivity obtained in the first cycle. The maximum productivities of lipids and total sugars were obtained in the first cycle at 22°C and an irradiance of 1251.1 μE m^{-2} s^{-1}. The frequency of renewal of the semicontinuous culture can be reduced to increase the productivities and diminishing the contamination of the culture due to other microorganisms.

Acknowledgements

The authors are grateful for the financial support provided by National Science and Technology Council (CONACYT-Mexico) (Grant 247402), the Secretary for Research and Posgraduated studies—Instituto Politécnico Nacional (SIP-IPN, Grants 20151343, 20152207, and 20161677). The authors would also like to thank Claudia Guerrero Barajas for her participation in reviewing the paper.

Author details

Luis C. Fernández Linares*, Kevin Á. González Falfán and Citlally Ramírez-López

*Address all correspondence to: lfernand36@gmail.com

Professional Interdisciplinary Biotechnology Unit of the National Polytechnic Institute, Mexico City, Mexico

References

[1] Hadj-Romdhane F, Jaouen P, Pruvost J, Grizeau D, Van Vooren G, Bourseau P. Development and validation of a minimal growth medium for recycling Chlorella vulgaris culture. Bioresour Technol. 2012 [cited 2016 Jul 14];123. Available from: http://www.ncbi.nlm.nih.gov/pubmed/22940343

[2] Mendoza JL, Granados MR, de Godos I, Acién FG, Molina E, Heaven S, et al. Oxygen transfer and evolution in microalgal culture in open raceways. Bioresour Technol. 2013;137:188–195.

[3] Aguirre A-M, Bassi A, Saxena P. Engineering challenges in biodiesel production from microalgae. Crit Rev Biotechnol [Internet]. 2012 [cited 2016 Apr 11];8551:1–16. Available from: http://www.ncbi.nlm.nih.gov/pubmed/22804334

[4] Fon Sing S, Isdepsky A, Borowitzka MA, Reza Moheimani N. Production of biofuels from microalgae. Mitig Adapt Strategies Glob Chang. 2013 [cited 2016 Apr 11];18(1): 47–72. Available from: http://link.springer.com/10.1007/s11027-011-9294-x

[5] Ramanna L, Guldhe A, Rawat I, Bux F. The optimization of biomass and lipid yields of Chlorella sorokiniana when using wastewater supplemented with different nitrogen sources. Bioresour Technol. 2014;168:127–135.

[6] Balat M. Potential alternatives to edible oils for biodiesel production—A review of current work. Energy Convers Manag. 2011; 52, 1479–1492.

[7] Harun R, Singh M, Forde GM, Danquah MK. Bioprocess engineering of microalgae to produce a variety of consumer products. Renew Sustain Energy Rev. 2010;14(3):1037–1047.

[8] Demirbas A, Fatih Demirbas M. Importance of algae oil as a source of biodiesel. Energy Convers Manag [Internet]. 2011 [cited 2016 Jul 18];52(1):163–170. Available from: http://dx.doi.org/10.1016/j.enconman.2010.06.055

[9] Yusuf C, Chisti Y. Biodiesel from microalgae. Biotechnol Adv [Internet]. 2007 [cited 2016 Apr 11];25(3):294–306. Available from: http://www.ncbi.nlm.nih.gov/pubmed/17350212

[10] Borowitzka MA, Moheimani NR. Open pond culture systems. In: Algae for Biofuels and Energy, (Ed). Borowitzka MA, Moheimani NR. Dordrecht: Springer Netherlands; 2013. 133–152. Available from: http://link.springer.com/10.1007/978-94-007-5479-9_8

[11] Molina Grima E, Acién Fernández FG, Camacho FG, Chisti Y. Photobioreactors: light regime, mass transfer, and scaleup. J Biotechnol 1999;70:231–247.

[12] International Energy Agency. Biorefinery Concept [Internet]. 2016 [cited 2016 Apr 4]. Available from: http://www.iea.org/topics/renewables/

[13] González-Delgado Á-D, Kafarov V. Microalgae based biorefinery: issues to consider. CT&F. 2011;4(4):5–22.

[14] IEA, Bioenergy Secretariat. Biorefineries: adding value to the sustainable utilisation of biomass. 2009. 1–16. Available from: http://www.ieabioenergy.com/wp-content/uploads/2013/10/Task-42-Booklet.pdf

[15] Posten C, Schaub G. Microalgae and terrestrial biomass as source for fuels: a process view. J Biotechnol. 2009;142(1):64–69.

[16] Clarens AF, Resurreccion EP, White MA, Colosi LM. Environmental life cycle comparison of algae to other bioenergy feedstocks. Environ Sci Technol [Internet]. 2010 [cited 2016 Apr 6];44(5):1813–1819. Available from: http://pubs.acs.org/doi/abs/10.1021/es902838n

[17] Gouveia L. From tiny microalgae to huge biorefineries. Oceanography [Internet]. 2014;2(1):2–9. Available from: http://www.esciencecentral.org/journals/from-tiny-microalgae-to-huge-biorefineries-2332-2632-2-120.php?aid=24524

[18] Milano J, Ong HC, Masjuki HH, Chong WT, Lam MK, Loh PK, et al. Microalgae biofuels as an alternative to fossil fuel for power generation. Renew Sustain Energy Rev 2016;58:180–197.

[19] Trivedi J, Aila M, Bangwal DP, Kaul S, Garg MO. Algae based biorefinery—How to make sense? Renew Sustain Energy Rev. 2015;47:295–307.

[20] Barsanti L, Gualteri P. Algae: Anatomy, Biochemistry and Biotechnology. CRC Press; Taylor & Francis Group, Boca Raton, FL, USA; 2014.

[21] Salis A, Nicolò M, Guglielmino S, Solinas V. Biodiesel from microalgae. In: Handbook of Hydrocarbon and Lipid Microbiology Protocols. McGenity, Terry J., Timmis, Kenneth N., Nogales, Balbina (Eds). Springer-Verlag; Berlin Germany, 2010; 2827–2839.

[22] Hernandez J-P, de-Bashan LE, Rodriguez DJ, Rodriguez Y, Bashan Y, et al. Growth promotion of the freshwater microalga Chlorella vulgaris by the nitrogen-fixing, plant growth-promoting bacterium Bacillus pumilus from arid zone soils. Eur J Soil Biol [Internet]. 2009 [cited 2016 Apr 11];45(1):88–93. Available from: http://linkinghub.elsevier.com/retrieve/pii/S1164556308001052

[23] Oncel S, Microalgae for a macroenergy world. Renew Sustain Energy Rev. 2013; 26:241–264.

[24] Hu Q, Sommerfeld M, Jarvis E, Ghirardi M, Posewitz M, Seibert M, et al. Microalgal triacylglycerols as feedstocks for biofuel production: perspectives and advances. Plant J [Internet]. 2008 [cited 2016 Apr 17];54(4):621–639. Available from: http://www.ncbi.nlm.nih.gov/pubmed/18476868

[25] Schenk PM, Thomas-Hall SR, Stephens E, Marx UC, Mussgnug JH, Posten C, et al. Second generation biofuels: high-efficiency microalgae for biodiesel production. BioEnergy Res [Internet]. 2008 [cited 2016 Apr 11];1(1):20–43. Available from: http://link.springer.com/10.1007/s12155-008-9008-8

[26] Das P, Aziz SS, Obbard JP. Two phase microalgae growth in the open system for enhanced lipid productivity. Renew Energy. 2011;36(9):2524–2528.

[27] Hadj-Romdhane F, Zheng X, Jaouen P, Pruvost J, Grizeau D, Croué JP, et al. The culture of Chlorella vulgaris in a recycled supernatant: effects on biomass production and medium quality. Bioresour Technol 2013;132:285–292.

[28] Mendoza JLL, Granados MRR, de Godos I, Acién FG, Molina E, Banks C, et al. Fluid-dynamic characterization of real-scale raceway reactors for microalgae production. Biomass and Bioenergy. 2013;54:267–275.

[29] Feng Y, Li C, Zhang D. Lipid production of *Chlorella vulgaris* cultured in artificial wastewater medium. Bioresour Technol. 2011;102(1):101–105.

[30] Mutanda T, Ramesh D, Karthikeyan S, Kumari S, Anandraj A, Bux F. Bioprospecting for hyper-lipid producing microalgal strains for sustainable biofuel production. Bioresour Technol [Internet]. 2011 [cited 2016 Apr 11];102(1):57–70. Available from: http://www.ncbi.nlm.nih.gov/pubmed/20624676

[31] He Q. Microalgae as Platforms for Recombinant Proteins. In: Handbook of Microalgal Culture. Biotechnology and Applied Phycology, A. Richmond (Ed.), Blackwell Science, Oxford, 2004; pp. 472.

[32] Suganya T, Varman M, Masjuki HH, Renganathan S. Macroalgae and microalgae as a potential source for commercial applications along with biofuels production: a biorefinery approach. Renew Sustain Energy Rev 2016;55:909–941.

[33] Richardson JW, Johnson MD, Outlaw JL. Economic comparison of open pond raceways to photo bio-reactors for profitable production of algae for transportation fuels in the Southwest. Algal Res [Internet]. 2012 [cited 2016 Jul 18];1(1):93–100. Available from: http://dx.doi.org/10.1016/j.algal.2012.04.001

[34] Acién FG, Fernández JM, Magán JJ, Molina E. Production cost of a real microalgae production plant and strategies to reduce it. Biotechnol Adv [Internet]. 2012;30(6):1344–1353. Available from: http://dx.doi.org/10.1016/j.biotechadv.2012.02.005

[35] Morales-Amaral M del M, Gómez-Serrano C, Acién FG, Fernández-Sevilla JM, Molina-Grima E. Outdoor production of *Scenedesmus* sp. in thin-layer and raceway reactors using centrate from anaerobic digestion as the sole nutrient source. Algal Res. 2015;12:99–108.

[36] Mata TM, Martins AA, Caetano NS. Microalgae for biodiesel production and other applications: a review. Renew Sustain Energy Rev. 2010;14(1):217–232.

[37] Pawlowski A, Mendoza JL, Guzmán JL, Berenguel M, Acién FG, Dormido S. Selective pH and dissolved oxygen control strategy for a raceway reactor within an event-based approach. Control Eng Pract [Internet]. 2015;44:209–218. Available from: http://www.scopus.com/inward/record.url?eid=2-s2.0-84939644511&partnerID=tZOtx3y1

[38] Ras M, Steyer J-PP, Bernard O. Temperature effect on microalgae: a crucial factor for outdoor production. Rev Environ Sci Biotechnol. 2013;12(2):153–164.

[39] Wen X, Du K, Wang Z, Peng X, Luo L, Tao H, et al. Effective cultivation of microalgae for biofuel production: a pilot-scale evaluation of a novel oleaginous microalga *Graesiella* sp. WBG-1. Biotechnol Biofuels [Internet]. 2016 [cited 2016 Aug 26];9(1):123. Available from: http://www.ncbi.nlm.nih.gov/pubmed/27303444

[40] Ho S-H, Chen C-NN, Lai Y-Y, Lu W-B, Chang J-S. Exploring the high lipid production potential of a thermotolerant microalga using statistical optimization and semi-continuous cultivation. Bioresour Technol [Internet]. 2014 [cited 2016 Jul 18];163:128–135. Available from: http://linkinghub.elsevier.com/retrieve/pii/S0960852414005306

[41] Fábregas J, Patiño M, Arredondo-Vega BO, Tobar JL, Otero A. Renewal rate and nutrient concentration as tools to modify productivity and biochemical composition of cyclostat cultures of the marine microalga Dunaliella tertiolecta. Appl Microbiol Biotechnol [Internet]. 1995 [cited 2016 Jul 18];44(3–4):287–292. Available from: http://link.springer.com/10.1007/BF00169918

[42] Zhou W, Min M, Li Y, Hu B, Ma X, Cheng Y, et al. A hetero-photoautotrophic two-stage cultivation process to improve wastewater nutrient removal and enhance algal lipid accumulation. Bioresour Technol 2012;110:448–455.

[43] Bao P, Huang H, Hu Z-Y, Häggblom MM, Zhu Y-G. Impact of temperature, CO_2 fixation and nitrate reduction on selenium reduction, by a paddy soil Clostridium strain. J Appl Microbiol [Internet]. 2013 [cited 2016 Jul 18];114(3):703–712. Available from: http://doi.wiley.com/10.1111/jam.12084

[44] Ho S-H, Chen C-Y, Chang J-S. Effect of light intensity and nitrogen starvation on CO_2 fixation and lipid/carbohydrate production of an indigenous microalga Scenedesmus obliquus CNW-N. Bioresour Technol 2012;113:244–252.

[45] Chiu SY, Kao CY, Tsai MT, Ong SC, Chen CH, Lin CS. Lipid accumulation and CO_2 utilization of Nannochloropsis oculata in response to CO_2 aeration. Bioresour Technol

[Internet]. 2009;100(2):833–838. Available from: http://dx.doi.org/10.1016/j.biortech.2008.06.061

[46] Radmann EM, Reinehr CO, Costa JAV. Optimization of the repeated batch cultivation of microalga Spirulina platensis in open raceway ponds. Aquaculture. 2007;265(1–4):118–126.

[47] Tebbani S, Titica M, Ifrim G, Caraman S. Control of the light-to-microalgae ratio in a photobioreactor. 2014 18th International Conference on System Theory, Control and Computing (ICSTCC) [Internet]. IEEE; 2014 [cited 2016 Jul 18]. 393–398. Available from: http://ieeexplore.ieee.org/lpdocs/epic03/wrapper.htm?arnumber=6982448

[48] Parsons TR, Maita Y, Lalli CM. A manual of chemical & biological methods for seawater analysis. Oxford: Pergamon Press, 1984; pp.135–139.

[49] Blanken W, Cuaresma M, Wijffels RH, Janssen M. Cultivation of microalgae on artificial light comes at a cost. Algal [Internet]. 2013;2(4):333–340. Available from: http://dx.doi.org/10.1016/j.algal.2013.09.004

[50] Griffiths M. Microalgal Cultivation Reactor Systems. In: Biotechnological Applications of Microalgae: Biodiesel and Value-Added Products. Bux F, (Ed). CRC Press; Taylor & Francis Group, Boca Raton, FL, USA; 2013; pp. 51–76.

[51] Sydney E, Novak A, Cesar de Carvalho C, Soccol C. Chapter – Respirometric Balance and Carbon Fixation of Industrially Important Algae. In: Biofuels from Algae. Pandey A, Lee D, Chisti Y, Soccol SR, (Eds). Elsevier, Burlington, MA, USA; 2014; pp. 67–84.

[52] UN-Water. Wastewater management—A UN-water analytical brief [Internet]. 2012 [cited 2016 Aug 28]. Available from: http://www.unwater.org/fileadmin/user_upload/unwater_new/docs/UN-Water_Analytical_Brief_Wastewater_Management.pdf

[53] Dibenedetto A. Production of aquatic biomass and extraction of Bio-oil. In: Biorefinery: From Biomass to Chemicals and Fuels, Aresta M, Dibenedetto A, Dumeignil F, (Eds). Walter de Gruyter GmbH & Co. Germany; 2015; pp. 81–100.

[54] Sudhir P, Murthy SDS. Effects of salt stress on basic processes of photosynthesis. Photosynthetica [Internet]. 2004 [cited 2016 Apr 11];42(4):481–486. Available from: http://link.springer.com/10.1007/S11099-005-0001-6

[55] Yen H-W, Hu I-C, Chen C-H, Chang J-S. Design of Photobioreactors for Algal Cultivation. In: Biofuels from Algae. Pandey A, Lee D, Chisti Y, Soccol SR, (Eds). Elsevier, Burlington, MA, USA; 2014; pp. 23–46.

[56] Duong VT, Li Y, Nowak E, Schenk PM. Microalgae isolation and selection for prospective biodiesel production. Energies [Internet]. 2012 [cited 2016 Apr 11];5:1835–1849. Available from: www.mdpi.com/journal/energies

[57] Hall P, Gifford J. Bioenergy Options for New Zealand [Internet]. Richardson M, editor. Scion, Energy Group, New Zealand, 2008 [cited 2016 Apr 11]. Available from: https://

www.niwa.co.nz/sites/niwa.co.nz/files/import/attachments/Situation_Analysis_-_Bioenergy_Options.pdf

[58] Karapatsia A, Penloglou G, Chatzidoukas C, Kiparissides C. An experimental investigation of *Stichococcus* sp.cultivation conditions for optimal co-production of carbohydrates, proteins and lipids following a biore fi nery concept. Biomass and Bioenergy [Internet]. 2016 [cited 2016 Apr 9]; Available from: http://dx.doi.org/10.1016/j.biombioe.2016.01.009

[59] Ho S-HH, Chen C-YY, Chang J-SS. Effect of light intensity and nitrogen starvation on CO_2 fixation and lipid/carbohydrate production of an indigenous microalga Scenedesmus obliquus CNW—N. Bioresour Technol [Internet]. 2012 [cited 2016 Jul 18];113:244–252. Available from: http://dx.doi.org/10.1016/j.biortech.2011.11.133

[60] Ho SH, Kondo A, Hasunuma T, Chang JS. Engineering strategies for improving the CO_2 fixation and carbohydrate productivity of Scenedesmus obliquus CNW-N used for bioethanol fermentation. Bioresour Technol [Internet]. 2013;143:163–171. Available from: http://dx.doi.org/10.1016/j.biortech.2013.05.043

[61] Ashokkumar V, Salam Z, Tiwari ON, Chinnasamy S, Mohammed S, Nasir F, et al. An integrated approach for biodiesel and bioethanol production from Scenedesmus bijugatus cultivated in a vertical tubular photobioreactor. Energy Convers Manag [Internet]. 2015;101:778–786. Available from: http://dx.doi.org/10.1016/j.enconman.2015.06.006

[62] Chen C, Chang Y, Chang H. Outdoor cultivation of Chlorella vulgaris FSP-E in vertical tubular-type photobioreactors for microalgal protein production. Algal [Internet]. 2016;13:264–270. Available from: http://dx.doi.org/10.1016/j.algal.2015.12.006

[63] Zhou X, Xia L, Ge H, Zhang D, Hu C. Feasibility of biodiesel production by microalgae *Chlorella* sp. (FACHB-1748) under outdoor conditions. Bioresour Technol [Internet]. 2013;138:131–135. Available from: http://dx.doi.org/10.1016/j.biortech.2013.03.169

[64] Wang SK, Hu YR, Wang F, Stiles AR, Liu CZ. Scale-up cultivation of Chlorella ellipsoidea from indoor to outdoor in bubble column bioreactors. Bioresour Technol 2014;156:117–122.

[65] Bhowmick G De, Subramanian G, Mishra S, Sen R. Raceway pond cultivation of a marine microalga of Indian origin for biomass and lipid production: a case study. Algal [Internet]. 2014;6(PB):201–209. Available from: http://dx.doi.org/10.1016/j.algal.2014.07.005

[66] Voltolina D, Sánchez-Saavedra M del P, Torres-Rodríguez LM. Outdoor mass microalgae production in Bahia Kino, Sonora, NW Mexico. Aquac Eng. 2008;38(2):93–96.

[67] Keeney DR, Nelson DW. Nitrogen-inorganic forms. Black CA, editor. Methods of Soil Analysis Part 2: Agronomy. Madison, Wisconsin, U.S.A.: American Society of Agronomy; 1982. 643–698.

[68] Ramírez-López C, Chairez I, Fernández-Linares L. A novel culture medium designed for the simultaneous enhancement of biomass and lipid production by Chlorella vulgaris UTEX 26. Bioresour Technol. 2016;212:207–216.

[69] Wellburn AR. The espectral determination of chlorophylls, a and b, as well as total carotenoids, using various solvents with spectrophotometers of different resolution. J Plant Physiol. 1994;144:307–313.

[70] DuBois M, Gilles KA, Hamilton JK, Rebers PA, Smith F. Colorimetric method for determination of sugars and related substances. Anal Chem [Internet]. 1956 [cited 2016 Jul 21];28(3):350–356. Available from: http://pubs.acs.org/doi/abs/10.1021/ac60111a017

[71] Bradford MM. A rapid and sensitive method for the quantitation of microgram quantities of protein utilizing the principle of protein-dye binding. Anal Biochem [Internet]. 1976 [cited 2016 Jul 21];72(1–2):248–254. Available from: http://linkinghub.elsevier.com/retrieve/pii/0003269776905273

[72] Griffiths MJ, van Hille RP, Harrison STL. The effect of nitrogen limitation on lipid productivity and cell composition in Chlorella vulgaris. Appl Microbiol Biotechnol [Internet]. 2014 [cited 2016 Aug 27];98(5):2345–2356. Available from: http://link.springer.com/10.1007/s00253-013-5442-4

[73] Millán-Oropeza A, Fernández-Linares L. Biomass and lipid production from Nannochloropsis oculata growth in raceway ponds operated in sequential batch mode under greenhouse conditions. Environ Sci Pollut Res [Internet]. 2016 [cited 2016 Jul 21];1–9. Available from: http://link.springer.com/10.1007/s11356-016-7013-6

[74] Hase R, Oikawa H, Sasao C, Morita M, Watanabe Y. Photosynthetic production of microalgal biomass in a raceway system under greenhouse conditions in Sendai City. J Biosci Bioeng [Internet]. 2000 [cited 2016 Jul 21];89(2):157–163. Available from: http://linkinghub.elsevier.com/retrieve/pii/S1389172300887307

[75] Singh SP, Singh P. Effect of temperature and light on the growth of algae species: a review. Renew Sustain Energy Rev. 2015;50:431–444.

[76] Hakalin NLS, Paz AP, Aranda DAG, Moraes LMP. Enhancement of cell growth and lipid content of a freshwater microalga *Scenedesmus* sp. by optimizing nitrogen, phosphorus and vitamin concentrations for biodiesel production. Nat Sci [Internet]. 2014 [cited 2016 Jul 21];6(12):1044–1054. Available from: http://www.scirp.org/journal/PaperDownload.aspx?DOI=10.4236/ns.2014.612095

[77] Grobbelaar JU. Algal nutrition-mineral nutrition. In: Handbook of Microalgal Culture. Biotechnology and Applied Phycology, A. Richmond (Ed.), Blackwell Science, Oxford, 2004; [cited 2016 Jul 21]; 95–115. Available from: http://doi.wiley.com/10.1002/9780470995280.ch6

[78] Abdel-Raouf N, Al-Homaidan AA, Ibraheem IBM. Microalgae and wastewater treatment. Saudi J Biol Sci. 2012;19(3):257–275.

[79] García LM. Elimination of CO2 with autochthonous microalgae. (Ph D Thesis). University of León, Spain; 2008.

[80] Park JBKBK, Craggs RJJ, Shilton ANN. Wastewater treatment high rate algal ponds for biofuel production. Bioresour Technol [Internet]. 2011 [cited 2016 Jul 21];102(1):35–42. Available from: http://dx.doi.org/10.1016/j.biortech.2010.06.158

[81] Chenl CY, Durbin EG. Effects of pH on the growth and carbon uptake of marine phytoplankton. Mar. Ecol. Prog. Ser. 1994;109: 83–94.

[82] Ugwu CUU, Aoyagi H, Uchiyama H. Photobioreactors for mass cultivation of algae. Bioresour Technol. 2008;99(10):4021–4028.

[83] Balat M, Balat H. Progress in biodiesel processing. Appl Energy [Internet]. 2010 [cited 2016 Jul 21];87(6):1815–1835. Available from: http://dx.doi.org/10.1016/j.apenergy.2010.01.012

[84] Andrade RCE, Vera BAL, Cárdenas LCH, Morales AED. Biomass production of the microalgae *Scenedesmus sp.* Using sewage from fishmongers. Rev Técnica la Fac Ing Univ del Zulia. 2009; 32(2):126–134.

[85] Ueda R, Hirayama S, Sugata K, Nakayama H. Process for the production of ethanol from microalgae. 1996; Patent US 5578472 A. https://www.google.com/patents/US5578472

[86] He Q, Yang H, Wu L, Hu C. Effect of light intensity on physiological changes, carbon allocation and neutral lipid accumulation in oleaginous microalgae. Bioresour Technol 2015;191:219–228.

[87] Toledo-Cervantes A, Morales M, Novelo E, Revah S. Carbon dioxide fixation and lipid storage by Scenedesmus obtusiusculus. Bioresour Technol. 2013;130:652–658.

[88] Liu J, Huang J, Sun Z, Zhong Y, Jiang Y, Chen F. Differential lipid and fatty acid profiles of photoautotrophic and heterotrophic Chlorella zofingiensis: assessment of algal oils for biodiesel production. Bioresour Technol [Internet]. 2011 [cited 2016 Jul 22];102(1): 106–110. Available from: http://dx.doi.org/10.1016/j.biortech.2010.06.017

4

Biomass Valorization: Agricultural Waste in Environmental Protection, Phytomedicine and Biofuel Production

Inyinbor Adejumoke Abosede,
Oluyori Abimbola Peter and
Adelani-Akande Tabitha Adunola

Abstract

Industrialization is a major promoter of any nation's economy; it is not without detrimental effects on our immediate environment. Human exposure to various pollutants discharged into the environment may lead to serious health challenges. In the same vein, discharge from the combustion of fossil fuel contributes a great deal to the environmental pollution. The resulting quest for a clean and sustainable environment has spurred myriads of research into advantageous utilization of waste biomass in industrial wastewater treatment and environmentally friendly/alternative energy. Similarly, diverse waste materials have been adopted as sources of important phytochemicals with different medicinal applications. This chapter focuses on the application of waste biomass in environmental remediation, curative medicine, and clean/alternative energy.

Keywords: waste, biomass, bioadsorption, phytochemicals, bioethanol

1. Introduction

Huge amounts of wastes generated during various agro-processes over time constitutes nuisance to the environment. Agro-wastes are cellulosic materials and depending on their sources, may possess high fixed carbon content, multifunctional group surface, and bioactive agents. These characteristics have been well explored and have led to their applications in environmental remediation, phytomedicine, and biofuel generation [1–3].

Biosorption, a sorption technique that utilizes biomass in sequestering various toxicants, is an attractive technique in wastewater treatment. It adds economic advantages to the operational simplicity and ability to remove very low concentration of toxicants offered by commercial activated carbon. Surface tailoring toward a specific pollutant is another unique advantage of biomass [4]. Multifunctional group surface and cellulosic backbone of biomass present them as suitable materials for pollutants uptake. Modification and surface treatment may enhance the sorption efficiency of raw biomass [5].

Similarly, waste biomasses have been explored for the beneficial phytochemicals, which they contain. These phytochemicals have been useful as antimicrobials, antioxidants, hepatoprotectives, and anticancer agents, which are generally preferred because of their minimal side effects on their consumers [1, 6]. Obtaining these useful phytochemicals from waste biomass is a very useful and timely development since threat to food security usually results whenever useful compounds of medicinal interest are identified in major food sources [7].

In the same vein, the use of agricultural waste in biofuel production has brought about provision of alternative energy and reduction in environmental pollution. The use of renewable resources such as cornstarch leads to competition with human food supply. In view of this, there is a drift in the use of agricultural waste for biofuel production [8]. Agro-wastes are lignocellulosic materials; their rich cellulose content qualifies them as feedstock in bioethanol production.

This chapter is aimed at explicitly depicting the potential of agricultural wastes in environmental remediation, curative medicine, and biofuel production with focus on research efforts, successes, and gaps.

1.1. Effective wastewater treatment: panacea to water pollution challenges

The advent of industrialization in developed economy brought to bear great environmental pollution challenges. The discharge of toxic wastewater and polluting gases increased exponentially with industrialization and economy [9]. The hydrosphere continuously receives wastewater with loads of toxicants. The effects of such discharge on the aquatic organism and aquatic ecosystem and by extension on human are enormous. Wastewater contains different categories of pollutants ranging from organics to inorganics. Organic pollutants such as dyes form a major part of wastewater discharged from dye-utilizing industries such as leather, textile, and paint industries.

Dyes are large organic molecules, which are nonbiodegradable and carcinogenic; they cause skin, eye, and respiratory tract irritations as well as the reduction of dissolved oxygen and the hindrance of sunlight penetration into water bodies [5]. Heavy metals, on the other hand, are inorganic pollutants with serious deleterious effects on plants and human [10, 11]. Effective treatment of wastewater containing the aforementioned toxicants before their discharge into the environment is thus very important.

Conventional methods of wastewater treatment exist, namely electrochemical treatment, ion exchange, precipitation, reverse osmosis, evaporation, solvent extraction, adsorption with activated charcoal, among others. Disadvantages associated with conventional methods of wastewater treatment may be a concern to the developed and the developing economy. For instance, toxic sludge generation in chemical precipitation subsequently results in soil and water pollution. Inability of conventional methods to remove low concentration of pollutants also calls for concern. In addition, cost of installation, operation, and maintenance is a great concern to developing economies. Adsorption using activated charcoal is a way out of some of these earlier mentioned challenges. However, activated charcoal is very expensive.

Naturally occurring low-cost adsorbents/biosorbents, such as clay and agricultural wastes, particularly, crop residues, may effectively address the challenges associated with conventional methods of wastewater treatment. Due to the basic composition of crop residues, they have been widely investigated as a potential tool for effective wastewater treatment.

1.1.1. Crop residues as a tool in environmental remediation: characteristics, uniqueness and versatility

Crop residues are the part of crop production processes not used as food for human. They are not the primary products usually consumed as food. These may come in the form of stalks, straw, leaves, roots, peels, husks, cobs, shells, nuts, and even waste woods. The composition of these crop residues may vary depending on their sources [12]. However, they are characterized by similar features and basic composition.

1.1.1.1. Polyols backbones

The basic components of crop residues are cellulose, hemicellulose, lignin, and some extractives [13]. While cellulose is a crystalline α, β (1–4) linked D-glucose polymer with high molecular weight, hemicellulose is an amorphous, shorter chain polymer of various sugars. These two components (cellulose and hemicellulose) contain numerous hydroxyl groups and thus result in a characteristic polyols backbone in crop residues. The polyols surface presents suitable sites for the uptake of pollutants. The deprotonation of hydroxyl group leaves on the residue surface a number of electron-rich oxygen atoms. These electron-rich species could act as Lewis bases, thus attracting electron-deficient species such as metal ions as depicted in **Figure 1**.

1.1.1.2. Multifunctional group surface

Aside the basic hydroxyl groups present in these crop residues, their surfaces may also carry other functional groups such as carbonyl, carboxyl, methoxyl, sulfhydryl, ether, acetamido, or amino groups. These functional groups, however, may vary depending on the crop residue source, variety, age, and environmental and cultivation parameters [14–16]. Common to these functional groups is the possession of lone pairs with which they chelate (bite) metal ions. Metal and dye sequestering by different parts of the residue can occur via various processes among which are the following:

Complexation: A process in which metal ions form complexes with organic molecules. This organic species must have an atom or atoms having lone pair of electrons to donate.

Chelation: A process in which organic molecules containing more than one functional group with donor electron pairs simultaneously donate these to a metal atom resulting in the formation of a ring structure involving the metal atom. Due to the multiple binding sites, chelates are more stable than complexes.

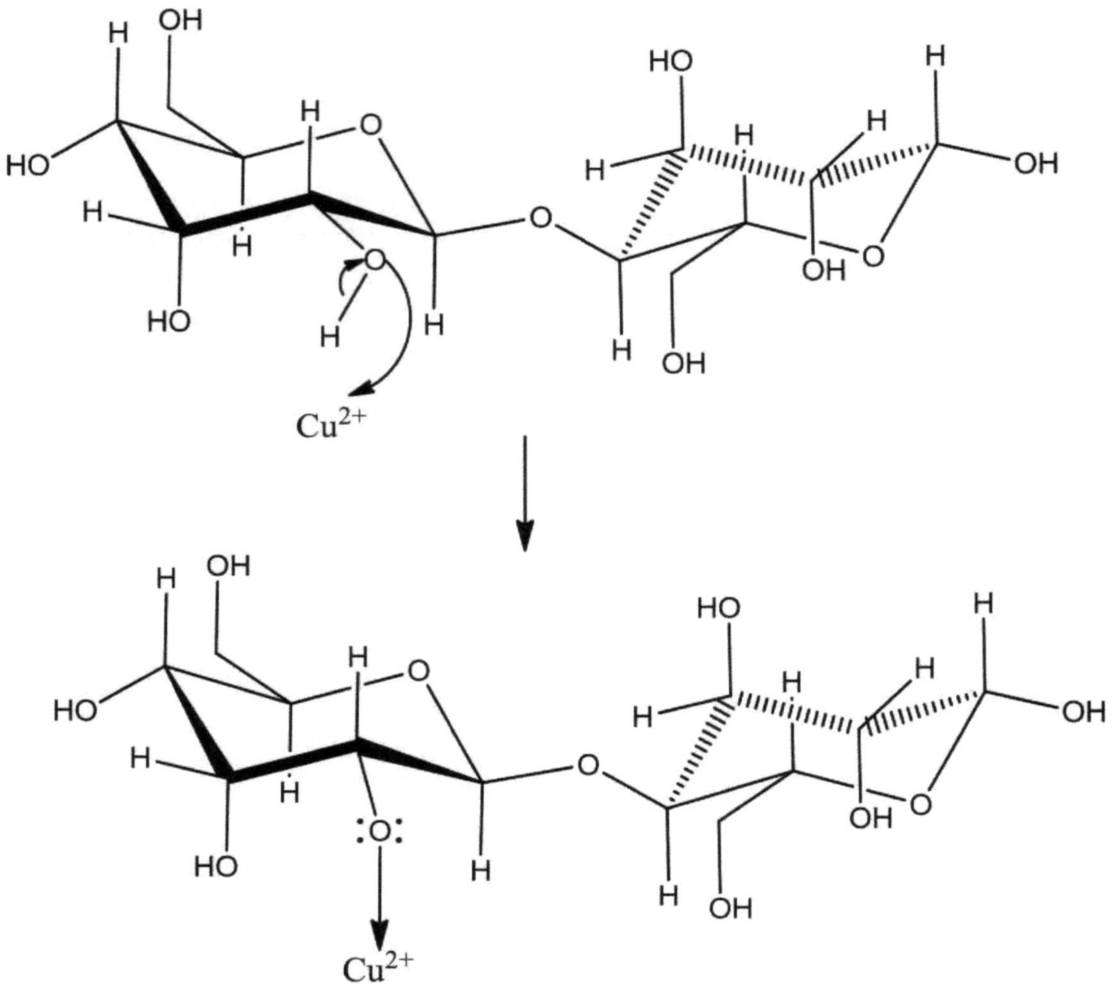

Figure 1. Possible dissociation in cellulosic chain and attraction between Lewis base and metalion. (Source: Authors).

Precipitation: Precipitation is a double decomposition reaction in which toxic metal ions are removed from solutions. This has been used severally in biosorption.

Reduction: Reduction is a process involving electron gain. Removal of toxic hexavalent chromium from aqueous solution by biosorption has been reported, and it is often associated with the simultaneous reduction of Cr (VI) to Cr (III).

Immobilization may be the result of more than one mechanism; for example, metal complexation may be followed by metal reduction or metal precipitation.

The high binding ability of the aforementioned functional groups with a wide range of pollutants via various mechanisms presents crop residues as excellent tools in environmental remediation.

1.1.1.3. Fibrous nature and possible porosity

The characteristics of crop residue are well understood and established by studying their surface morphology and porosity. While some crop residue depicts vivid threadlike surface fibers, pores scarcely appear in some, and some others may possess naturally wide-open pores and hollows. Fibrous character is synonymous with wood waste [17], and a range of water pollutants can be locked up within the crop residue fibers. Pores and hollows also serve as mediums for transporting various types of pollutants into the bioadsorbents. Mesopores (pores with size range between 2 and 50 μm) are more useful in the transport and trapping of large molecules such as dyes into the bioadsorbents [4]. Scanning electron microscopy studies (SEM) of two biomass (*Irvingia gabonensis* nut and *Raphia hookerie* epicarp) as investigated by Inyinbor and colleagues reveal the fibrous nature of *I. gabonensis* biomass, while several moderately large pores were observed in *R. hookerie* biomass (**Figure 2a** and **b**).

1.1.1.4. Carbon content and versatility

Crop residues are generally characterized by high carbon content. Hence, they have found great application in diverse fields. Their use in environmental remediation spans through utilizing raw bioadsorbents, modified biomass, as well as biomass-generated activated carbon for pollutant uptake. The high fixed-carbon content in raw biomass significantly aids pollutant uptake [18].

Pollutants may be chemisorbed via its interactions with surface functional group(s), locked in the fibers, or trapped within the biomass pores. Via one or a combination of these mechanisms, biomasses sequester pollutants from aqueous solutions. Although raw biomasses are effective in pollutant uptake, surface modification, as well as functionalization, increases the sorption capacity. Thermal and acid treatment increase sorbent porosity and subsequently its sorption capacity [4, 5]. Commonly used activating agents are KOH, NaOH, H_3PO_4, H_2SO_4, $ZnCl_2$, among others [19]. Highly porous-activated carbons have been prepared from various crop

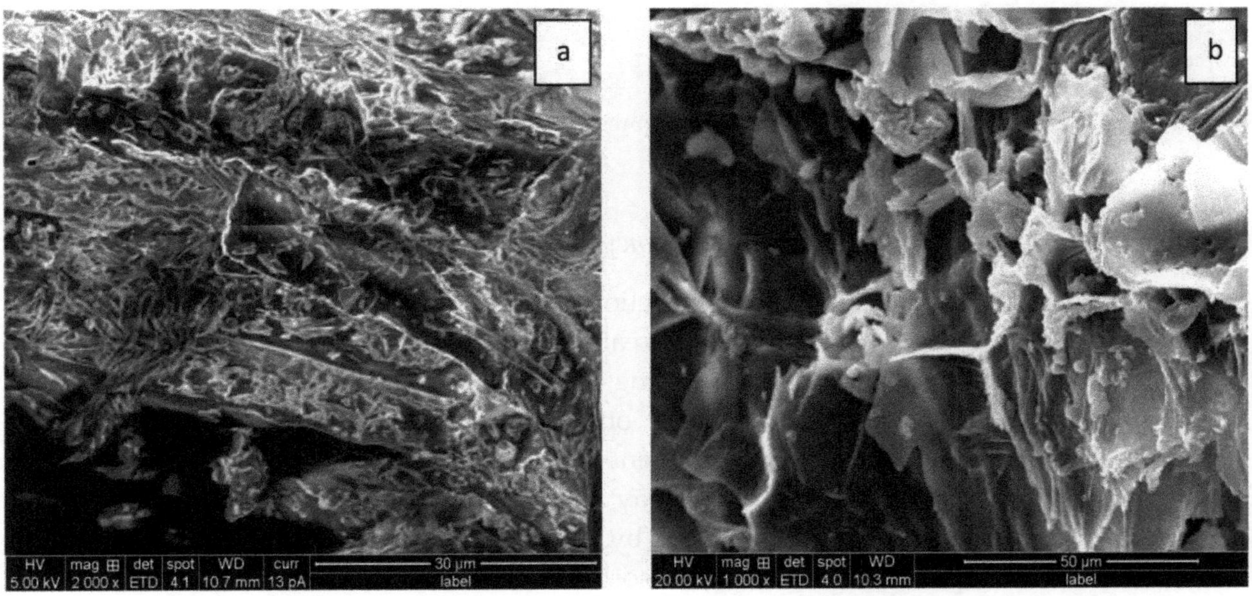

Figure 2. (a) Fibrous surface of *I. gabonensis* bioadsorbents (Source: Authors); (b) moderately large pores on the surface of *R. hookerie* bioadsorbent (Source: Authors).

residues using different activating agents. Prepared porous-activated carbons have been applied to the uptake of dyes and heavy metals and a few of previously reported work is summarized in **Table 1**.

Crop residue	Treatment	Surface area (m^2/g)	Pollutants	q_{max} (mg/g)	References
Corn cob	Thermal/H_3PO_4	600–700	Methylene blue	0.81–28.65	[20]
Palm stem	Thermal/K_2SO_4	—	Methylene blue	149	[21]
Rambutan peel	Microwave induced	971	Acid yellow	215.05	[22]
	KOH activation				
Olive mill	Thermal/KOH	1641	Cr (III)		[23]
Pineapple waste	$ZnCl_2$/Thermal	914.67	Methylene blue	288.34	[24]
Grape waste	$ZnCl_2$/Thermal	1455	Methylene blue	417	[25]
			Metanil yellow	386	
Silkworm cocoon	$(NH_4)_2HPO_4$/KOH	1153–2797	Congo red	512 g/kg	[26]
	Activation				
Durian seed	Thermal/KOH	980.62	Methyl red	384.62	[27]
Tomatoe waste	$ZnCl_2$/Thermal	1093	Methylene blue	400	[28]
			Metanil yellow	385	
Peanut shells	Microwave irradiation	370.10	Direct black	110.6	[29]
	Pyrolysis		Reactive red	284.5	
Potato peels	H_3PO_4/Thermal	1.04	Co^{2+}	373–480	[30]
Guava seed	Carbonization and physical activation	1000	Pb^{2+}	96	[31]
Tropical almond				112	

Table 1. Activated carbon preparation from different crop residue and their applications in dyes and heavy metal uptake.

1.1.2. Pretreatment methods and biosorbent efficiency

Surface treatment, modification, and/or functionalization enhance biosorption efficiency of bioadsorbents. The desired feature can be created in bioadsorbents via designed treatment and/or modification. Biomass treatment using concentrated acid may blow up the fibers subsequently leading to mesopore creation on biomass surface [5]. Mesopores are large enough to transport dyes into the bioadsorbents. The release of adventitious water from the biomass framework via thermal treatment may also result in generation of pores [4]. Texture and porosity of biomass can also be altered by microwave irradiation. Hollows, pores, and cavities serve as transport medium for the adsorbate into bioadsorbents [32].

The possibility of surface modification and/or functionalization gives additional advantage to crop residue. The surface of raw bioadsorbents may be tailored toward specific pollutants via the introduction of specific functional group(s) that has(ve) affinity for the desired pollutant.

Surface modification may involve the grafting of functional group(s) onto the biomass frameworks, which have a high tendency to reduce sorbent porosity and subsequently its surface area [33, 34]. Interestingly, surface modified biomaterial usually generates adsorbents with far superior adsorption capacity. Sections 1.21 and 1.22 of this chapter introduced the possibility of pollutants uptake onto biomass via surface functional groups. Introduction of such functional groups with unique metal/dye binding characteristics enhances biomass sorption capacity. Metal ions can be complexed by thiol, amino, carboxylates, nitrogen, and oxygen-based moieties. Bonding via electrostatic interactions can also occur between metal and unprotonated carboxyl, oxygen, and sulphate groups. **Table 2** compares some reported literature works that reveal increase in sorption potential of modified biomass over their unmodified counterparts.

Crop residue	Pollutants	q_{max} (mg/g)	References
Citrus sinensis	Reactive yellow 42		[35]
Raw		36.36	
Immobolized		51.47	
Chemically treated		106.36	
Citrus sinensis	Reactive red 45		[35]
Raw		18.28	
Immobolized		39.06	
Chemically treated		31.45	
Irvingia gabonensis	Rhodamine B		[5]
Raw		212.77	
Acid treated		232.00	
Lagerstoemia speciosa bark	Cr (VI)		[36]
Raw		20.41	
HNO_3 modified		24.39	
Phytolacca americana L	Pb(II)		[37]
Raw		13.19	
HNO_3 modified		14.51	
Charred dika nut waste	Rhodamine B		[4]
Raw		52.90	
Chitosan coated		217.39	
Eucalyptus sheathiana bark	Zn (II)		[38]
Raw		128.21	
NaOH modified		250.00	

Table 2. Comparing the efficiency of raw and modified biomass.

1.1.3. Crop residue in dyes and heavy metal biosorption

Crop residues, based on their versatility earlier described, have been widely used in the uptake of diverse range of pollutants. Pollutants of interest here are dyes and heavy metals. This section, therefore, presents some recent works reported in literatures using crop residues in dye and heavy metal sorption.

The uptake of Cu (II) ions using almond shell has been reported. Binary metal mixture viz Cu (II) and Pb (II) was used in this study. The presence of lead was reported to affect copper uptake due to competition for sorption sites. The affinity of Pb (II) for almond shell was reported to be higher than that of Cu (II). While equilibrium sorption fitted best into the Sips model, maximum sorption was obtained to be 9.0 and 13.7 mg/g for Cu (II) and Pb (II), respectively [39]. Water chestnut peel was used in the uptake of Rhodamine B dye from aqueous solution with consideration of various sorption operational parameters. Heterogeneous sorption controlled by both liquid film and intraparticle diffusions was reported for the sorption mechanism. Adsorption data fitted best into the Freundlich isotherm model [40]. The use of *Trichoderma reesei* in the uptake of Co^{2+}, Cu^{2+}, Ni^{2+}, Pb^{2+}, and Zn^{2+} ions has been reported with consideration of various sorption operational parameters. While the uptake of the metal ions onto the bioadsorbents obeyed the Pseudo second kinetics models, biosorption process was affirmed to be spontaneous and feasible [41].

The uptake of Pb (II) from aqueous solution onto peanut shells has been reported. Decrease in bioadsorbents efficiency was observed with increased temperature. Equilibrium sorption data fitted best into the Langmuir isotherm model with maximum monolayer sorption capacity of 39 mg/g [42]. Batch sorption of Cd^{2+} onto Arundo donax reed leaves has been reported. Cd^{2+} removal was reported to be pH-dependent, and fast sorption kinetics was observed. The maximum monolayer sorption capacity was reported to be 27.90 mg/g. Equilibrium sorption data was best described by the Freundlich isotherm model, while the kinetic data best obeyed the pseudo second-order kinetics model [43]. The uptake of Cr (VI) using cork powder has been investigated and reported. Sorption parameters, such as contact time, pH, and concentration of Cr (VI), showed significant effects on pollutant uptake, and maximum uptake was observed between pH 2 and 3 [44]. The use of loquat (*Eriobotrya japonica*) leaves waste in Cd (II) uptake was investigated and reported. Uptake of Cd (II) was reported to be dependent on solution pH, initial cadmium ion concentrations, biosorbent dose, contact time, and temperature. Uptake of Cd (II) onto loquat leaves occurred via endothermic process; thus, bioadsorption was more favored at a high temperature. Equilibrium sorption data fitted well into both the Langmuir and Temkin isotherm models [45]. Removal of Malachite green using pea shells (*Pisum sativum L*) has also been studied and reported. Optimum operational parameter was established, and equilibrium sorption data was tested with some isotherm models. Freundlich isotherm model best described the sorption process, and pseudo second-order kinetics best explains the kinetics data [46]. Chir pine sawdust in the bioadsorption of Congo red and basic violet from aqueous has been reported. Equilibrium was attained between 45 and 60 min. While the equilibrium sorption data for the uptake of basic violet obeyed the Langmuir isotherm model, Freundlich isotherm model best described the uptake of Congo red onto the bioadsorbents. Maximum monolayer sorption capacity was 11.3 and 5.8 gkg for basic violet and Congo red, respectively. Chemisorption was reported for the uptake of Congo red onto Chir pine sawdust, while physical sorption ruled the uptake of basic violet onto Chir pine sawdust [47].

The use of water hyacinth leaves in the uptake of amaranth (acid red 27) anionic dye has also been reported. The effects of various operational parameters were reported, and the dye uptake depended greatly on these effects. Dye uptake was reported to increase with increased contact time and dye concentration. Sorption kinetic data was best explained by pseudo second-order kinetic model, while the equilibrium sorption data fitted best into the Langmuir isotherm model with 70 mg/g [48]. *Nauclea diderrichii* seed biomass with low surface area was used for the uptake of toxic Cr (III), and it was found to be effective. Fast kinetics was reported for this study, with data fitting best into the pseudo second-order kinetics. Equilibrium sorption data was reported to fit best into the Freundlich isotherm model, and the maximum monolayer sorption capacity was 483.81 mg/g [49].

The effectiveness of oat hull in the removal of malachite green was investigated and reported. Optimum sorption condition was established, and the pseudo second-order kinetic models best described sorption kinetic data. High temperature favored the uptake of malachite green, while the Freundlich isotherm model best described the sorption equilibrium data. The maximum monolayer sorption capacity was reported to be 83 mg/g [50]. The use of *R. hookerie* fruit epicarp in the removal of Rhodamine has been investigated and reported. The prepared bioadsorbents were found to be effective in the removal of the pollutants. Freundlich isotherm best described the equilibrium sorption data, and maximum monolayer sorption capacity was reported to be 666.67 mg/g [2].

1.1.4. Conventional methods versus biosorption

The preference for adsorption over other wastewater treatment techniques relates to its simplicity of operations and ability to remove pollutants of very low concentration. Biosorption, however, adopts the aforementioned characteristics as well as abundance, availability, and economic advantages. In order to attain zero pollution outflows by most industries, combinations of methods are employed. While adsorption may act effectively in the removal of residue concentration as well as protects resins, ion exchange resins are very efficient in the combat of highly metal-loaded wastewater [51]. The application of activated carbon, ion exchange resins, chelating resin, and bioadsorbents by batch adsorption in removal of metal ions from real and aqueous effluents is summarized in **Table 3**.

In conclusion, the uptake of dyes and heavy metals onto various crop residues has been widely investigated. Amazing results of their efficiency were recorded. Freundlich isotherms that suggest a multilayer sorption commonly describe the sorption process. With the understanding that sorption onto crop residue may either be via functional group or available pores or the combination of the two, multilayer sorption may be expected often in pollutant uptake using plant residue.

Optimum operational parameters that can give insight into further applications were well studied and reported. However, only a few reports exist for applications of biomass to real

Techniques	Pollutants	Percentage (%)	References
Ion exchange resin (mix)			[52]
	Cr	68.18–about 85%	
	Cu	82.09–87.25%	
	Fe	58.33–about 90%	
	Ni	50.00–75.58%	
Chelatin resin			[52]
	Cr	6.67–69.23%	
	Cu	54.48–85.19%	
	Fe	25.00–87.50%	
	Ni	11.11–39.54%	
Activated carbon			[52]
	Cr	46.67–about 85%	
	Cu	40.74–82.35%	
	Fe	58.33–83.33%	
	Ni	5.26–11.11%	
Ion exchange resin (INDION225H)			[51]
	Cu^{2+}	9.01–14.49%	
	50–100 mg/L		
Eucalyptus bark			[38]
	Zn^{2+}	33.75–40.12%	
	20–70 mg/L		
Olive stone			[53]
	Cu^{2+}	6.61–19.87%	
	10–300 mg/L		
Pine bark			[53]
	Cu^{2+}	35.25–76.00%	
	10–300 mg/L		
Peanut shell			[42]
	Pb^{2+}	65.73%	
	100 mg/L		

Table 3. Comparison of pollutant removal techniques and percentage removal.

industrial effluent. Result-oriented applications of crop residues to real industrial effluents present a solution to the lingering challenge of water pollution. Real industrial effluents are a composite of various ions and other materials such as starch and salt. Thus, pilot studies using binary/ternary effluents as well as dye-metal mixture will assist in understanding the behavior of crop residue in real industrial effluent applications.

1.2. Biomass valorization in phytomedicine

The use of plant materials as sources of beneficial nutritional component dates back to prehistoric time. Noah obtained alcohol from vine as recorded in the Bible; Hippocrates encouraged the consumption of medicinal food [54], and the list continues.

Centuries later, several plants have been identified as sources of phytochemicals with potent medicinal value. Initially, plant extracts were used directly, but in these modern times, bioactive extracts are fractionated, and pure compounds isolated therefrom are then investigated for their inherent pharmacological properties. This, of course, is preceded by the characterization of the isolates using spectroscopic tools such as 1D and 2D-Nuclear Magnetic Resonance, Infrared, and mass spectrometry.

Despite the advent of several synthetic drugs via improved scientific methods of drug development, natural products have continued to maintain relevance by consistently providing new compounds with interesting structural motifs, which serve as sources of new drug leads.

The consequence of this intense plant utilization is the generation of a huge amount of waste, which eventually constitutes environmental pollution among other problems. In line with waste valorization, which promotes environmental sustainability, several scientists have engaged in intense search into possible economically beneficial application of plant wastes.

In the same vein, the discovery of beneficial phytochemicals in waste plant materials would be a welcome development for scientists. This is because there would not be any threat to food security as is the case when an edible food source is found to supply uniquely valuable phytochemicals [55].

This section is, therefore, dedicated to the interesting medicinal application of phytochemicals obtained from waste biomasses.

1.2.1. Direct application of plant parts

Conventionally, the active components in different plant parts are first extracted with the aid of a suitable solvent, for example, water and ethanol. The extract solution is then concentrated to obtain the crude extract, which is then applied to proffer solution to different health challenges.

However, these same plant parts can be used directly for different purposes without conventional extraction. An example of such usage is the use of banana leaves, plantain leaves, and *Thaumatococcus daniellii* in food packaging. Particularly, these leaves have been employed in the packaging of bean pudding. Although the initial reason for employing these leaves as packaging materials was their surface area and their ability to withstand heat, it has also been observed that these leaves perform additional functions such as flavour enhancement and possible supply of medicinal compounds to the consumer [56]. Quantitative phytochemical analysis of the leaves of *T. daniellii* prior to and after usage revealed that some of these phytochemicals (saponins, flavonoids, and tanins) have leached either into the bean pudding or the water used for steaming [57].

It will be interesting to initiate a research that will reveal the exact position of these leached phytochemicals as the presence of saponins in food samples could constitute a natural preservative, which will prevent the growth of food spoilage organisms such as *Escherichia coli*. Saponins have been earlier implicated in antimicrobial activity [58]. Similarly, the presence of flavonoids in bean pudding would be advantageous since they are potent scavengers of free radicals in human systems.

Furthermore, agro-wastes can be utilized in biotreatment procedure for safe long-term storage of cereal products. An example is the inhibition of *Aspergillus parasiticus* and aflatoxin production during the storage of wheat, rice, and maize by powdered neem (*Azadiratcha indica*) and kika leaves (*Acia nilotica*). The agro-wastes (leaves) in their powdered forms were applied to the cereals at three concentrations 5, 10, and 20% out of which 20% neem gave the best overall result. Neem leave powder is, therefore, a safe and cheap natural additive for the improvement of the storage of selected cereal products [59]. Similarly, pomegranate peels powder at 1–3% concentration have been shown to provide high storage stability in prepared beef sausage samples, and the presence of antimicrobial/antioxidant phenolics explains this important activity [60].

1.2.2. Extraction of phytochemicals from waste biomass

In several homes, various plant parts such as lemon grass and tea have been boiled in water in order to obtain tonics, which are drunk afterward to achieve different objectives. This domestic procedure is known as extraction.

Extraction, in this context, is the process of separating active constituents or secondary metabolites of interest from waste plant matrix by the use of extractants (solvents). Different methods of extraction that have been utilized in the extraction of phytochemicals include the following:

1.2.2.1. Maceration and percolation

These are cold traditional methods of extraction in which water or organic solvents are made to permeate pulverized biomass with the aim of transferring the phytochemicals in the biomass into the solvent. The extract solution is afterward concentrated using an appropriate technique, for example, rotary evaporator to obtain the crude extract. While maceration emphasizes soaking the biomass in the choice solvent to achieve the extraction, percolation involves the permeation of the solvent through the biomass in the direction of gravity and filtration in order to obtain the extract solution. These traditional methods of extraction have been employed in the extraction of phytochemicals from sweet potato peels, leading to extracts and isolates with antioxidant and antifungal properties [1].

Although these methods may not be as fast as other methods discussed in this section, they do preserve phytochemicals which are thermolabile by nature, thus retaining the pharmacological activities of interest.

1.2.2.2. Hydrodistillation/assisted hydrodistillation

Hydrodistillation (HD) is a method, which is employed for the extraction of fragrances/flavors from plants. In principle, the vapors of the volatile constituents are conveyed by steam to a condenser, where they are returned to liquid state and form two layers: aqueous and oil-rich layers. The two methods via which hydrodistillation is executed are steam distillation and Clevenger distillation [61]. This important extraction method yielded the essential oils from tangerine peels, which were successfully adopted for fish preservation [62].

In recent times, the conventional HD has been assisted and improved with ohmic and microwave heating processes. Assisting hydrodistillation with ohmic heating reduced the extraction time by over 50%, that is, from 1 h to 24.75 min [63], while the GC-MS analysis conducted did not show any difference in the chemical constitution of the oil. Although a shorter extraction time was achieved using the microwave-assisted hydrodistillation [64], the ohmic-assisted HD was the greener method of extraction since its energy consumption is relatively low.

1.2.2.3. Sohxlet extraction

This is a hot method of extraction in which the soxhlet apparatus is employed to achieve the extraction process. The soxhlet apparatus is depicted in **Figure 3**.

This method was successfully utilized in our laboratory to extract constituents of *Ipomoea batatas* peel-waste [65]. Although this method has good speed and high yield as advantages, the possibility of thermal decomposition of the active principles remains a subject of huge concern and constitutes a limitation to this application.

1.2.2.4. Pressurized fluid extraction

One of the emerging methods that have been employed for extractions from waste biomass is pressurized fluid extraction (PFE). We will focus on two PFE methods in this section: supercritical extraction and subcritical extraction.

Supercritical extraction is an extraction method in which the conventional solvent has been replaced by a supercritical fluid: a supercritical fluid is a liquid substance at a temperature and

Figure 3. Three soxhlet apparatus on heating mantles during the hot extraction of the phytochemicals in water melon rind (Source: Authors).

pressure above its critical point—a point where distinct liquid and gas phases do not exist. A common example of this supercritical fluid is CO_2, which is sometimes modified by cosolvents such as ethanol and methanol [66]. This environment friendly method was successfully applied to citrus peels, a common by-product in citrus processing industries [66]. Fatty acid esters, phenols, coumarin derivatives, and terpene derivatives were among the useful compounds that were isolated in the process.

Although more expensive to set up, the advantage of this method is its environmental friendliness and extremely easy solvent removal.

The cost cut by improved percentage yield achieved by PFEs has not been compared with the cost of setting up pressurized fluid extraction apparatus. Despite this gap, the safety guaranteed by the absence of obnoxious chemicals continues to make PFE a preferred alternative.

In the same vein, subcritical water (pressurized hot water) has also been employed to achieve environment friendly extractions. Subcritical water is water under 0.2–21 MPa pressures and temperatures higher than its boiling point but lower than its critical temperature. Subcritical water was first reported in 1994 [67], and several other applications of subcritical water has been reported since then. Recently, this method was applied to several agro-waste biomasses in Canada. These waste biomasses are potato peel, barley hull, lentil husk, and flaxseed hull as contained in a recent review by Saldana and Valdivieso-Ramírez [68].

1.2.3. Effect of extraction method on bioactivity

The different extraction conditions associated with different extraction methods have considerable effect on the pharmacological potency of various phytochemicals. Such conditions include the following:

- Extraction pressure
- Extraction time
- Extraction solvent
- Extraction temperature

Trabelsi et al. studied the effect of high pressure (300 and 500 MPa) and time (3 and 10 min) on the antioxidant/antimicrobial activity of citrus peel extracts [66]. While the antimicrobial activity was not at all influenced by high pressure, the highest Total Phenolic Content (TPC) and antioxidant capacities were observed at 300 MPa and 3 min. Extraction at higher temperature and longer time probably drew out more "nonantioxidant" matter or decomposed some antioxidant constituents in the extracts, thus reducing the TPC per gram.

Furthermore, extraction solvent and extraction time have significant effect on the antimicrobial activity of bioactive extracts from watermelon seeds as shown by a recent study from the School of Science, Landmark University, Nigeria [58]. Cold methanol extract was more active against *Staphylococcus aureus* compared with the hot methanol extract. Thermal decomposition might have affected some of the phytochemicals that are responsible for the observed activity.

Furthermore, the cold chloroform extract that did not inhibit the growth of *S. aureus* at all was, however, active against *Pseudomonas aeruginosa.* This simply shows that the choice of a solvent in phytochemical extraction greatly influences the nature of compounds extracted and ultimately the biological activity of the extract. The effect of different extraction methods on percentage yield and bioactivity is vividly displayed in **Table 4**.

Generally, the introduction of pressure, higher temperature, and supercritical fluids led to the achievement of better yields and improved/preserved biological activity.

1.2.4. Medicinal products from waste biomass

Different kinds of plant biomass have been employed in the search for medicinal phytochemicals. Selected examples are shown in **Table 5**.

In line with the principles of green chemistry, waste biomass that otherwise would have been discarded have been exploited as sources of beneficial/medicinal phytochemicals by several research scientists.

1.2.4.1. Antioxidants from waste biomass

Antioxidants, as the name implies, are compounds that oppose or prevent oxidation. They are compounds that scavenge free radicals, which have been known to cause unwanted oxidation of important molecules in the body.

$$R + ArOH \rightarrow RH + ArO \tag{1}$$

The unwanted free radical (R) is reduced by the antioxidant compound (ArOH), resulting in the harmless RH and ArO which is a more stable radical.

Furthermore, a high consumption of antioxidants has been correlated with lower incidence of cancer [82]. During renovation of cultivations, the roots of *Rugosa rose* are usually treated as wastes. Olech et al. studied, among other parameters, the antioxidant potential of this postharvest residue with the aim of proposing new uses for it. Interestingly, these roots displayed significant antiradical activity compared with ascorbic acid. Hence, these roots could be explored as a source of antioxidant supplements in food and pharmaceutical industries [83].

In the same vein, tomato-processing by-products (TPP) have also been shown to contain antioxidants. Particularly, the presence of simple polyphenols in TPP suggests that it could be safely used as a functional ingredient in the production of antioxidant-rich foods [84].

Similarly, Oluyori et al. recently reported a successive extraction of sweet potato peels using ethyl acetate and methanol, respectively. While the extracts did not show significant antimicrobial activity against the selected microorganisms, the DPPH assay and the TPC determination revealed the methanolic extract as the extract with higher antioxidant capacity [6].

These and many other research efforts continue to prove that there is, indeed, a vast amount of wealth in agro wastes, which are usually discarded.

S/No	Extraction method	Process condition	Percentage of yield (%)	Remarks	Reference (s)
1	Room temperature water extraction	25 C; 3 h; 1:20; w/v	<30.35%	% yield of CPS by microwave-assisted extraction was the highest and the antitumor activity same extract was the best	[69]
	Hot water extraction	78 C; 3 h; 1:20; w/v	<30.35%		
	Microwave-assisted extraction	5 min; 280 W; 70°C; 1:20; w/v	30.35%		
	Ultrasound-assisted extraction	13 min; 250 W; 45 C	<30.35%		
	Cellulase-assisted extraction	55 C; 80 min; pH 5.0	32.50%		
2	Ultra-high pressure extraction (UPE)	60% ethanol; 400 MPa; 2 min; 30°C; 1:5 solid to liquid ratio	4.395% chlorogenic acid	UPE showed higher extraction yield, shorter extraction time, lower energy consumption, and higher purity of extracts	[70]
	Heat reflux extraction	60 C; 3 h	3.471% chlorogenic acid		
	Ultrasonic extraction	250 W; 30°C; 40 min; 42 KHz	3.601% chlorogenic acid		
	Soxhlet extraction	60 C; 2 h	3.793% chlorogenic acid		
3	Ultra-high hydrostatic pressure (UHP)	60% ethanol; 70:1 ml/g solvent: herb; 500 Mpa hydrostatic pressure; 3 min	26.12%	High efficiency at room temperature was observed for UHP; flavanoids and salidroside were used to monitor the extraction yield	[71]
	Reflux extraction	80 C; 120 min; 4:1 ml/g	13.83%		
	Soxhlet extraction	80°C; 240 min;	17.00%		
4	Supercritical fluid extraction (SFE)	250 bar and 40°C		SFE gave the best extraction yield and the most active extract	[72]
	Soxhlet extraction	Ethanol; 6 h	10.00%		
	Steam distillation	4 h	0.60%		
	Marceration	40 °C; 48 h; material: solvent 1:10	8.90%		

S/No	Extraction method	Process condition	Percentage of yield (%)	Remarks	Reference (s)
5	Soxhlet extraction	720 min; 200 ml solvent	9.58%	% yield of scopoletin was used to determine the extraction efficiency	[73]
	Reflux extraction	480 min; 100 ml solvent	10.95%		
	Ultrasound-assisted extraction	20 min; 60 W; 50 ml solvent	28.55%		
	Microwave-assisted extraction	1 min; 560 W; 50 month = 1 solvent	45.10%		
	Supercritical fluid extraction	60 min; 6500 psig;	7.99%		
6	Steam distillation	403–423 K; 24 h; atm pressure	2.32%	% yield of essential oil from Kabosu (*Citrus sphaerocarpa*) peel was used to measure the extraction efficiency	[74]
	Supercritical fluid extraction	313–352 K; 10–30 Mpa; 0.19–0.21 g/s CO_2 flow rate	1.55%		
	Cold press	100 ml distilled water; centrifuged at 10,000 G for 15 min/frozen at 253 K	0.12%		
7	Supercritical fluid extraction	600 bar/40°C	88.20 mg/100 g oil	Extraction of tocopherols	[75]
		600 bar/80°C	85.57 mg/100 g oil		
	n-hexane extraction	—	62.38 mg/100 g oil		

Table 4. Different extraction methods: process conditions and effects on percentage(%) of yield.

1.2.4.2. Antimicrobial compounds from waste biomass

The emergence of multidrug-resistant bacteria has continued to be a major concern across the globe. This is the reason for the increased attention given to the search for new antimicrobial agents.

Generally, many people after consumption discard the seeds of watermelon. However, several scientists have now shown that inherent in these seeds are phytochemicals with varying activities against *Staphylococcus* sp., *E. coli, Proteus* sp., *Klebsiella* sp., and *P. aeruginosa* [57]. Hence, these seeds could serve as renewable sources of new antimicrobial agents.

A similar study focused on the valorization of sweet potato peels. It was interesting to have isolated beneficial triterpenoids from sweet potato peels that we normally discard. Although the isolates (an example is shown in **Figure 4**) from the peels of *I. batatas* Lam exhibited low

S/No	Biomass type	Plant source	Medicinal phytochemical of interest	Application	References
1	Seed	Sour cherry seed kernel	Alpha-tocopherol, tocotrienols, tocopherol-like compounds	Antioxidants	[76]
2		Germinated fenugreek seed	Vitexin, Isovitexin	Antioxidants	[77]
3	Stem bark	*Mitrephora celebica*	Ent-tradyloban-19-oic acid and ent-kaur-16-en-19-oic acid	Antimicrobial	[78]
4	Leaves	Leaves of *X. laavigata*	Essential oils (γ-muurolene, δ-cadinene, α-copaene, germacrene-D, bicyclogermacrene and (E)-cariophyllene	Antitumor activity	[79]
		Leaves of *Clinacanthus nutaus lindau*	13(2)-hydroxy-(13(2)-R)-phaephytin b, 13(2)-hydroxy-(13(2)-R)-phaephytin a, 13(2)-hydroxy-(13(2)-S)-phaephytin a	Anti-herpes simplex activity, Anti-HSV-IF activity	[80]
5	Flowers	Elder flower: *Sambucus nigra*	Naringenin	Nutraceutical application (modulation of glucose metabolism)	[81]
6	Peels	Sweet potato peels	3-Friedelanol; Stigmasterol; Urs-13(18)-ene-3β-yl acetate	Antifungal	[1]

Table 5. Different plant biomass sources and the phytochemicals of interest.

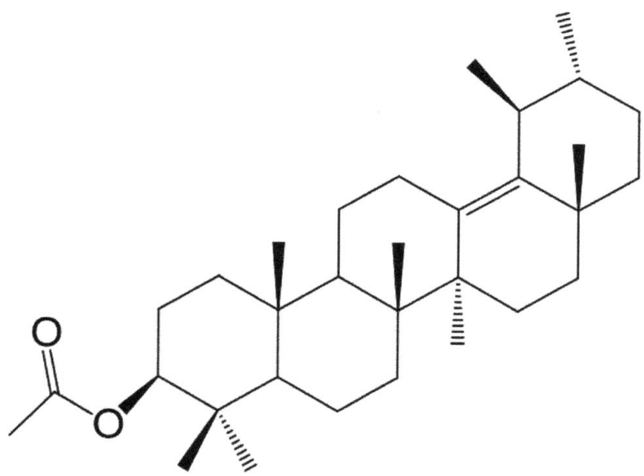

Figure 4. A natural antifungal compound from sweet potato peel: Urs-13(18)ene-3β-yl acetate (Source: Authors).

antibacterial activity, the same isolates were more active against two fungi: *Tricophyton metagrophytes* and *Sporothrix schenkii* (IC_{50}, 50 μg/ml) [1].

Similarly, extracts from drumstick peel recently exhibited significant activity against two human pathogenic bacteria: *S. aureus* and *E.coli*. This extract, which was nontoxic to red blood cells, also showed some photocatalytic property [85].

1.2.4.3. Anticancer compounds from waste biomass

Cancer is the second leading cause of death in the United States after heart disease. In fact, it has been said to be responsible for one out of every four deaths [86]. Although there are current cures for cancer, the high cost of treatment is a challenge to many cancer patients. In view of the increasing cost of cancer treatment, the presence of anticancer compounds in waste biomass is a blessing to many, especially, people in the developing world [87]. The roots of *R. rose*, which showed some antiradical activity as earlier mentioned, also exhibited significant anticancer potential. About 90% inhibition of cervical (HeLa) and breast (T47D) cancer cell lines was observed in the presence of the rugosa root extracts, showing that the roots that have hitherto been ignored are an important phytochemical source [83]. Similarly, Oluyori et al. recently carried out an anticancer guided fractionation/isolation of an alcoholic extract from sweet potato peels [88]. An anticancer fraction was obtained, and a glucocerebroside {N-((E)-3,4-dihydroxy-1-(((2R,3R,4S,5S,6R)-3,4,5-trihydroxy-6-(hydroxymethyl)tetrahydro-2H-pyran-2-yl)oxy)octadec-8-en-2-yl)-2-hydroxytetracosanamide} was isolated and characterized as the main anticancer compound. Our results suggest that the consumption of sweet potato peels may supply the body with phytochemicals that can prevent the onset or the development of cancer [89]. These findings among others project agro-wastes as potential sources of economically important compounds with multifaceted medicinal applications.

1.3. Harnessing agro-wastes for bioethanol production

Conversion of agricultural waste into fuel has become popular in the past few decades occasioned by the increasing demand for fossil fuel. Campbell and Laherrere [89] had predicted the reduction in availability of "cheap fuel" years ago. In a bid to forestall imminent fall in supply, many countries ventured into research seeking alternative energy sources. This led to the production of biofuel from various sources including cornstarch, sugar cane, and now from cellulosic biomass. Biofuels have several advantages over fossil fuel including being environment friendly, thus reducing pollution. There is also a reduction in the release of greenhouse gases and consequent global warming [90]. Biofuels are produced from renewable sources such as rind of various fruits, corn stover, "weeds" including switch grass, winter rye, and faba bean to mention a few [91]. This section elaborates the extent of research on bioethanol production from agricultural wastes, exposes possible challenges in bioethanol production from agro-wastes as well as exposes gaps from various previous research works efforts thus opening new areas of investigations to researchers.

1.3.1. A drift to biofuel as an alternative energy source

Biofuels are any kind of fuel made from living things or from the wastes they produce [92]. Biofuels can also be referred to as renewable energy produced from organic matter by converting its complex carbohydrates into energy. The carbon dioxide (CO_2) emitted *during biofuel production* is not larger than that consumed during photosynthesis. In order words, the process is CO_2 neutral and beneficial to the environment as opposed to combustion of fossil fuels that leads to increase in greenhouse gases, which causes ozone layer depletion. Biofuels include biogas, biodiesel, and bioethanol, but bioethanol has enjoyed vast investment and is popular as a transport fuel. It is presently the most widely used in the transport industry, and it is receiving attention at the international and national fronts [93]. Countries at the forefront of bioethanol production include the United States of America (USA) and Brazil. Blending ethanol with gasoline is advantageous because it brings about reduction in consumption of fossil fuel, and *in recent time, bioethanol has* provided a means of powering engines without using fossil fuels. Currently, there are vehicles that can run solely on bioethanol as well as in blends with gasoline, for example, E5, E10, and E5 implies that 5 parts of bioethanol is blended with 95 parts of gasoline. When gasoline is blended with bioethanol, it has a higher octane number, which implies that the engine has increased thermal efficiency and works more efficiently.

1.3.2. Second-generation bioethanol

For the production of bioethanol, sugars are fermented using yeast and other microorganisms. Commercial quantities of bioethanol have been produced from corn, sugar cane, sweet sorghum, and cassava [94]. Bioethanol produced in this manner is referred to as first-generation ethanol. In a bid to find feedstock that will not compete with (human) food and (animal) feed and also meet up with global demand for bioethanol production, second-generation bioethanol production was introduced [93]. This process exploits the use of lignocellulosic biomass such as agricultural wastes as feedstock because they are renewable, available in abundance, and relatively cheap. Utilization of such agricultural waste involves pretreatment of such waste, then hydrolysis of the pretreated material, and finally, fermentation of monomeric sugars released to yield bioethanol.

Lignocellulosic materials *currently in use* include agricultural wastes/residues; forest residues such as fallen branches, leaves, twigs, and sawdust, municipal solid wastes including paper and cardboard products as well as industrial wastes and energy crops. Energy crops include herbaceous crops such as switch grass, fast-growing hybrids of poplar and leucaena trees [95]. Among all these sources of lignocellulose, agricultural wastes/residues are the most abundant and provide higher quantities of cellulose. Agricultural wastes are parts of the plant that are left after the main crop has been obtained. Examples are corn stover, sugar cane baggase, spent sugar beets pulp, sweet sorghum straw, cotton seed hairs, oil palm frond, and banana stem. All these can be converted into bioethanol to provide energy to drive various engines. These materials are of great advantage because the crop is grown for dual purpose: first, as a source of food or feed and second, to provide lignocellulosic biomass, which will serve as feedstock for bioethanol production. This is of great advantage to the farmer who now obtains economic value from both the crop and its waste.

1.3.3. Pretreatment technologies

As stated earlier, agricultural wastes undergo pretreatment, and hydrolysis before the fermentation of sugars can be carried out. Pretreatment is a process of solubilization and separation of one or more components of a biomass. In this case, the lignocellulosic material breaks down thereby making cellulose more susceptible to enzymatic attack [95]. The cell wall of agricultural waste consists of cellulose, hemicelluloses, and lignin. Of these three, cellulose, which is the highest constituent, is responsible for the mechanical strength of the cell wall. The glucose monomers of cellulose are connected by β-1,4 glycosidic bonds; hemicellulose is a branched heteropolymer of D-xylose, L-arabinose, D-mannose, D-glucose, D-galactose, and D-glucuronic acid, while lignin is composed of three major phenolic components, namely p-coumaryl alcohol, coniferyl alcohol, and sinapyl alcohol and other extractable components.

Various types of pretreatment ranging from physical, chemical to biological pretreatment have been described [96, 97]. Pretreatment is aimed at removing lignin and hemicellulose, reducing the crystallinity of cellulose, and increasing porosity of the lignocellulosic materials. Physical pretreatment could be a combination of mechanical methods whereby the waste materials are comminuted by chipping, grinding, and milling, leading to reduced cellulose crystallinity. Pyrolysis is another form of physical pretreatment that involves treating the agricultural waste at temperatures above 300°C. Another form of physical pretreatment of agricultural waste steam is explosion that involves heating the biomass to a desired temperature for a specified period of time [98]. Chemical pretreatment involves treating lignocellulosic material with chemicals, for example, ozonolysis necessitates using ozone to degrade lignin and hemicellulose. Acid [99] and alkaline pretreatments are other forms of chemical pretreatment using concentrated acids such as H_2SO_4 and bases, respectively, to treat lignocellulosic materials. Apart from the aforementioned chemical pretreatment methods, lignin degradation could be catalyzed by a process known as oxidative delignification in the presence of hydrogen peroxide. Combinations of physical, chemical, and biological pretreatment methods can be used on lignocellulosic agricultural waste [100, 101]. Examples of such methods are steam explosion (a physical pretreatment method) and ammonia fiber explosion (AFEX that is a chemical pretreatment method); ultrasonic (physical) pretreatment of rice hull followed by biological pretreatment with *Pleurotus ostreatus* [102].

Biological pretreatment is highly advocated as it can prevent the formation of inhibitory by-products that can adversely affect subsequent stages of hydrolysis and fermentation. Biological pretreatment is also said to be cost-effective, environment friendly compared with chemical methods. Brown rot, white rot, and soft rot fungi are currently used for biological pretreatment [103, 104].

Advantages of biological pretreatment include mild conditions to undertake the process and low energy requirements, ecofriendly, less hazardous to health, more economical, and more effective compared with other chemical and physiochemical methods of pretreatment [105]. White rot fungi has been recognized as efficient for lignin degradation, and their activity is said to enhance the subsequent enzymatic hydrolysis. These organisms degrade the lignin seal and thus give access to cellulases attack on cellulose, but this process is often slow, spanning from 28 days to about 120 days [104]. Apart from longer pretreatment period, biological pretreatment also requires a large space and carefully controlled growth conditions; sometimes the organisms used act on lignin as well as hemicellulose and cellulose [93, 103]. Agricultural wastes are abundant feedstock for bioethanol production; however, there is need to focus more on pretreatment preferably biological pretreatment in order to exploit the benefit of this feedstock for biofuel production.

1.3.3.1. Hydrolysis

After pretreatment, hydrolysis of cellulose is carried out; this involves the conversion of cellulose (and hemicelluloses) to monomeric sugars using chemicals or enzymes [97]. Cellulases are often used for enzymatic hydrolysis of the pretreated agricultural waste [103]. Enzymatic hydrolysis leads to the release of fermentable sugars from the pretreated waste. This process of enzymatic hydrolysis is dependent on some factors including pH of the reaction, temperature, substrates, time, and enzyme activity.

1.3.3.2. Fermentation

When hydrolysis and fermentation are carried out separately, it is called Single Hydrolysis and Fermentation (SHF), but when enzymatic hydrolysis and fermentation are carried out concurrently, the process is termed Simultaneous Saccharification and Fermentation (SSF). In another process known as simultaneous saccharification and cofermentation (SSCF), enzymatic hydrolysis of both glucose and hemicellulose is carried out followed by fermentation of the sugars (glucose and xylose) produced in the same vessel [104].

Consolidated bioprocessing, on the other hand, integrates cellulase production with hydrolysis and fermentation of lignocellulosic material for biofuel production. It has the advantage of potentially reduced production cost, higher energy input, and higher conversion efficiency. A major challenge has, however, been finding a suitable consortium of organism to drive this process. Organisms such as *Saccharomyces cerevisiae, Clostridium thermocellum,* and *C. phytofermentans* are, however, being genetically engineered to perform this purpose [3]. Agricultural residues such as winter rye straw, oilseed rape straw, and faba bean straw have been shown to yield 66, 70, and 52% (of theoretical calculation) ethanol after pretreatment by wet oxidation and fermentation by *S. cerevisiae* [106].

One of the challenges faced in the fermentation of monomeric sugars to ethanol is that ethanol produced in the medium is inhibitory to some organisms used for fermentation. In addition, certain products such as furfural and hydroxylmethylfurfural will limit or totally stop the activities of organisms used for fermentation. Apart from these, other factors such as high temperature, osmotic stress, and contamination by other organisms can adversely affect the process. Eventually, the ethanol produced by fermentation is made to undergo distillation [97].

Bioethanol yield of 11.11% was obtained when preheated rice husk was treated with *Trichophyton soudanense* that was fermented with palm wine yeast. But without fungal treatment, the heat-treated rice husk did not give up to 5% bioethanol yield [107]. Also, the highest cellulose and lignin yields recorded from rice husk in their study were 44.50 and 28.90% obtained from rice husk pretreated with *Trichophyton mentagrophyte* and a coculture of *T. soudanense* and *T. mentagrohyte,* respectively, while the highest hemicellulose value of 30.50% was obtained from preheated rice husk treated with *T. soudanense* .

Agro-waste such as coffee mucilage has also been reported to produce bioethanol. Unlike some agro-wastes that require pretreatment, it was established that coffee mucilage could provide up to 60 g/L of reducing sugar, which could be directly converted to bioethanol using *S. cerevisiae* Y2034 [108]. Pretreatment of wild cassava peels (*Manihot glaziovii*) with enzymes and a combination of enzyme with alkali followed by simultaneous saccharification and fermentation (SSF) was reported. Results revealed that the latter treatment led to higher ethanol production of about 95% theoretical yield [109].

Banana waste composed of approximately 14% of lignin, 14.8% of hemicellulose, and 13.2% of cellulose was reported to yield 7.45% v/v ethanol [95, 104]. Paulova et al. [110] also reported increased ethanol production as conversion of cellulose to glucose increased; in the same vein, as the quantities of furfural and hydroxymethyl-furfural reduced, ethanol yield increased. For example, 78% conversion of cellulose to glucose brought about 0.34 g/g yield of ethanol in the presence of 88 mg/l furfural and 6 mg/l hydromethyl-furfural. On the other hand, 64% cellulose to glucose conversion brought resulted in an ethanol yield of 0.04%, while 158 mg/g and 10 mg/g of furfural and hydromethyl-furfural were detected.

In conclusion, utilizing agricultural waste as energy source in the form of bioethanol is a viable way of converting waste to wealth. These wastes are available in abundance and renewable hence continuity in production of bioethanol is assured. Nevertheless, there is a need for further research on how to minimize production cost especially at the pretreatment stage. In the same vein, future research should focus on how to scale-up second-generation bioethanol production so that commercial production of bioethanol will not depend on food crops only.

Author details

Inyinbor Adejumoke Abosede[1]*, Oluyori Abimbola Peter[1] and Adelani-Akande Tabitha Adunola[2]

*Address all correspondence to: inyinbor.adejumoke@landmarkuniversity.edu.ng

1 Department of Physical Sciences, Landmark University, Omu Aran, Nigeria

2 Department of Biological Sciences, Landmark University, Omu Aran, Nigeria

References

[1] Oluyori A.P., Shaw A.K., Preeti R., Sammajay R., Atolani O., Olatunji G.A. and Fabiyi O. A. (2015) Natural antifungal compounds from the peels of *Ipomoea batatas Lam*; Natural Product Research; 30(18):2125–2129.

[2] Inyinbor, A.A., Adekola, F.A. and Olatunji, G.A.(2016) Kinetic and thermodynamic modeling of liquid phase adsorption of Rhodamine B dye onto *Raphia hookerie* fruit epicarp; Water Resources and Industry; 15:14–27.

[3] Joshi B., Bhatt M.R., Sharma D., Joshi S., Malla R., Sreerama L. (2011) Lignocellulosic ethanol production: current practices and recent developments; Biotechnology and Molecular Biology Review; 6(8):172–182.

[4] Inyinbor, A.A., Adekola, F.A. and Olatunji, G.A. (2016) Liquid phase adsorptions of Rhodamine B dye onto raw and chitosan supported mesoporous adsorbents: isotherms and kinetics studies; Applied Water Science, 1–11, http://link.springer.com/article/10.1007/s13201-016-0405-4

[5] Inyinbor, A.A., Adekola, F.A. and Olatunji, G.A. (2015) Adsorption of Rhodamine B dye from aqueous solution on *Irvingia gabonensis* biomass: kinetics and thermodynamics studies; South Africa Journal of Chemistry; 68:115–125.

[6] Oluyori A.P. and Olatunji G.A. (2016) Phytochemical analysis, antimicrobial and antioxidant activity of Peels' extracts from *Ipomoea batatas Lam*; Journal of International Research in Medical and Pharmaceutical Sciences; 6(4):157–164.

[7] Concannon, J.A. and DeMeo, T. (1997) Golden seal: facing a hidden crisis; Endangered Species Bulletin; 22(6):10–12.

[8] Imamoglu, E., and Sukan, F.V. (2014) The effects of single and combined cellulosic agrowaste substrates on bioethanol production; Fuel; 134:477–484.

[9] Chen, S. (2015) Environmental pollution emissions, regional productivity growth and ecological economic development in China; China Economic Review; 35:171–182.

[10] Anastopoulos, I. and Kyzas, G.Z. (2015) Progress in batch biosorption of heavy metals onto algae; Journal of Molecular Liquids; 209:77–86.

[11] Dada, A.O., Adekola, F.A. and Odebunmi, E.O. (2015) A novel zerovalent manganese for removal of copper ions: synthesis, characterization and adsorption studies; Applied Water Science; http://link.springer.com/article/10.1007/s13201-015-0360-5.

[12] Johnson, M.F.J., Barbour, N.W. and Weyers, S.L. (2007) Chemical composition of crop biomass impacts its decomposition; Soil Science Society of America Journal; 71:155–162.

[13] Patel, A., Sindhu, D.K., Arora, N., Singh, R.P., Pruthi, V. and Pruthi, P.A. (2015) Biodiesel production from non-edible lignocellulosic biomass of Cassia fistula L. fruit pulp using oleaginous yeast *Rhodosporidium kratochvilovae* HIMPA1; Bioresource Technology; 197:91–98.

[14] Bhatnagar, A., Sillanpaa, M. and Witek-Krowiak, A. (2015) Agricultural waste peels as versatile biomass for water purification—a review; Chemical Engineering Journal; 270:244–271.

[15] Jeong, C.Y., Dodla, S.K., Wang, J.J. (2016) Fundamental and molecular composition characteristics of biochars produced from sugarcane and rice crop residues and by-products; Chemosphere; 142:4–13.

[16] Sindhu, R., Gnansounou, E., Binod, P. and Pandey, A. (2016) Bioconversion of sugarcane crop residue for value added products: an overview, Renewable Energy; http://dx.doi.org/10.1016/j.renene.2016.02.057.

[17] Cretescu, I., Lupascu, T., Buciscanu, I., Balau-Mindru, T. and Soreanu, G. (2016) Low-cost sorbents for the removal of acid dyes from aqueous solutions; Process Safety and Environmental Protection; http://dx.doi.org/10.1016/j.psep.2016.05.016.

[18] Bello, O.S. Auta, M. and Ayodele, O.B. (2013) Ackee apple (*Blighiasapida*) seeds: a novel adsorbent for the removal of Congo red dye from aqueous solutions; Chemistry and Ecology; 21(1):58–71.

[19] Gao, Y., Yue, Q. and Gao, B. (2016) Insights into properties of activated carbons prepared from different raw precursors by pyrophosphoric acid activation; Journal of Environmental Sciences; 41:235–243.

[20] El-Sayed, G.O., Yehia, M.M., Asaad, A.A. (2014) Assessment of activated carbon prepared from corncob by chemical activation with phosphoric acid; Water Resources and Industry; 7–8:66–75.

[21] AlOthman, Z.A., Habila, M.A., Ali, R., Ghafar, A.A., Hassouna, M.S.E. (2014) Valorization of two waste streams into activated carbon and studying its adsorption kinetics, equilibrium isotherms and thermodynamics for methylene blue removal; Arabian Journal of Chemistry; 7:1148–1158.

[22] Njoku, V.O., Foo, K.Y., Asif, M. and Hameed, B.H. (2014) Preparation of activated carbons from rambutan (*Nephelium lappaceum*) peel by microwave-induced KOH activation for acid yellow 17 dye adsorption; Chemical Engineering Journal; 250:198–204.

[23] Bautista-Toledo, M.I., Rivera-Utrilla, J., Ocampo-Pẽrez, R., Carrasco-Marin, F. and Sănchez-Polo, M. (2014) Cooperative adsorption of bisphenol-A and chromium(III) ions from water on activated carbons prepared from olive-mill waste; Carbon; 7 (3):338–350.

[24] Mahamad, M.N., Zaini, M.A.A. and Zakaria, Z.A. (2015) Preparation and characterization of activated carbon from pineapple waste biomass for dye removal; International Biodeterioration & Biodegradation; 102:274–280.

[25] Sayğılı, H., Güzel, F. and Önal, Y. (2015) Conversion of grape industrial processing waste to activated carbon sorbent and its performance in cationic and anionic dyes adsorption; Journal of Cleaner Production; 93:84–93.

[26] Li, J., Ng, D.H.L., Song, P., Kong, C., Song, Y. and Yang, P. (2015) Preparation and characterization of high-surface area activated carbon fibers from silkworm cocoon waste for Congo red adsorption; Biomass and Bioenergy; 7(5):189–200.

[27] Ahmad, M.A., Ahmad, N. and Bello, O.S. (2015) Modified durian seed as adsorbent for the removal of methyl red dye from aqueous solutions; Applied Water Science; 5:407–423.

[28] Sayğılı, H. and Güzel, F. (2016) High surface area mesoporous activated carbon from tomato processing solid waste by zinc chloride activation: process optimization, characterization and dyes adsorption; Journal of Cleaner Production; 113:995–1004.

[29] Georgin, J., Dotto, G.L., Mazutti, M.A. and Foletto, E.L. (2016) Preparation of activated carbon from peanut shell by conventional pyrolysis and microwave irradiation-pyrolysis to remove organic dyes from aqueous solutions; Journal of Environmental Chemical Engineering; 4:266–275.

[30] Kyzas, G.Z., Deliyanni, E.A., and Matis, K.A. (2016) Activated carbons produced by pyrolysis of waste potato peels: cobalt ions removal by adsorption; Colloids and Surfaces A: Physicochemical and Engineering Aspects; 490:74–83.

[31] Largitte, L., Brudey, T., Tant, T., Dumesnil, P.C. and Lodewyckx, P. (2016) Comparison of the adsorption of lead by activated carbons from three lignocellulosic precursors; Microporous and Mesoporous Materials; 219:265–275.

[32] Inyinbor, A.A., Adekola, F.A. and Olatunji, G.A. (2016) Kinetics and isothermal modeling of liquid phase adsorption of rhodamine B onto urea modified *Raphia hookerie* epicarp; Applied Water Science (In Press), http://link.springer.com/article/10.1007/s13201-016-0471-7

[33] Xu, X., Gao, B., Tan, X., Yue, Q., Zhong, Q. and Li, Q. (2011) Characteristics of amine-crosslinked wheat straw and its adsorption mechanisms for phosphate and chromium (VI) removal from aqueous solution; Carbohydrate Polymers; 84:1054–1060.

[34] Kumar, R., Barakat, M.A., Daza, Y.A., Woodcock, H.L. and Kuhn, J.N. (2013) EDTA functionalized silica for the removal of Cu(II), Zn(II) and NI(II) from aqueous solution; Journal of Colloid and Interface Science; 408:200–205.

[35] Asgher, M, and Bhatti, H.N. (2010) Mechanistic and kinetic evaluation of biosorption of reactive azo dyes by free, immobilized and chemically treated Citrus sinensis waste biomass; Ecological Engineering; 36:1660–1665.

[36] Srivastava, S., Agrawal, S.B. and Mondal, M.K. (2015) Biosorption isotherms and kinetics on removal of Cr(VI) using nativeand chemically modified Lagerstroemia speciosa bark; Ecological Engineering; 85:56–66.

[37] Wang, G., Zhang, S., Yao, P., Chen, Y., Xu, X., Li, T., Gong, G. (2015) Removal of Pb(II) from aqueous solutions by *Phytolacca americana* L. biomass as a low cost biosorbent; http://dx.doi.org/10.1016/j.arabjc.2015.06.011

[38] Afroze, S., Sen, K.T. Ang, H.M. (2016) Adsorption removal of zinc (II) from aqueous phase by raw and base modified *Eucalyptus sheathiana* bark: kinetics, mechanism and equilibrium study; Process Safety and Environmental Protection; 102:336–352

[39] Ronda, A., Martìn-Lara, M.A., Dionisio, E., Blázquez, G. and Calero, M. (2013) Effect of lead in biosorption of copper by almond shell; Journal of the Taiwan Institute of Chemical Engineers; 44:466–473.

[40] Khan, T.A., Nazir, M. and Khan, E.A. (2013) Adsorptive removal of rhodamine B from textile wastewater using water chestnut (*Trapa natans* L.) peel: adsorption dynamics and kinetic studies; Toxicological and Environmental Chemistry; 95(6):919–931.

[41] Ghaedi, M., Hajati, S., Karimi, F., Barazesh, B. and Ghezelbash, G. (2013) Equilibrium, kinetic and isotherm of some metal ion biosorption; Journal of Industrial and Engineering Chemistry; 19:987–992.

[42] Taşar Ş., Kaya, F. and Özer, A. (2014) Biosorption of lead(II) ions from aqueous solution by peanut shells: equilibrium, thermodynamic and kinetic studies; Journal of Environmental Chemical Engineering; 2:1018–1026.

[43] Ammari, T.G. (2014) Utilization of a natural ecosystem bio-waste; leaves of Arundo donax reed, as a raw material of low-cost eco-biosorbent for cadmium removal from aqueous phase; Ecological Engineering; 71:466–473.

[44] Sfaksi, Z., Azzouz, N., Abdelwahab, A. (2014) Removal of Cr(VI) from water by cork waste; Arabian Journal of Chemistry; 7:37–42.

[45] Awwad A.M. and Salem, N.M. (2014) Kinetics and thermodynamics of Cd(II) biosorption onto loquat (*Eriobotrya japonica*) leaves; Journal of Saudi Chemical Society; 18:486–493.

[46] Khan, T.A., Rahman, R., Ali, I., Khan, E.A. and Mukhlif, A.A. (2014): Removal of malachite green from aqueous solution using waste pea shells as low-cost adsorbent; adsorption isotherms and dynamics; Toxicological and Environmental Chemistry; http://dx.doi.org/10.1080/02772248.2014.969268.

[47] Khan, T.A., Sharma, S., Khan, E.A. and Mukhlif, A.A. (2014) Removal of Congo red and basic violet 1 by chir pine (*Pinus roxburghii*) sawdust, a saw mill waste: batch and column studies; Toxicological and Environmental Chemistry; 96(4):555–568.

[48] Guerrero-Coronilla, I., Morales-Barrera, L. and Cristiani-Urbina, E. (2015) Kinetic, isotherm and thermodynamic studies of amaranth dye biosorption from aqueous solution onto water hyacinth leaves; Journal of Environmental Management; 152:99–108.

[49] Omorogie, M.O., Babalola, J.O., Unuabonah, E.I., Song, W. and Gong, J.R. (2016) Efficient chromium abstraction from aqueous solution using a low-cost biosorbent: *Nauclea diderrichii* seed biomass waste; Journal of Saudi Chemical Society; 20:49–57

[50] Banerjee, S., Sharma, G.C., Gautam, R.K., Chattopadhyaya, M.C., Upadhyay, S.N. and Sharma, Y.C. (2016) Removal of Malachite Green, a hazardous dye from aqueous solutions using *Avena sativa* (oat) hull as a potential adsorbent; Journal of Molecular Liquids; 213:162–172.

[51] Thakare, Y.N. and Jana, A.K. (2015) Performance of high density ion exchange resin (INDION225H) for removal of Cu(II) from waste water; Journal of Environmental Chemical Engineering; 3:1393–1398.

[52] Charles, J., Bradu, C., Morin-Crini, N., Sancey, B., Winterton, P., Torri, G., Badot, P. and Crini, G. (2016) Pollutant removal from industrial discharge water using individual and combined effects of adsorption and ion-exchange processes: chemical abatement; Journal of Saudi Chemical Society; 20:185–194.

[53] Blázquez, G., Martìn-Lara, M.A., Dionisio-Ruiz, E., Tenorio, G. and Calero, M. (2011) Evaluation and comparison of the biosorption process of copper ions onto olive stone and pine bark; Journal of Industrial and Engineering Chemistry; 17:824–833.

[54] Chishti, H. (1988). The Traditional Healer's Handbook. Vermont: Healing Arts Press. pp. 11. ISBN 0892814381.

[55] Concannon J.A. and DeMeo T.E. (1997). Golden seal: facing a hidden crisis; Endangered Species Bull; 22(6):10–12.

[56] Ojekale, A.B, Makinde, S.C. and Osileye, O. (2007). Phytochemistry and anti-microbial evaluation of *Thaumatococcus danielli,* Benn. (Benth.) leaves; Nigerian Food Journal; 25 (2):176–183.

[57] Dosumu, O.O. and Akinnuoye, G.A. 2014. Effect of steaming of beans pudding on the phytochemical composition of *Thaumatococcus Daniellii* Wrapper; NIFOJ; 32(1):110–116.

[58] Adunola A.T., Chidimma A.L., Olatunde D.S. and Peter O.A. (2015); Antibacterial activity of watermelon (*Citrullus lanatus*) seed against selected microorganisms; African Journal of Biotechnology; 14(14):1224–1229.

[59] Sultana, B., Naseer R., and Nigam P.. (2015) Utilization of agro-wastes to inhibit aflatoxins synthesis by *Aspergillus parasiticus*: a biotreatment of three cereals for safe long-term storage; Bioresource Technology; 197:443–450.

[60] El-Nashi H.B., Abdel Karim A.F., Abdel Rahman N.R. and Abd El-Razik M.M.. 2015 Quality characteristics of beef sausage containing pomegranate peels during refrigerated; Annals of Agricultural Sciences; 60(2):403–412.

[61] Gavahian, M., Farhoosh, R., Farahnaky, A., Javidnia, K. and Shahidi, F. (2015) Comparison of extraction parameters and extracted essential oils from Mentha piperita L. using hydrodistillation and steamdistillation; International Food Research Journal; 22(1):283–288.

[62] Qi H. and Kaijun, X. (2016) The effects of tangerine peel (*Citri reticulatae* pericarpium) essential oils as glazing layer on freshness preservation of bream (*Megalobrama amblycephala*) during superchilling storage; Original Research Article; Food Control; 69:339–345.

[63] Gavahian, M., Farahnaky, A., Javidnia, K. and Majzoobi, M. (2012) Comparison of ohmic-assisted hydrodistillation with traditional hydrodistillation for the extraction of essential oils from *Thymus vulgaris* L; Innovative Food Science and Emerging Technologies; 14:85–91.

[64] Gavahian, M., Farahnaky, A., Farhoosh, R., Javidnia, K. and Shahidi, F. (2015) Extraction of essential oils from *Mentha piperita* using advanced techniques: microwave versus ohmic assisted hydrodistillation; Original Research Article; Food and Bioproducts Processing; 94:50–58.

[65] Oluyori AP, Olatunji GA, and Atolani O. 2013. Alkyd resin from *Ipomoea batatas* Lam; Pakistan Journal of Scientific and Industrial Research; 56:57–58.

[66] Trabelsi, D., Aydi, A., Zibetti, G.W., Porta, D., Sconagmiglio, M., Cricchio, V., Langa, E., Abderrabba, M. and Mainar, A.M., Supercritical extraction from Citrus *Aurantium amara* peels using CO_2 with ethanol as co-solvent; The Journal of Supercritical Fluids; http://dx.doi.org/doi:10.1016/j.supflu.2016.07.003.

[67] Kubatov, A., Lagade, A.J.M., Miller, D.J., and Hawthorn, S.B. (2000) Selective extraction of oxygenates from savory and peppermint using subcritical water; Flavour and Fragrance Journal; 16:64–73.

[68] Saldana, M.D.A. and Valdivieso-Ramírez, C.S. (2015) Pressurized fluid systems: phytochemical production from biomass; Journal of Supercritical Fluids; 96:228–244.

[69] Zhu, Z., Dong, F., Liu, X., Qian, L., Yang, Y., Liu, F., Chen, L., Wang, T., Wang, Z. and Zhang, Y. (2016) Effects of extraction methods on the yield, chemical structure and anti-tumor activity of polysaccharides from *Cordyceps gunnii* mycelia; Carbohydrate Polymers; 140:461–471.

[70] Hu, W., Guo, T., Jiang, W., Dong, G., Chen, D., Yang, S. and Li, H. (2015) Effects of ultrahigh pressure extraction on yield and antioxidant activity of chlorogenic acid and cynaroside extracted from flower buds of *Lonicera japonica*; Chinese Journal of Natural Medicines; 13(6):445–453.

[71] Zhang, S., Bi, H. and Liu, C. (2007) Extraction of bio-active components from *Rhodiola sachalinensis* under ultrahigh hydrostatic pressure; Separation and Purification Technology; 57(2):277–282.

[72] Kotnik, P., Škerget, M. and Knez, Ž. (2006) Supercritical fluid extraction of chamomile flower heads: comparison with conventional extraction, kinetics and scale-up; The Journal of Supercritical Fluids; Selected Papers from the 3rd International Meeting on High Pressure Chemical Engineering, Erlangen, Germany, May 10–12, 2006; vol.43(2):192–198.

[73] Tatke, P. and Rajan, M. (2014) Comparison of conventional and novel extraction techniques for the extraction of scopoletin from *Convolvulus Pluricaulis*; Indian Journal of Pharmaceutical Education and Research; 48(1):27–31.

[74] Takuya, S., Masahiro, T., Hideo I., Teruaki, M., Yukihiro, K. Chiho, S., Yoshito, S., Munehiro, H., Armando, T., Mitsuru, S, Junshi S. and Motonobu G. (2013) Supercritical CO2 extraction of essential oil from Kabosu (*Citrus sphaerocarpa* Tanaka) peel; Flavour; 2:18. DOI: 10.1186/2044-7248-2-18.

[75] Abdalbasit A.M., Bertrand M. and Maznah I. (2011) Comparison of supercritical fluid and hexane extraction methods in extracting Kenaf (*Hibiscus cannabinus*) seed oil lipids; Journal of the American Oil Chemists' Society; 88(7):931–935.

[76] Bak, I., Lekli, I., Juhasz, B., Varga, E., Varga, B., Gesztelyi, R., Szendrei, L. and Tosaki, A. (2010) Isolation and analysis of bioactive constituents of sour cherry (*Prunus cerasus*) seed kernel: an emerging functional food; Journal of Medicinal Food; 13(4):905–10. DOI: 10.1089/jmf.2009.0188.

[77] Swati, K., Suchandra, C. Saroj, G. (2014) Bioactive constituents of germinated fenugreek seeds with strong antioxidant potential; Journal of Functional Foods; 6:270–279, DOI:10.1016/j.jff.2013.10.016.

[78] Zgoda-Pols, J.R., Freyer, A.J., Killmer, L.B. and Porter, J.R. (2002) Antimicrobial diterpenes from the stem bark of *Mitrephora celebica*; Fitoterapia; 73(5):434.

[79] Quintans, J.S., Soares, B.M., Ferraz, R.P., Oliveira, A.C., da Silva, T.B., Menezes, L.R., Sampaio, M.F., Prata, A.P., Moraes, M.O., Pessoa, C., Antoniolli, A.R., Costa, E.V. and Bezerra, D.P. (2013) Chemical constituents and anticancer effects of the essential oil from leaves of *Xylopia laevigata*; Planta Medica; 79(2):123–30.

[80] Sakdarat, S., Shuyprom, A., Pientong, C., Ekalaksananan, T. and Thongchai S. (2009) Bioactive constituents from the leaves of *Clinacanthus nutans Lindau*; Bioorganic and Medicinal Chemistry; 17(5):1857–60.

[81] Bhattacharya, S., Christensen, K.B., Olsen, L.C., Christensen, L.P., Grevsen, K., Færgeman, N.J., Kristiansen, K., Young, J.F. and Oksbjerg, N. (2013) Bioactive components from flowers of *Sambucus nigra* L. increase glucose uptake in primary porcine myotube cultures and reduce fat accumulation in *Caenorhabditis elegans*; Journal of Agricultural and Food Chemistry; 61(46):11033–40.

[82] Casquete, R., Castro, S.M., Martín, A., Ruiz-Moyano, S., Saraiva, J.A., Córdoba, G.M. and Teixeira, P. (2015) Evaluation of the effect of high pressure on total phenolic content, antioxidant and antimicrobial activity of citrus peels; Innovative Food Science & Emerging Technologies; 31:37–44.

[83] Olech, M., Nowak, R., Nowacka, N., Pecio, L., Oleszek, W., Los, R., Malm, A. and Rzymowska, J. (2015) Evaluation of rose roots, a post-harvest plantation residue as a source of phytochemicals with radical scavenging, cytotoxic, and antimicrobial activity; Industrial Crops and Products;69:129–136.

[84] Kalogeropoulos, N., Chiou, A., Pyriochou, V., Peristeraki, A. and Karathanos, V.T. (2012) Bioactive phytochemicals in industrial tomatoes and their processing by-products; LWT —Food Science and Technology; 49(2):213–216.

[85] Surendra T.V., Roopan S.M., Arasu M.V., Al-Dhabi, N.A., Sridharan M., (2016) Phenolic compounds in drumstick peel for the evaluation of antibacterial, hemolytic and photocatalytic activities; Journal of Photochemistry and Photobiology, B: Biology; 161:463–471.

[86] Cancer facts and figures 2015, American Cancer Society.

[87] Pinar, K., Jeffrey, S., Mustaqeem, A.S., Holly, V.H. and Nilay, D.S. (2011) Impact of new drugs and biologics on colorectal cancer treatment and costs. Impact of new drugs and biologics on colorectal cancer treatment and costs; Journal of Oncology Practice; 7(3):30–37.

[88] Oluyori, A.P., Shaw A.K., Olatunji G.A., Rastogi P., Sanjeev, M., Datta D., Arora, A., Sama A. and Reddy, P.S. Sweet potato peels and cancer prevention; Nutrition and Cancer; (In Press), 68(8):1330–1337.

[89] Campbell, C.J. and Laherrere, J.H. 1998. The end of cheap oil; Scientific America; 78–84.

[90] Vimmerstedt, LJ, Bush, BW, Hsu, DD, Inman, D and Peterson, S.O. 2015. Maturation of biomass-to-biofuels conversion technology pathways for rapid expansion of biofuels production: a system dynamics perspective; Biofuel, Bioproduct and Biorefining; 9 (2):158–176.

[91] Khan, A.M., Khaliq, S. and Sadiq R. 2015. Investigation of waste banana peels and radish leaves for their biofuels potential; Bulletin Chemical Society of Ethiopia; 29(2):239–245.

[92] Igwe, C., Agbaeze, E., Obike, I. and Sonde, U. (2012) Characterization and determination of ethanol fuel composite qualities of *Saccharum officinarum, Pennisetum purpureum* and *Costusafer*; Asian Journal of Plant Science and Research; 2(5):643–649.

[93] Sarkar, N., Ghosh, S., Bannerjee, S. and Aikat, K. (2012) Bioethanol production from agricultural wastes: an overview; Renewable Energy; 37:19–27.

[94] Ohmain, I. (2010) Emerging bio-ethanol projects in Nigeria: their opportunities and challenges; Energy Policy; 38:7161–7168.

[95] Gupta, A. and Verma, P. (2015) Sustainable bio-ethanol production from agro-residues: a review. Renewable and Sustainable Energy Reviews; 41:550–567.

[96] Sun, Y. and Cheng, J. (2002) Hydrolysis of lignocellulosic materials for ethanol production: a review; Bioresource Technology; 83:1–11.

[97] Hendrick A.T. and Zeeman G. (2009) Pretreatments to enhance the digestibility of lignocellulosic biomass; Bioresource Technology; 100:10–18.

[98] Ballesteros M., Oliva J.M., Negro M.J., Manzanares P. and Ballesteros I. (2004) Ethanol from lignocellulosic materials by a simultaneous saccharification and fermentation process (SFS) with *Kluyveromyces marxianus* CECT 10875; Process Biochemistry; 39:1843–1848.

[99] Lloyd T.A., Wyman C.E. (2005) Combined sugar yields for dilute sulfuric acid pretreatment of corn stover followed by enzymatic hydrolysis of the remaining solids; Bioresource Technology; 96:1967–1977.

[100] Ma F., Yang N., Xu C., Yu H., Wu J. and Zhang X. (2010) Combination of biological pretreatment with mild acid pretreatment for enzymatic hydrolysis and ethanol production from water hyacinth; Bioresource Technology; 101:9600–9604.

[101] Yu H., Zhang X., Song L., Ke J., Xu C., Du W. and Zhang J. (2010) Evaluation of white-rot fungi-assisted alkaline/oxidative pretreatment of corn straw undergoing enzymatic hydrolysis by cellulase; Journal of Bioscience and Bioengineering; 110(6):660–664.

[102] Yu J., Zhang J. He J., Liu Z. and Yu Z. (2009) Combinations of mild physical or chemical pretreatment with biological pretreatment for enzymatic hydrolysis of rice hull; Bioresourse Technology; 100:903–908.

[103] Yu H., Guo G., Zhang X., Yan K. and Xu C. (2009) The effect of biological pretreatment with the selective white-rot fungus *Echinodontium taxodii* on enzymatic hydrolysis of softwoods and hardwoods; Bioresoure Technology; 100:5170–5175.

[104] Anwar, Z., Gulfraz, M. and Irshad, M. (2014) Agro-industrial lignocellulosic biomass a key to unlock the future bio-energy: a brief review; Journal of Radiation Research and Applied Sciences; 7:163–173.

[105] Saritha M., Arora A., Lata. (2012) Biological pretreatment of lignocellulosic substrates for enhanced delignification and enzymatic digestibility; Indian Journal Microbiology; 52(2):122–130, https://www.ncbi.nlm.nih.gov/pmc/articles/PMC3386436/.

[106] Petersonn A., Thomsen M.H., Hauggaard-Nielsen H. and Thomsen A. 2007. Potential bioethanol and biogas production using lignocellulosic biomass from winter rye, oilseed rape and faba bean; Biomass and Bioenergy; 31:812–819.

[107] Ezeonu, C.S., Onwurah, I.N.E., Ubani, C.S., Ejikeme, C.M. and Ogodo, A.C. (2016) Achievements in the life sciences *Trichophyton Soudanense* and *Trichophyton*; Achievements in the Life Sciences; 10:72–79.

[108] Yadira, P.B., Sergio, S., Fernando, S.E.L., Sebastian, P.J. and Eapen, D. (2014) Bioethanol production from coffee mucilage; Energy Procedia; 57:950–956.

[109] Moshi, A.P., Temu, S.G., Nges,I.A., Malmo, G., Hosea, K.M.M., Elisante, E. and Mattiasson, B. (2015), Combined production of bioethanol and biogas from peels of wild Cassava *Manihot glaziovii*; Chemical Engineering Journal; 279:297–306.

[110] Paulova, L., Patakova, P., Rychtera, M. and Melzoch, K. (2014) High solid fed-batch SSF with delayed inoculation for improved production of bioethanol from wheat straw; Fuel; 122:294–300.

5

Biomass as Raw Material for Production of High-Value Products

Sibel Irmak

Abstract

Industrial production of a wide range of value-added products heavily relies on fossil resources. Lignocellulosic biomass materials are receiving increased attention as a renewable, economical, and abundant alternative to fossil resources for the production of various value-added products. Biomass feedstocks utilized for these productions include energy crops, agricultural biomass residues, forest biomass, and food-based biomass wastes. Various conversion technologies are used for production value-added products from biomass. Efficiencies of conversion technologies highly depend on the types of biomass used as raw materials that differ in contents and compositions of cellulose, hemicellulose, and lignin structures in biomass. In some conversion technologies, such as chemical, biochemical, and hydrothermal conversion techniques, biomass materials must be first broken down into smaller molecular weight components (e.g., oligosaccharides and monosaccharides) in order to be efficiently converted into target products. In this matter, pretreatment and hydrolysis play critical roles on the yield of the product(s). The chapter describes lignocellulosic materials that are used for production of top value-added products and conversion technologies to produce products in high yields. Future developments in the conversion of lignocellulosic biomass into value-added products are directly correlated to improvements of conversion technologies and selection the right types of biomass in the process.

Keywords: biomass types, value-added products, biofuels, bioproducts, conversion methods

1. Introduction

Being a nonedible portion of the plant, lignocellulosic biomass materials are attractively growing the attention as sustainable and renewable energy sources. Biomass materials can be used for producing a wide range of value-added products, including biofuels (ethanol, hydrogen, etc.), bioproducts products (sugar and sugar alcohols, etc.), and industrially important chemicals (e.g., solvents) [1]. Conversion can be performed using a variety of methods, including chemical, biochemical, and thermochemical processes. Each method offers several advantages or disadvantages for high yielding of a certain product.

Biomass can be derived from forestry wastes such as residues of the trees and shrubs, energy crops like sorghum, miscanthus, kenaf, switchgrass, corn, sugarcane, and any agricultural residues such as corn stovers, wheat straw, etc. The diversity in the chemical composition of biomass (cellulose, hemicellulose, and lignin constituents) can affect the conversion technologies employed for production of high-value products.

2. Top value-added products from biomass

2.1. Biofuels

Ethanol is a renewable fuel made from corn, sugarcane, sweet sorghum, and other variety of carbohydrate containing sources. In the United States, the ethanol fuel industry is largely based on corn. The production consists of the fermentation of sugars to produce a dilute alcohol solution, which is followed by distillation and dehydration to produce fuel-grade ethanol. Ethanol is blended with gasoline in various amounts for use in vehicles (E10, E15, and E85). It was reported that the use of 10% ethanol (E10) blends decreases greenhouse gas emissions by 12–19% compared to petroleum-based fuels [2].

Butanol is also a biomass-based renewable fuel that can be produced by alcoholic fermentation of biomass feedstocks [3]. Butanol is commonly produced using fossil fuels, but it can also be produced from biomass. Butanol can blend with gasoline better than ethanol since it has longer hydrocarbon chain thus, has lower polarity. Butanol has an energy density similar to that of gasoline [4]. However, because of limitations and difficulties in its production it is not considered a viable biofuels like ethanol. Biobutanol is produced by ABE (acetone-butanol-ethanol) fermentation process using various substrates. ABE fermentation suffers from limitations and difficulties including low butanol titer, low butanol yield, high substrate cost (grain and molasses), end-product inhibition, and high product recovery cost by distillation [5].

Methanol is another alcoholic fuel that has a high octane rating, easily distributable, and has low volatility. Methanol has chemical and physical fuel properties similar to ethanol. Methanol is mostly produced by the catalytic conversion of syngas from fossil sources but it can be also produced from lignocellulosic biomass materials [6]. Methanol can be used directly as fuel or fuel in fuel cells or as a feedstock for the gasoline additive methyl tertiary butyl ether.

Biodiesel is another most widely used liquid biofuel; however, its production does not rely on lignocellulosic fraction of biomass. Biodiesel is produced from vegetable oil or animal fat with an alcohol and a catalyst through transesterification reaction.

Producer gas is the mixture of gases produced by the gasification of biomass materials. It is composed of CO, H_2, CO_2, and low molecular weight hydrocarbons such as CH_4. Producer gas can be utilized as a fuel gas for heat or for electricity generation [7]. *Syngas* (synthesis gas) is a mixture of CO and H_2. Syngas has the ability to replace natural gas as a more thermally efficient liquid fuel. Electricity can be generated from the power provided by the combustion of syngas. It can also be used as a fuel source or as an intermediate for the production of other chemicals [8, 9].

Hydrogen is considered as a promising gas fuel. Hydrogen has the highest specific energy content of all conventional fuels. Its main use as fuel is in fuel cells. Fuel cells are being considered an attractive option for power generation, because of their high efficiency with no pollution [10]. Its only waste or byproduct is pure water, while hydrocarbon fuels produce massive amounts of carbon dioxide, a greenhouse gas. The development of less expensive and convenient methods for hydrogen production is challenging issue that limits its use. Among other various conversion methods, aqueous-phase reforming (APR) technique is a promising method for high-yielding hydrogen gas production [11, 12].

2.2. Bioproducts and industrially important chemicals

Nonedible lignocellulosic biomass materials are attracting increasing attention as renewable, economical, and abundant resources to reduce dependency on petroleum resources and minimize energy and material feedstock costs. In addition to energy and fuels, biomass can be used to create valuable carbon-based chemicals and materials, known as bioproducts. These products are sugars and sugar alcohols, glycerine, furfurals, cellulose fiber and derivatives, carbonaceous materials, resins, bioplastics, etc.

Activated carbons have been used for many applications including wastewater treatment (as adsorbent and filter) [13], catalyst [14], catalyst support [15], storage material [16], etc., and can be prepared from lignocellulosic biomass. *Mesoporous carbons* play important roles as catalysts supports as well as adsorbents, membranes, supercapacitors, chemical sensors, etc. [17, 18]. Mesoporous carbons can be prepared from lignin and enhance its mesoporosity by physical and chemical activation methods.

Furfural is a natural precursor to furan-based chemicals. It has been considered an important building block for the production of nonpetroleum-derived chemicals, a new generation of bioplastics, and potential biofuels or fuel additives. Furfural and its derivatives have been used to make jet and diesel fuel range alkanes, to serve as gasoline blend stock, and to develop a new generation of biofuels and bioplastics [19, 20].

Sugar alcohols are important products in the food industry. For instance, xylitol is a pentose sugar alcohol used as a sugar substitute in the food industry because of its low caloric and anticarcinogenic properties [21]. In addition, xylitol is a building block for a variety of commodity chemicals.

Addition to sugar alcohols, following C5 and C6 sugar-derived *platform chemicals* can be transformed into new families of useful molecules: 1,4-diacids (succinic acid, fumaric acid, and malic acid), 2,5-furan dicarboxylic acid, 3-hydroxy propionic acid, aspartic acid, glucaric acid, glutamic acid, itaconic acid, levulinic acid, 3-hydroxybutyrolactone, and glycerol [22].

Biomass feedstocks are converted into plenty of *other intermediate platforms, building blocks, secondary chemicals, and products* that are used in industry, transportation, textiles, food supply, environment, housing, etc. For instance, 1,3-butadiene is the building block for the production of rubbers, which are used in the production of tires for light vehicles. Ethyl lactate is a biodegradable solvent produced by the esterification of ethanol and lactic acid, which are produced from biomass. It is used in industrial applications to replace volatile organic petroleum-derived compounds. Lactic acid, which is mostly produced by microbial fermentation of carbohydrates, is used in many applications, including in food, pharmaceuticals, polymers, etc. Succinic acid, which is a dicarboxylic acid, is precursor for the synthesis of high-value products such as commodity chemicals, polymers, surfactants, and solvents.

Lignocellulosic biomass-based *polymers* have been considered as good candidates for sustainable development as well as ecofriendly environment. A wide range of polymers can be prepared from biomass-derivatives by different reaction routes [22]. For instance, C5 and C6 sugars and their derivatives can be either incorporated in polymer backbone or be used as pendant groups to prepare glycopolymers that mimics structural and functional responsibilities of glycoproteins [23].

3. Parameters affecting the product yield

3.1. Pretreatments

Despite their potential, the complex and rigid structures of biomass materials limit their use in many applications. Biomass materials must first be broken down into components with smaller molecular weights (e.g., oligosaccharides and monosaccharides) in order to be efficiently converted into a range of products. The goal of pretreatment is to make the cellulose accessible to breakdown process (hydrolysis) for conversion to fuels or value-added products. Various pretreatment techniques change the physical and chemical structure of lignocellulosic biomass and improve hydrolysis rates. Pretreatment makes biomass accessible to deconstruction by altering structural features of biomass such as removing lignin and reducing cellulose crystallinity, thereby increasing porosity. Successful production of biofuels and other bioproducts from lignocellulosic biomass depends on the pretreatment and deconstruction methods applied as well as the physical and chemical properties of the biomass. An efficient pretreatment method followed by solubilization in aqueous media without using toxic and hazardous chemicals is necessary to obtain reduced molecular weight of carbohydrates from biomass to produce various biofuels and bioproducts other value-added products.

In *ammonia fiber explosion* (AFEX) pretreatment method, liquid ammonia is mixed with biomass under moderate pressure (0.7–2.8 MPa) and temperature (70–200°C) before rapidly releasing the pressure. The temperature of the reaction, residence time, and ammonia concentration are main parameters that affect the dissolution [24]. The sudden release of pressure in the system explosively ruptures the biomass fibers and eventually breaks down the cellulose, hemicellulose, and lignin polymers into smaller components, and increases the size and the number of micropores in the cell wall.

Steam explosion is one of the most common and widely employed physicochemical pretreatments for lignocellulosic biomass. Biomass is treated with high-pressure (0.69–4.83 MPa) saturated steam at high temperature (160–260°C) for a short time (several seconds to a few minutes) and then the pressure is rapidly released. This treatment destroys the fibril structure and makes the materials undergo an explosive decompression [25]. After the treatment, hemicellulose is degraded, part of lignin is solubilized, and cellulose binding is reduced. Partial degradation of hemicellulose and formation of toxic compounds are the main drawbacks of steam explosion process. Toxic and/or inhibitory compounds can affect and reduce the performance of fermentations steps [26]. Substances that may act as inhibitors of microorganisms include phenolic compounds and other aromatics, aliphatic acids, furan aldehydes, inorganic ions, and bioalcohols or other fermentation products [27].

Alkali pretreatment involves the use of bases, such as sodium, potassium, calcium, and ammonium hydroxides, for pretreatment of biomass. Sodium hydroxide is the most common base used in this application. Alkali pretreatment can be carried out at ambient conditions, but pretreatment times take hours or days rather than minutes or seconds [28]. Alkali pretreatment disrupts the lignin structure, causes partial decrystallization of cellulose, removes acetyl, and the various uronic acid substitutions on hemicellulose and increases the accessibility of enzymes to cellulose and hemicellulose [29]. A neutralizing step to remove lignin and inhibitors (salts, phenolic acids, furfural, and aldehydes) is required before enzymatic hydrolysis. Compared with acid processes, alkaline processes cause less sugar degradation, and many of the caustic salts can be recovered and/or regenerated [28].

Acid pretreatment is based on mixing biomass with concentrated or diluted acids at temperatures between 130°C and 210°C. The use of concentrated acid can count as a treatment method rather than pretreatment as explained in the hydrolysis section in the detail. The most commonly used acid in the acid pretreatment is dilute sulfuric acid (H_2SO_4). Other acids such as hydrochloric acid (HCl), phosphoric acid (H_3PO_4), and nitric acid (HNO_3) can also be used. As an alternative to these inorganic acids, organic acids such as maleic acid and fumaric acid can also be used for dilute acid pretreatment [30]. The application time could take from a few minutes to hours depending on the type of acid, concentration, and temperature used in the process. This pretreatment method solubilizes hemicellulose fraction and releases monomeric sugars and soluble oligomers from the cell wall matrix into the hydrolysate. After removing hemicellulose, increase in porosity improves enzymatic digestibility. However, the sugars released from hemicellulose can be further degraded to furfural and hydroxymethyl furfural, which takes role as inhibitors in fermentation process.

Other common pretreatment methods used for biomass are *oxidative delignifications*. These methods are (a) lignin biodegradation catalyzed by the peroxidase enzyme in the presence of H_2O_2, (b) lignin degradation caused by ozone, which attacks and cleaves aromatic ring structures, while hemicellulose and cellulose are hardly decomposed (*ozonolysis*) [25], (c) the internal lignin and hemicellulose bonds in presence of an organic are broken by using methanol, ethanol, acetone, ethylene glycol, triethylene glycol, or tetrahydrofurfuryl alcohol or aqueous organic solvent mixture in presence of inorganic acid catalysts such as HCl and H_2SO_4 or organic acid catalysts such as oxalic, acetylsalicylic, and salicylic acids (*organosolv process*) [28], and (d) biomass is treated with oxygen or air as an oxidizer in combination with water at elevated temperature (e.g., 195°C) and pressure (e.g., 1.2 MPa) (*wet oxidation*). Wet oxidation mainly solubilizes hemicellulose and removes lignin from biomass structure. Lignin is decomposed into carbon dioxide, water, and carboxylic acids. In this method, the formation of strong inhibitors such as furfural and 5-hydroxymethylfurfural is very low.

There are also some methods that use *ultrasound, microwave, and radiofrequency radiations* as pretreatments methods. These pretreatments are used in conjunction with other methods. Ultrasound produces sonochemical and mechanoacoustic effects, which affect the chemical and physical composition of biomass. The mechanoacoustic alters the surface structure of the biomass while the sonochemical production of oxidizing radicals is effective on biomass components [31]. It was reported that sonication of biomass-water mixture under 20 kHz ± 50 Hz ultrasound for 8 min did not change the hydrolysis that was performed in subcritical water after sonication. Although hydrolyzed solid biomass fraction remained the same, the molar mass of polysaccharides fractions in the hydrolysates notably decreased because of degradation. The amount of monosaccharides in sonicated hydrolysates was high that confirming the positive effect of sonication on degradation of polysaccharides. For instance, hydrolysis of kenaf biomass at 250°C subcritical water treatment resulted in 10% of more xylose release in the sonicated hydrolysate compared to nonsonicated one [32].

On the other hand, microwave treatment has enhanced the surface disruption and the breaking of lignin structures in switchgrass and improved enzymatic saccharification 53% more compared to conventional heating [33]. Microwave pretreatment has positive effect on solubilization of switchgrass in subcritical water. Hydrolysis percentages and total organic carbons released into solution are higher in microwave-treated biomass samples. When microwave pretreatment is applied at higher temperature, solubilization significantly increases (**Table 1**).

Biomass[a]	**Hydrolysis**[b] **(%)**	**TOC (mg L^{-1})**
Untreated	49.01 ± 0.9	1132 ± 13
MW—120°C	57.09 ± 0.6	1541 ± 14
MW—150°C	62.91 ± 0.7	1679 ± 17

[a]MW: microwave-treated biomass.
[b]Determined by weighing microwave-treated biomass samples before and after solubilization in subcritical water treatment.

Table 1. Solubilization of microwave pretreated samples in subcritical water.

Microwave pretreatment was used for solubilization of lignocellulosic biomass in combination with acid and alkali treatments followed by enzymatic hydrolysis [33–35]. Chimentão et al. [36] investigated hydrolysis of dilute acid-pretreated cellulose in a conventional oven and under microwave heating. Although the method was called "mild hydrothermal conditions," the hydrolysis process was accelerated using acids (sulfuric and oxalic acids).

Radio frequency (RF) heating is another promising dielectric heating technology, which is used as an initial breakdown of the lignocellulosic matrix. Dielectric heating transforms electromagnetic energy into heat that is effective on breakdown of biomass structure. The electromagnetic field could generate nonthermal effects, which can also accelerate the destruction of the crystallinity structure [37]. RF heating prevents uncontrolled heating and overheating that protects the product from degradation. RF has large penetration depth (10–30 m) and higher energy efficiency than microwave [38–40]. Efficiency of pretreatment highly depends on temperature, frequency, and type of product/biomass (water content, chemical composition, etc.). Radio frequency-assisted dielectric heating was usually combined with alkaline pretreatment for destruction of biomass materials [41, 42].

Biological pretreatments are attractive approaches for delignification of lignocellulosic biomass. Advantages of these processes are mild operation conditions, environment friendly, and low energy requirement [43]. The formations of toxic materials (furfural, hydroxymethylfurfural, etc.) are low. However, these pretreatments are time consuming [44]. The microbial treatment includes fungi, such as white-rot fungi, brown-rot fungi, and soft-rot fungi, actinomycetes, and bacteria to degrade recalcitrant polymeric structures in biomass [45]. Brown rots are mainly effective on polysaccharides with minimal lignin degradation, whereas white and soft rots attack carbohydrates and lignin. Most of the brown-rot fungi degrade cellulose and hemicellulose more rapidly than lignin in lignocellulosic biomass. White-rot fungi are the most effective for biological pretreatment of lignocellulosic materials since they are able to efficiently degrade all components of plant cell walls, both carbohydrates and lignin. Enzymes such as peroxidases and laccase take roles on lignin degradation by white-rot fungi. Bacteria and actinomycetes are not as efficient as white- and brown-rot fungi in pretreatment of biomass. Lignocellulosic biomass materials such as wheat, corn and rice straws, corn stover, corn stalks, beech wood chips, pine wood chips, and switchgrass were successfully pretreated by these methods using white-rot fungi, brown-rot fungi, and soft-rot fungi [46–48].

Many applications have combined one pretreatment with other one for effective breakdown of biomass structure.

3.2. Hydrolysis of biomass materials

The major hydrolysis processes typically used for the solubilization of biomass require either use of toxic, corrosive, and hazardous chemicals (e.g., acid and alkali treatments) or longer retention times (e.g., enzymatic hydrolysis), which collectively make the process environmentally unsafe and/or expensive. Mineral acids are commonly used to dissolve hemicelluloses, whereas lignin is typically dissolved by alkaline or organosolv pretreatments [45, 49]. Recovery of the chemical catalyst is often crucial to the success of these processes [24]. On the other hand, generally harsh conditions (e.g., high temperatures and high acid concentrations) are needed to release glucose from biomass complex structures. Pyrolysis and other side reactions at higher temperatures become very important, and the amount of undesirable byproducts (tars) increases as the temperature is increased above 220°C [50].

Concentrated acid hydrolysis has been applied to breakdown lignocellulosic efficiently. The hydrolysis reaction for cellulose conversion into sugars is principally the degradation of chemical bonds in cellulose, involving the hydrolytic cleavage of beta-1,4-glycosidic bond, which is catalyzed by H_3O^+ ions of an acid. The reaction rate depends on the H_3O^+ ion concentration, the reaction temperature, and the chemical environment of the glycosidic bond [51]. The acid hydrolysis process usually employs sulfuric acid and hydrochloric acid at concentrations of 1–10% using a moderate temperature (in the range of 100–150°C) [52]. A two-step sulfuric acid hydrolysis is a widely used technique for releasing sugars from biomass. Biomass is first treated with concentrated sulfuric acid at a low temperature and then hydrolyzed with diluted sulfuric acid at an elevated temperature. Concentrated acid recrystallizes cellulose to less crystallized oligosaccharides followed by less concentrated and higher reaction temperature for converting recrystallized oligosaccharides to monosaccharides. Advantages of concentrated acid hydrolysis process are higher conversion from polysaccharides to monosaccharides with minimum formation of reaction byproducts with careful control of reaction conditions. The use of concentrated acid for biomass hydrolysis has also several drawbacks such as energy consumption, equipment corrosion, handling of nonsafe chemicals, an added necessary step of acid neutralization, the formation of byproducts that create an inhibitory effect in the fermentation [53, 25], and other negative environmental impacts. On the other hand, *hydrothermal treatments* (subcritical and supercritical water) are alternative ways to hydrolyze lignocellulosic biomass in an environmentally friendly manner by only operating temperature and pressure conditions. The main shortcoming of these applications is their very high investment cost. **Figure 1** shows morphological changes occur in biomass structure (kenaf) after subcritical water hydrolysis treatment [32].

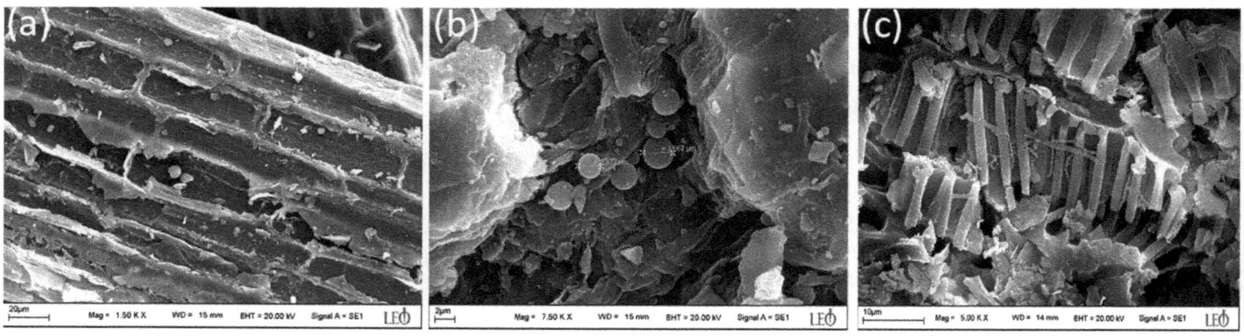

Figure 1. SEM images of untreated (a), 200°C (b) and 250°C (c) subcritical water treated kenaf samples.

Before treatment biomass exhibits rigid and highly ordered fibrils. The cell walls are visible creating a "brickwork-like" appearance to surface (a). The compacted outer layer is partially removed at 200°C and a range of discrete droplet morphologies that contain lignin is appeared on the cell. The 250°C-subcritical water treatment reduces and degrades lignocellulosic structure, leaving highly degraded solids. The maximum solubilization yield of wheat straw and kenaf biomass were found to be 70–75% in subcritical water medium (under 250°C and 27.58 MPa carbon dioxide pressure conditions) [11, 32]. However, the hydrolysates obtained in this process contained high molecular weight polysaccharides that were difficult to utilize for producing value-added products such as gas biofuel hydrogen.

3.3. Chemical composition of biomass materials

Chemical composition of biomass and structures of biopolymers (cellulose, hemicellulose, and lignin) are two important factors affecting the yield of the biofuels, bioproducts, and chemicals produced from biomass. Composition of lignin, cellulose, and hemicellulose in biomass materials significantly differ among biomass species. For instance, some biomass materials such as hardwoods contain more cellulose in their structures, whereas others such as straws and leaves have more of hemicelluloses. It is known that lignin content of herbaceous plants such as grasses is very low compared to softwoods, which are known to have highest amount of lignin in their structures [54]. On the other hand, polymerization degree and(or) structures of biopolymers can also considerably varies among biomass species. For instance, the chemical structure of lignin is based on syringyl (S), guaiacyl (G), and p-hydroxyphenyl (H) units. Softwood lignins are mainly composed of residues derived from guaiacyl units (lignin type G), whereas hardwood lignins contain both syringyl and guaiacyl units with minor amount of p-hydroxyphenyl (lignin type GS). Lignins from grasses are composed of the three basic precursors (lignin type HGS) [55, 56]. Hemicellulose fractions of softwoods mainly have D-mannose-derived structures such as galactoglucomannans, whereas hemicelluloses in hardwoods have D-xylose-derived structures such as arabinoglucuronoxylan. Xylan is a polypentose hemicellulose structure in biomass materials that displays a wide range of compositions, molecular sizes, and structures depending upon its source [57, 58]. This diversity among biomass materials can significantly affect the yield of value-added products directly produced from biomass as raw materials. On the other hand, this diversity can also affect solubilization efficiency of the biomass materials and, therefore, contents and compositions of the biomass components in the hydrolysates. The differences in the hydrolysates will considerably change the yield of the target compounds and byproducts produced from these biomass hydrolysates. For instance, molecular weight distribution of carbohydrates in the hydrolysates can significantly affect the method employed for biofuel or useful chemicals production. The more degraded organics containing hydrolysates are preferable for the production of certain various value-added products from biomass. For example, high-yielding hydrogen gas production from biomass hydrolysates requires reduced molecular weight oxygenated compounds containing biomass hydrolysates as feeds in aqueous-phase reforming gasification process [59].

3.4. Conversion methods

3.4.1. Chemical

Sugars released from biomass can be hydrogenated to C5-6 polyols (*sugar alcohols*) by using a chemical reducing agent such as sodium borohydride ($NaBH_4$). Biomass hydrolysates can also be utilized for direct production of derivatives, such as *furfural, hydroxymethyl furfural,* and/or *levulinic acid*. It is also possible to produce C2-3 *glycols* from biomass hydrolysates by hydrogenolysis. Sugar containing hydrolysates are further upgraded via oxidation or halogenation reactions [60]. Sugars obtained from biomass hydrolysis can be converted to N-heterocyclic components, pyrones, and aromatics that can be further converted into a *variety of chemical intermediates*.

Furfural is a valuable compound for a variety of chemical applications and it serves as a precursor for the synthesis of many fine chemicals and biofuels. It is produced industrially by acid-catalyzed hydrolysis and dehydration of pentoses (mainly xylose) in lignocellulosic feedstocks (sugarcane bagasse, corn cobs, sunflower stalk, etc.) at temperatures ranging from 153 to 240°C [61]. During the initial stage, the hemicellulose is hydrolyzed to xylans, which generate pentose carbohydrates to be further converted into furfural. Commercially, furfural is produced using sulfuric acid as a homogeneous catalyst. Significant quantities of steam are used in the process in order to strip the furfural and to avoid its further degradation.

Xylitol is currently produced through chemical reduction of xylose derived from birch wood chips and sugarcane bagasse hemicellulose hydrolysate. The chemical process adapted for xylitol production from xylan-rich biomass demands high production costs in terms of temperature and pressure input, as well as the formation of byproducts that require expensive separation and purification steps [62].

3.4.2. Thermochemical

Thermochemical conversion processes are combustion, pyrolysis, gasification, and liquefaction.

Combustion of biomass is the least complex conversion method to transform biomass into energy. Combustion process takes places at 800–1000°C. Complete combustion involves the production of heat as a result of the oxidation of carbon and hydrogen-rich biomass to CO_2 and H_2O. The biomass with high moisture content is not suitable for this process, predrying may be necessary in some cases. The high-pressure steam formed in the process can be utilized as hot air, hot water, steam, or electricity to produce hot gases. Combustion systems for electricity and heat production are similar to most fossil-fuel fired power plants. Combustion technologies can produce high NOx emissions.

Pyrolysis is thermal degradation of lignocellulosic biomass in the absence of oxygen around 500°C. The products are combination of solid (biochar) and gaseous fractions and liquid biooil (biocrude). These products can be used directly or after processing as fuel. Biochar and biooil can also be utilized for production of chemicals and value-added products [63]. Pyrolysis is a part of gasification and combustion, which consists of a thermal degradation of the initial solid biomass into gases and liquids without an oxidizing agent. Thermal decomposition of organic components in biomass starts at 350–550°C and goes up to 700–800°C in the absence of air/oxygen [64]. Pyrolysis can be fast or slow process depending on time taken for processing the feed into pyrolysis products. In slow pyrolysis, primarily biochar is produced at lower temperatures. At higher temperatures biooils are produced through fast pyrolysis. In slow pyrolysis, biomass is typically heated at slow heating rates (up to 10–20°C min^{-1} with sufficient time allowed for repolymerization reactions to maximize the solid yields while fast pyrolysis uses much higher heating rates >10–200°C s^{-1}), higher processing temperature and short vapor residence times (less than 2 s) that produce 50–70 wt% biooil yield (dry biomass basis). Fast pyrolysis usually requires a finely ground biomass feed (<1 mm); however, slow hydrolysis accepts a wide range of particle sizes (5–50 mm) [65].

Gasification is the conversion of biomass into a combustible gas mixture at 700–1600°C. The produced gas is rich in CO, H_2, CH_4, and CO_2. This combustible gas mixture can be used for different purposes after cleaning. The clean gas can be used directly as an engine fuel or upgraded to liquid fuels or converted into chemical feedstocks by different methods including biological fermentation or catalytic upgrading through the Fischer-Tropsch process [7, 19].

Lignocellulosic biomass can also be converted into liquid and gaseous fuels by *hydrothermal gasification processes.* Hydrothermal gasification processes are applied in either gaseous phase (steam reforming), in supercritical water, or in liquid phase (aqueous phase reforming). Steam reforming needs the lignocellulosic material to be dried prior to gasification. The temperatures applied in this conversion are also high, i.e., 800°C or above and considerable tar and char are formed. The thermodynamic critical points for water are 374.3°C and 221.2 bar for temperature and pressure, respectively. Effective gasification of lignocellulosics in supercritical water needs to go quite higher values than the critical points, e.g., above 600°C for complete gasification [66]. APR gasification process is a promising process since conversion reactions take place at moderate temperatures (225–265°C, and at pressures of 27–54 bar) in presence of a reforming catalyst (such as precious metal, Pt). The water-gas shift reaction ($CO + H_2O \leftrightarrow H_2 + CO_2$) is thermodynamically favored at processing temperatures. Biomass hydrolysates, which are composed of biomass-derived mixed oxygenated hydrocarbons, are successfully converted into gaseous products that are rich in hydrogen [12, 59, 67].

Hydrothermal liquefaction of biomass is the thermochemical conversion of biomass into liquid fuels at 280–370°C and 10–25 MPa in aqueous medium [68]. In this process, water is an important reactant and catalyst. It is still in liquid state and has a range of promising properties at the liquefaction conditions [68]. In this process, lignocellulosic biomass is break down into soluble low molecular weight products that undergo further condensation, cyclization, and polymerization reactions forming biooil that is soluble in organic solvents. These biooils are most commonly used as a replacement for petroleum fuels but they can be also served as feedstock for production of a wide range of value-added products. For instance, lignin-derived phenolics can be converted into aromatic chemicals by hydrogenation [69], and carbohydrates are utilized for catalytic production of hydrogen or consumed in fermentation [70].

Various value-added products can be produced or synthesis from biomass or biomass-derived compounds by thermochemical methods. For instance, sugars can be reduced into sugar alcohols by thermochemical reduction. In industrial reaction, sorbitol production by thermochemical reduction of glucose is mainly carried out discontinuously in stirred tank reactors at 100–180°C under 5–15 MPa of H_2 pressure in the presence of a catalyst, usually raney-type nickel, or ruthenium catalysts. **Table 2** shows thermochemical sugar alcohol production from glucose, simple biomass model compound, at various processing temperatures under 2.0 MPa H_2 pressure in the presence of carbon supported ruthenium catalyst. The reaction time was 60 min [71].

The results showed that 80°C hydrogenation temperature is not enough for complete reduction of glucose solution. Further increase of the process temperature causes complete reduction of all glucose; however, the contents and compositions of reduced products (sugar alcohols) significantly change as temperature increased. The main reduction product of glucose is sorbitol and its highest yield is at 100°C at the processing conditions described above.

Temperature (°C)	Sugar alcohols (% concentration)				
	Glucose	Mannitol	Sorbitol	Xylitol	Other products [b]
80	8.8	0.9	90.3	n.d.	n.d.
100	n.d.[a]	2.2	97.3	n.d.	n.d.
120	n.d.	5.9	88.6	n.d.	5.5
140	n.d.	16.5	59.7	2.7	21.1
160	n.d.	21.8	41.0	3.6	33.6
180	n.d.	23.4	20.7	4.1	51.8
200	n.d.	19.7	12.1	2.1	66.1

[a]n.d., not detected.
[b]Total of other reduced products (glycerol, erythritol, etc.).

Table 2. Thermochemical sugar alcohol production from reduction of glucose at various temperatures.

Other types of value-added products that can be produced by a thermochemical method are carbonaceous materials. Nonsolubilized residue containing mostly lignin and lignin destruction products can be utilized for production of carbon-based materials such as activated carbon and mesoporous carbons, etc. [72].

3.4.3. Biochemical

Biochemical conversion can also be used for breaking down biomass into sugars that can then be converted into biofuels (gaseous or liquid fuels) and bioproducts through the use of microorganisms and enzymes. This process is usually used for treating high moisture containing organic wastes. Biochemical processes are relatively slow processes that require more time for conversion of biomass into useful compounds [73]. The most popular biochemical technologies are *anaerobic digestion* and *fermentation* [74]. Biomass from various sources is biochemically degraded by *anaerobic digestion* in highly controlled, oxygen-free condition circumstances to produce biogas containing mostly methane and carbon dioxide. Biogas can be burned directly for heat or steam or converted to electricity. It can also be upgraded to biomethane or reformed into hydrogen fuel. In anaerobic digestion, bacteria are used to hydrolyze carbohydrates into sugars that are digestible by other bacteria. Methane and carbon dioxide are produced from those digestible components.

Fermentation is used commercially on a large scale to produce ethanol from carbohydrate containing materials. Structural carbohydrates in biomass are broken down into sugars by using enzymes. Then, released sugars are transformed into alcohols, organic acids, or hydrocarbons by microorganisms in fermentation. The intermediate sugars can also be utilized to obtain other value-added chemicals. Conversions in this process take place at 25–70°C under atmospheric pressure conditions.

Not only biofuels but also a wide range of value-added products such as food-based sugar alcohols (e.g., xylitol) can be produced by combined processes such as thermochemical process combined with a chemical or biochemical method [62].

4. Comparison of various types of biomass materials for high-yielding value-added products

4.1. Energy crops

Energy crops are specifically grown for its fuel value or to produce bioenergy. These plants usually require low cost and low maintenance to grow and they are utilized to make biofuels or directly exploited for their energy contents. Energy crops can be food crops (corn, sugarcane, sugar beet, sweet sorghum, etc.) or nonfood crops (poplar trees, switchgrass, miscanthus, kenaf, etc.). A major focus among them is nonfood energy crops.

Perennial grasses (C4 species) such as *switchgrass* and *miscanthus* are promising potential lignocellulosic biomass sources for biofuel production. They require less fertilizer, water, and energy for production. Since these grasses regrow from their roots, they do not need to be replanted each year. Switchgrass and miscanthus are being developed as bioenergy crops because they have high yield potential and desirable agronomic traits. Switchgrass is a native prairie plant that grows about 10 feet tall. Annual dry matter yields are 15–30 t ha^{-1}. It exhibits adaptation to relatively wet and fertile soils. Switchgrass is very tolerant of poor soils, flooding, and drought. It greatly reduces soil erosion. Giant miscanthus is a perennial warm-season (C4) sterile hybrid grass. It is native to subtropical and tropical regions of Africa and southern Asia, which has also been used as an ornamental plant in many countries. Giant miscanthus can grow best in the soils that are well drained and have a pH between 5.5 and 7.5 under medium to high fertility [75]. Giant miscanthus has produced more than double the biomass of upland switchgrass per acre. Yield estimates from 10–15 tons per acre. Harvestable stems of miscanthus are usually more than 9 feet long [76].

Kenaf (*Hibiscus cannabinus* L.), is a warm season annual herbaceous crop that belongs to Malvaceae family and grows best in the tropics and subtropics where the mean daily temperatures during the growing season exceed 20°C. Kenaf is a good candidate for bioenergy because it grows fast without special care and produces large amount of biomass. Under good conditions, kenaf will grow to a height of 5–6 m in 6–8 months and produce up to 30 t ha^{-1} y^{-1} of dry stem material [77].

Corn is one of the most important crops in the world, and the United States is the largest corn producer in the world. The total production of corn in the United States for the year 2014–2015 was reported to be 14.215 billion bushels [78]. A major use of corn in the United States is ethanol production (30.5%). Other uses beyond animal feed include the production of high fructose corn syrup (3.6%), sweeteners (2.1%), starch (1.8%), cereals (1.5%), beverage alcohol (1.0%), etc. [78]. The amount of cropland devoted to growing corn has considerably increased in recent years and now, the United States has excess corn production. Ethanol production capacity from corn has reached about 15.25 billion gallons in 2014/15, exceeding projected capacity demand of 15.1 billion gallons per year [79]. Most ethanol in the United States is made from corn because it is a relatively low-cost source of starch that can easily be converted into sugar, fermented, and distilled into ethanol. Today, more than 10% of vehicle fuel comes from ethanol. Producing ethanol increases the cost of corn, which leads to *higher* prices for other commodities such as dairy, meat, and livestock feed that farmers rely on. Since ethanol use has increased incredibly since 2000, the demand and the price of corn have increased. This has forced the farmers to use more pristine lands to plant corn. This land conversion and the associated use of agricultural fertilizers and pesticides have caused significant increases in greenhouse gas emissions. Even more emissions were released when ethanol consumed as fuels. Because ethanol has less energy content than gasoline, vehicles have to consume more fuel and thus produce more emissions.

Sugarcane (*Saccharum* spp.) is an efficient tropical/subtropical grass with nitrogen-fixing symbionts, thus needing little fertilizer. Brazil is the largest producer (31 million tons year^{-1}) and produces 25% of the world's sugarcane [80]. Commercial average of sugar cane yield (Australia, Colombia, and South Africa) and total dry matter (biomass) production are reported to be 84 and 39 t ha^{-1} yr^{-1}, respectively [81]. A cycle of sugarcane planting and harvesting is about 12 months. The average sugarcane yield from four cycles is about 60 t ha $^{-1}$ yr^{-1} [82]. The sugarcane industry's waste product, bagasse, can also be used to produce biofuels. Bagasse is the fibrous residue left over when sugarcane is squeezed for its juice. It was reported that energy balance of sugarcane-based ethanol is seven times greater than that of corn-based ethanol. Energy balance defined as the difference between the energy consumed to produce ethanol from crop and the energy released when it is consumed as biofuel. Only certain percent of corn starch can be converted into ethanol that required an initial breakdown process to release sugars from starch first. This step significantly increases the operation costs compared to sugar-based ethanol [83].

Poplar trees are already known to be good candidates for bioenergy because of their fast growth rates (up to 4 m annually and matures in about 6 years) and large biomass production. Poplar can grow on 2–3 year rotations, growing back after it is harvested, with seven or more cycles possible before replanting. After harvest, the tree resprouts from the same root stalk, a very efficient way to produce biomass. Hybrid poplars have been planted for their potential as an energy crop due to their high yield rates and adaptability to many growing conditions. Hybrid poplar growth and biomass yield could vary significantly depending on climatic conditions, soil characteristics, and species genotype [84, 85]. Poplar plantations have many environmentally desirable applications, including use as buffer strips to decrease erosion and nitrate in run-off from highly erodible fields, treatment, and removal of toxic materials from landfills and other soil contaminations, and an excellent sink of atmospheric CO_2 [86]. Poplar woods do not need to be stored like other bioenergy crops such as grasses; it can be harvested throughout the year. Poplar requires less water and other inputs than crops such as corn. The higher pest resistance, increased site adaptability, and development of poplar genotypes with improved yield make poplar trees important biomass materials for biofuel production.

Sweet sorghum (*Sorghum bicolor* L. Moench) is a widely adapted sugar crop that can be produced at less cost than corn [87]. It is a highly productive and versatile crop that can be cultivated in temperate climates. Sweet sorghum can be grown in many regions of the United States. Under favorable growing conditions, sorghum can produce high biomass yields with low rates of nitrogen fertilizer. Individual stalks can be over 10 feet tall. Two fractions of sweet sorghum can be used for biofuel production. The stalk and seed are used directly for biomass energy, and their high sugar content allows them to be fermented to make ethanol. Rather than producing starch, sweet sorghum carbohydrates are stored in the stalk as sugar, with sugar concentrations of 8–20% [88]. As the plant matures, carbohydrates are translocated from the stalks to the head to make starch in the seeds. Conversion of these sugar components into ethanol requires less energy than starch for same reason discussed for sugarcane. Sweet sorghum has the potential to produce up to 6000 L ha^{-1} of ethanol equivalent to corn grain yields of approximately 20 Mg ha^{-1} [89]. The main drawback in sweet sorghum is; since sugars in the sweet sorghum are not stable and rapidly degrade it is required to squeeze the juice out immediately for conversion. This reduces flexibility in harvesting and can increase transportation costs [90].

4.2. Agricultural biomass residues

Agricultural residues from well-established production chains are important sources of biomass that can provide a substantial amount of biomass for production of a wide range of value-added products. Since these residues are a natural byproduct of the food crop, they can be used as promising low-cost feedstocks without increasing the amount of land used for agriculture. Agricultural biomass includes *bagasse, straw, stem, stalk, leaves, husk, shell, peel, pulp, stubble*, etc. is produced annually worldwide and is not utilized to any significant extent.

Cereal straw is a typical example of an agriculture byproduct. Rice straw (includes stems, leaf blades, leaf sheaths, and the remains of the panicle after threshing) and wheat straw (includes nongrain portion of the wheat plant; stems, leaves, and chaff) are two of the most abundant cereal-based lignocellulosic wastes in the world. Wheat is the second most important grain that is cultivated in the United States, following only corn. The United States is a major wheat-producing country ranked third in production volume of wheat in the world. Total wheat production in United States was about 55 million tons in 2015. A good wheat crop yields between 1 and 1.2 tons of straw per acre on a dry matter basis. The annual global rice straw production is 731 million tons and Asia alone produces 667.6 million tons [1]. One ton of rice paddy produces approximately 290 kg rice straw. Rice husk is one of the major byproducts from the rice milling process and constitutes about 20% of paddy by weight. One ton of rice paddy produces roughly 220 kg rice husk [2].

High-yielding corn production activities generate large amounts of corn stover that is roughly 80% of all agricultural residues produced in the United States [91]. It has been estimated that corn in the United States produces 1.7 times more residue than other cereals. Corn stover contains 27.5% stalk, 8.2% cob, and 7.0% husk [92].

Biomass	Advantages and disadvantages
Switchgrass and miscanthus	Nonedible Require less fertilizer, water, and energy for production Do not need to be replanted each year Miscanthus produces more biomass than switchgrass
Kenaf	Nonedible Grows fast in the tropics and subtropics Does not need special care for production produces large amount of biomass
Corn	Edible Mostly used for ethanol production because of relatively low-cost source of starch Large amount of cropland devoted to corn Excess corn production in the United States
Sugarcane	Edible Grows in tropics and subtropics Needs little fertilizer Sugarcane juice is directly used in conversion techniques (no pretreatment needed) Waste product, bagasse, is an important feedstock for value-added products
Sweet sorghum	Needs little fertilizer Sugars in the juice can be directly utilized for value-added products (no pretreatment needed) Required to squeeze the juice out immediately (not stable) Limited flexibility in harvesting and transportation costs
Poplar trees	Nonedible Grow fast Produce large amount of biomass Can be harvested throughout the year Have many environmentally desirable applications such as reducing erosion
Cereal straw	Nonedible Natural byproduct of the food crops rice, wheat, corn, etc. Produce large amounts of biomass
Corn stover	Nonedible Natural byproduct of the food crop, corn Produce large amounts of biomass (more than cereals)
Forest biomass	Nonedible Largest source of lignocellulosic biomass High costs of harvesting and transportation Widely used in combustion process but not in gasification, pyrolysis, and fermentation
Food biomass wastes	Edible/nonedible Have low values Released in large amounts from food industries

Table 3. Comparison of biomass materials for production of value-added products.

4.3. Forest biomass

Forests are the largest source of lignocellulosic biomass that can be substitute for fossil fuels in the production of energy and other value-added products. Since it is a nonfood type of biomass, it is a promising feedstock for these conversions [93]. Forest biomass includes material left on logging sites (*trunks, crowns, and branches*), *unused wood* from forests and *wood manufacturing, and processing residues*. The wastes from wood industries are sawdust, off-cuts, trims, and shavings. Forest biomass utilization has some challenges that restrict fully production of value-added products from forests. The main issue is high costs of harvesting and transportation of these biomass materials. Second, forest biomass has been widely used in combustion process, however, other potential conversion technologies such as gasification, pyrolysis, and fermentation are not applied to forest biomass [94].

4.4. Food biomass wastes

The food industry produces a large number of residues and byproducts that can be used as biomass energy sources.

Bagasse is a promising biomass material that is produced during sugarcane processing. After sugarcane is milled for juice extraction, bagasse is obtained as a residue, which is about 25% of the total weight and contains 60–80% carbohydrates [95].

Pomace is also important food waste-based biomass candidate from wine industry that remains after the grapes are pressed and consists of skins, seeds, and a small percentage of stems. It is the most abundant wine-making waste representing about the 20% (w/w) of grapes used for the production of wine. It is rich in carbohydrates [96].

Nuts hulls, peanuts shells, corn stover, rice hull/husk, and other grain biomass residues can also be considered food biomass wastes. On the other hand, *corn bran*, a byproduct of the corn milling process, can also be counted as lignocellulosic biomass. A large starch producer in the United States releases approximately 9×10^5 tons of corn bran per year [97]. However, this byproduct currently has low value and is often used for animal feed alone or combined with corn germ cake or meal.

Table 3 summarizes advantages and disadvantages of biomass materials for utilization as raw materials for production of value-added products.

Author details

Sibel Irmak

Address all correspondence to: sibel.irmak@unl.edu

Biological Systems Engineering, Industrial Agricultural Products Center, University of Nebraska-Lincoln, Nebraska, USA

References

[1] Clark J, Deswarte F. Introduction to Chemicals from Biomass. 2nd ed. Chichester: Wiley; 2015. pp. 114–281. DOI: 10.1002/9781118714478

[2] Saini JK, Saini R, Tewari L. Lignocellulosic agriculture wastes as biomass feedstocks for second-generation bioethanol production: concepts and recent developments. 3 Biotech. 2015;5:337–353. DOI: 10.1007/s13205-014-0246-5

[3] Hansen AC, Kyritsis DC, Lee CF. Characteristics of biofuels and renewable fuel standards. In: Vertes AA, Blaschek HP, Yukawa H, Qureshi N, editors. Biomass to biofuels-strategies for global industries. New York: John Wiley; 2009. DOI: 10.1002/9780470750025.ch1

[4] Jin C, Yao M, Liu H, Lee CF, Ji J. Progress in the production and application of n-butanol as a biofuel. Renew Sustain Energy Rev. 2011;15:4080–4106. DOI: 10.1016/j.rser.2011.06.001

[5] Sabra W, Groeger C, Sharma PN, Zeng AP. Improved *n*-butanol production by a non-acetone producing *Clostridium pasteurianum* DSMZ 525 in mixed substrate fermentation. Appl Microbiol Biotechnol. 2014;98:4267–4276. DOI: 10.1007/s00253-014-5588-8

[6] Yin X, Leung DYC. Characteristics of the synthesis of methanol using biomass-derived syngas. Energ Fuels. 2005;19:305–310. DOI: 10.1021/ef0498622

[7] Wang L, Weller CL, Jones DD, Hanna MA. Contemporary issues in thermal gasification of biomass and its application to electricity and fuel production. Biomass Bioenergy. 2008;32:573–581. DOI: 10.1016/j.biombioe.2007.12.007

[8] De María R, Díaz I, Rodríguez M, Sáiz A. Industrial methanol from syngas: kinetic study and process simulation. Int J Chem Reac Eng. 2013;11:469–477. DOI: 10.1515/ijcre-2013-0061

[9] Khandan N, Kazemeini M, Aghaziarati M. Direct production of dimethyl ether from synthesis gas utilizing bifunctional catalysts. Appl Petrochem Res. 2012;1:21–27. DOI: 10.1007/s13203-011-0002-2

[10] Dresselhaus MS, Thomas IL. Alternative energy technologies. Nature. 2001;414:332–337. DOI: 10.1038/35104599

[11] Meryemoglu B, Hesenov A, Irmak S, Atanur OM, Erbatur O. Aqueous-phase reforming of biomass using various types of supported precious metal and raney-nickel catalysts for hydrogen production. Int J Hydrogen Energy. 2010;35:12580–12587. DOI: 10.1016/j.ijhydene.2010.08.046

[12] Cortright RD, Davda RR, Dumesic JA. Hydrogen from catalytic reforming of biomass-derived hydrocarbons in liquid water. Nature. 2002;418:964–967. DOI: 10.1038/nature01009

[13] Baccar R, Bouzid J, Feki M, Montiel A. Preparation of activated carbon from Tunisian olive-waste cakes and its application for adsorption of heavy metal ions. J Hazard Mater. 2009;162:1522–1529. DOI: 10.1016/j.jhazmat.2008.06.041

[14] Zhou F, Lu C, Yao Y, Sun L, Gong F, Li D, Pei K, Lu W, Chen W. Activated carbon fibers as an effective metal-free catalyst for peracetic acid activation: implications for the removal of organic pollutants. Chem Eng J. 2015;281:953–960. DOI: 10.1016/j.cej.2015.07.034

[15] Tsyntsarski B, Stoycheva I, Tsoncheva T, Genova I, Dimitrov M, Petrova B, Paneva D, Cherkezova-Zheleva Z, Budinova T, Kolev H, Gomis-Berenguer A, Ania CO, Mitov I, Petrov N. Activated carbons from waste biomass and low rank coals as catalyst supports for hydrogen production by methanol decomposition. Fuel Process Technol. 2015;137:139–147. DOI: 10.1016/j.fuproc.2015.04.016

[16] Ramesh T, Rajalakshmi N, Dhathathreyan KS. Activated carbons derived from tamarind seeds for hydrogen storage. J Energ Storage. 2015;4:89–95.

[17] Saha D, Deng S. Adsorption equilibrium and kinetics of CO_2, CH_4, N_2O, and NH_3 on ordered mesoporous carbon. J Colloid Interface Sci. 2010;345:402–409. DOI: 10.1016/j.jcis.2010.01.076

[18] Lee J, Yoon S, Hyeon T, Oh SM, Bum Kim K. Synthesis of a new mesoporous carbon and its application to electrochemical double-layer capacitors. Chem Commun. 1999;21:2177–2178. DOI: 10.1039/A906872D

[19] Huber GW, Iborra S, Corma A. Synthesis of transportation fuels from biomass: chemistry, catalysts, and engineering. Chem Rev. 2006;106:4044–4098. DOI: 10.1021/cr068360d

[20] Xing R, Subrahmanyam AV, Olcay H, Qi W, van Walsum GP, Pendse H, Huber GW. Production of jet-fuel-range alkanes from hemicellulose-derived aqueous solutions. Green Chem. 2010;12:1933–1946. DOI: 10.1039/C0GC00263A

[21] Ko BS, Kim J, Kim JH. Production of xylitol from D-xylose by a xylitol dehydrogenase gene-disrupted mutant of *Candida tropicalis*. Appl Environ Microbiol. 2006;72:4207–4213. DOI: 10.1128/AEM.02699-05

[22] Werpy TA, Petersen G. Top value added chemicals from biomass. Volume I. Results of screening for potential candidates from sugars and synthesis gas. N. R. E, Laboratory Report DOE/GO-102004–1992. U.S. Department of Energy (DOE), 2004.

[23] Godula K, Bertozzi CR. Synthesis of glycopolymers for microarray applications via ligation of reducing sugars to a poly(acryloyl hydrazide) scaffold. J Am Chem Soc. 2010;132:9963–9965.

[24] Mosier N, Wyman C, Dale BE, Elander R, Lee YY, Holtzapple M, Ladisch M. Features of promising technologies for pretreatment of lignocellulosic biomass. Bioresour Technol. 2005;96:673–686. DOI: 10.1016/j.biortech.2004.06.025

[25] Sun Y, Cheng J. Hydrolysis of lignocellulosic materials for ethanol production: a review. Bioresour Technol. 2002;83:1–11. DOI: 10.1016/S0960-8524(01)00212-7

[26] Olivia JM, Sáez F, Ballesteros I, Gónzalez A, Negro MJ, Manzanares P, Ballesteros M. Effect of lignocellulosic degradation compounds from steam explosion pretreatment on ethanol fermentation by thermotolerant yeast Kluyverinyces narxianus. Appl Microbiol Biotechnol. 2003;105:141–153. DOI: 10.1385/ABAB:105:1-3:141

[27] Jönsson LJ, Alriksson B, Nilvebrant NO. Bioconversion of lignocellulose: inhibitors and detoxification. Biotechnol Biofuels. 2013;6:16. DOI: 10.1186/1754-6834-6-16

[28] Kumar P, Barrett DM, Delwiche MJ, Stroeve P. Methods for pretreatment of lignocellulosic biomass for efficient hydrolysis and biofuel production. Ind Eng Chem Res. 2009:48:3713–3729. DOI: 10.1021/ie801542g

[29] Chang VS, Holtzapple MT. Fundamental factors affecting biomass enzymatic reactivity. Appl Biochem Biotechnol. 2000;84–86:5–37. DOI: 10.1385/ABAB:84-86:1-9:5

[30] Kootstra AMJ, Beeftink HH, Scott EL, Sanders JPM. Comparison of dilute mineral and organic acid pre-treatment for enzymatic hydrolysis of wheat straw. Biochem Eng J. 2009;46:126–131. DOI: 10.1016/j.bej.2009.04.020

[31] Bussemaker MJ, Zhang D. Effect of ultrasound on lignocellulosic biomass as a pretreatment for biorefinery and biofuel applications. Ind Eng Chem Res. 2013;52:3563–3580. DOI: 10.1021/ie3022785

[32] Ozturk I, Irmak S, Hesenov A, Erbatur O. Hydrolysis of kenaf (Hibiscus cannabinus L.) stems by catalytical thermal treatment in subcritical water. Biomass Bioenerg. 2010;34:1578–1585. DOI: 10.1016/j.biombioe.2010.06.005

[33] Hu Z, Wen Z. Enhancing enzymatic digestibility of switchgrass by microwave-assisted alkali pretreatment. Biochem. Eng J. 2008;38:369–378. DOI: 10.1016/j.bej.2007.08.001

[34] Zhu S, Wu Y, Yu Z, Zhang X, Wang C, Yu F, Jin S, Zhao Y, Tu S, Xue Y. Simultaneous saccharification and fermentation of microwave/alkali pre-treated rice straw to ethanol. Biosyst Eng. 2005;92:229–235. DOI: 10.1016/j.biosystemseng.2005.06.012

[35] Marx S, Ndaba B, Chiyanzu I, Schabort C. Fuel ethanol production from sweet sorghum bagasse using microwave irradiation. Biomass Bioenerg. 2014;65:145–150. DOI: 10.1016/j.biombioe.2013.11.019

[36] Chimentão RJ, Lorente E, Gispert-Guirado F, Medina F, López F. Hydrolysis of dilute acid-pretreated cellulose under mild hydrothermal conditions. Carbohydr Polym. 2014;111:116–124. DOI: 10.1016/j.carbpol.2014.04.001

[37] Hu Z, Wang Y, Wen Z. Alkali (NaOH) pretreatment of switchgrass by radio frequency-based dielectric heating. Appl Biochem Biotechnol. 2008;148:71–81. DOI: 10.1007/s12010-007-8083-1

[38] Balakrishnan PA, Vedaraman N, Sunder VJ, Muralidharan C, Saminathan G. Radio frequency heating - a prospective leather drying system for future. Drying Technol. 2004;22:1969–1982. DOI: 10.1081/DRT-200032738

[39] Izadifar M, Baik OD, Mittal GS. Radio frequency-assisted extraction of podophyllotoxin: prototyping of packed bed extraction reactors and experimental observations. Chem Eng Process. 2009;48:1437–1444. DOI: 10.1016/j.cep.2009.07.011

[40] Piyasena P, Dussault C, Koutchma T, Ramaswamy HS, Awuah GB. Radio frequency heating of foods: principles, applications and related properties - a review. Crit Rev Food Sci Nutr. 2003;43:587–606. DOI: 10.1080/10408690390251129

[41] Wang X, Taylor S, Wang Y. Improvement of radio frequency (RF) heating-assisted alkaline pretreatment on four categories of lignocellulosic biomass. Bioprocess Biosys Eng. 2016; 39:1539–1551. DOI: 10.1007/s00449-016-1629-2

[42] Iroba KL, Tabi LG, Dumonceaux T, Baik OD. Effect of alkaline pretreatment on chemical composition of lignocellulosic biomass using radio frequency heating. Biosyst Eng. 2013;116:385–398. DOI: 10.1016/j.biosystemseng.2013.09.004

[43] Okano K, Kitagaw M, Sasaki Y, Watanabe T. Conversion of Japanese red cedar (*Cryptomeria japonica*) into a feed for ruminants by white-rot basidiomycetes. Animal Feed Sci Technol. 2005;120:235–243. DOI: 10.1016/j.anifeedsci.2005.02.023

[44] Cardona CA, Sanchez OJ. Fuel ethanol production: process design trends and integration opportunities. Bioresour Technol. 2007;98:2415–2457. DOI: 10.1016/j.biortech.2007.01.002

[45] Galbe M, Zacchi G. Pretreatment of lignocellulosic materials for efficient bioethanol production. Adv Biochem Eng Biotechnol. 2007;108:41–65. DOI: 10.1007/10_2007_070

[46] Xu C, Ma F, Zhang X, Chen S. Biological pretreatment of corn stover by Irpex lacteus for enzymatic hydrolysis. J Agric Food Chem. 2010;58:10893–10898. DOI: 10.1021/jf1021187

[47] Canam T, Town JR, Tsang A, McAllister TA, Dumonceaux TJ. Biological pretreatment with a cellobiose dehydrogenase-deficient strain of *Trametes versicolor* enhances the biofuel potential of canola straw. Bioresour Technol. 2011;102:10020–10027. DOI: 10.1016/j.biortech.2011.08.045

[48] Hatakka A. Biodegradation of lignin. In: Hofrichter M, Steinbuchel A, editors. Biopolymers: Biology, Chemistry, Biotechnology, Applications, Vol. 1 Lignin, Humic Substances and Coal. Weinheim: Wiley-VCH; 2001. PP. 129–180. DOI: 10.1002/1521-3773(20020603)41:11<1963::AID-ANIE1963>3.0.CO;2-F

[49] Alvira P, Tomás-Pejó E, Ballesteros M, Negro M. Pretreatment technologies for an efficient bioethanol production process based on enzymatic hydrolysis: a review. Bioresour Technol. 2010;101:4851–4861. DOI: 10.1016/j.biortech.2009.11.093

[50] Brennan AH, Hoagland W, Schell DJ. High temperature acid-hydrolysis of biomass using an engineering scale plug flow reactor: results of low solids testing. Biotechnol Bioeng Symp. 1986;17:53.

[51] Klemm D, Philipp B, Heinze T, Heinze U, Wagenknecht W. General considerations on structure and reactivity of cellulose. Comprehensive Cellulose Chemistry: Fundamentals and Analytical Methods, Vol. 1. Verlag GmbH & Co. KGaA: Wiley-VCH; 1998. pp. 83–86. DOI: 10.1002/3527601929.ch2a

[52] Wingren A, Galbe M, Zacchi G. Techno-economic evaluation of producing ethanol from softwood: comparison of SSF and SHF and identification of bottlenecks. Biotechnol Prog. 2003;19:1109–1117. DOI: 10.1021/bp0340180

[53] Galbe M, Zacchi, G. A review of the production of ethanol from softwood. Appl Microbiol Biotechnol. 2002;59:618–628. DOI: 10.1007/s00253-002-1058-9

[54] Jorgensen H, Kristensen JB, Felby C. Enzymatic conversion of lignocellulose into fermentable sugars: challenges and opportunities. Biofuels Bioprod Biorefin. 2007;1:119–134. DOI: 10.1002/bbb.4

[55] Capanema EA, Balakshin MY, Kadla JF. Quantitative characterization of a hardwood milled wood lignin by NMR spectroscopy. J Agric Food Chem. 2005;53:9639–9649. DOI: 10.1021/jf0515330

[56] Lewis NG, Yamamoto E. Lignin: occurrence, biogenesis and biodegradation. Ann Rev PlantPhysiolPlantMolBiol.1990;41:455–496.DOI:10.1146/annurev.pp.41.060190.002323

[57] Ebringerova A. Hemicelluloses as biopolymeric raw materials. Papier Proceedings Paper: Cellulose-chemists roundtable discussion on environmental considerations during production, processing and use of dissolving pulps. Baden, Germany, 1992). 1992;46:726–733.

[58] Puls J. Chemistry and biochemistry of hemicelluloses. Relationship between hemicellulose structure and enzymes required for hydrolysis. Macromol Symposia. 1997;120:183–196. DOI: 10.1002/masy.19971200119

[59] Irmak S, Öztürk İ. Hydrogen rich gas production by thermocatalytic decomposition of kenaf biomass. Int J Hydrogen Energy. 2010;35:5312–5317. DOI: 10.1016/j.ijhydene.2010.03.081

[60] de Wit D, Maat L, Kieboom APG. Carbohydrates as industrial raw materials. Ind Crops Products. 1993;2:1–12. DOI: 10.1016/0926-6690(93)90004-S

[61] Agirrezabal-Telleria I, Gandarias I, Arias PL. Production of furfural from pentosan-rich biomass: analysis of process parameters during simultaneous furfural stripping. Bioresour Technol. 2013;143:258–264. DOI: 10.1016/j.biortech.2013.05.082

[62] Rafiqul ISM, Sakinah AMM. Processes for the production of xylitol-a review. Food Rev Int. 2013;29:127–156. DOI: 10.1080/87559129.2012.714434

[63] Yin R, Liu R, Mei JY, Fei W, Sun X. Characterization of bio-oil and bio-char obtained from sweet sorghum bagasse fast pyrolysis with fractional condensers. Fuel. 2013;112:96–104. DOI: 10.1016/j.fuel.2013.04.090

[64] Fisher T, Hajaligol M, Waymack B, Kellogg D. Pyrolysis behaviour and kinetics of biomass derived materials. J Appl Pyrolysis. 2002;62:331–349. DOI: 10.1016/S0165-2370(01)00129-2

[65] Chhiti Y, Kemiha M. Thermal conversion of biomass, pyrolysis and gasification: a review. Int J Eng Sci. 2013;2:75–85. ISSN: 2319 – 1813 ISBN: 2319-1805

[66] Xu X, Matsumura Y, Stenberg J, Antal MJ Jr. Carbon-catalyzed gasification of organic feedstocks in supercritical water. Ind Eng Chem Res. 1996;35:2522–2530. DOI: 10.1021/ie950672b

[67] Irmak S, Kurtulus M, Hasanoğlu (Hesenov) A, Erbatur O. Gasification efficiencies of cellulose, hemicellulose and lignin fractions of biomass in aqueous media by using Pt on activated carbon catalyst. Biomass Bioenerg. 2013;49:102–108. DOI: 10.1016/j.biombioe.2012.12.016

[68] Behrendt F, Neubauer Y, Oevermann M, Wilmes B, Zobel N. Direct liquefaction of biomass - review. Chem Eng Technol. 2008;31:667–677. DOI: 10.1002/ceat.200800077

[69] Bu Q, Lei H, Zacher AH, Wang L, Ren S, Liang J, Wei Y, Liu Y, Tang J, Zhang Q, et al. A review of catalytic hydrodeoxygenation of lignin-derived phenols from biomass pyrolysis. Bioresour Technol. 2012;124:470–477. DOI: 10.1016/j.biortech.2012.08.089

[70] Wei Z, Zeng G, Huang F, Kosa M, Sun Q, Meng X, Huang D, Ragauskas AJ. Microbial lipid production by oleaginous Rhodococci cultured in lignocellulosic autohydrolysates. Appl Microbiol Biotechnol. 2015;99:7369–7377. DOI: 10.1007/s00253-015-6752-5

[71] Irmak S, Meryemoglu B, Hasanoglu H, Erbatur O. Does reduced or non-reduced biomass feed produce more gas in aqueous-phase reforming process? Fuel. 2015;139:160–163. DOI: 10.1016/j.fuel.2014.08.028

[72] Meryemoglu B, Irmak S, Hasanoglu A. Production of activated carbon materials from kenaf biomass to be used as catalyst support in aqueous-phase reforming process. Fuel Proc Technol. 2016;151:59–63. DOI: 10.1016/j.fuproc.2016.05.040

[73] Balan V, da Costa Sousa L, Chundawat SP, Vismeh R, Jones AD, Dale BE. Mushroom spent straw: a potential substrate for an ethanol-based biorefinery. J Ind Microbiol Biotechnol. 2008;35:293–301. DOI: 10.1007/s10295-007-0294-5

[74] McKendry P. Energy production from biomass (part 2): conversion technologies. Bioresour Technol. 2002;83:47–54. DOI: 10.1016/S0960-8524(01)00119-5

[75] Allen C, Kaiser J, Cordsiemon R. Fact Sheet for Planting and Managing Giant Miscanthus in Missouri for the Biomass Crop Assistance Program (BCAP); 2011. USDA-Natural Resources Conservation Service, Plant Materials Center. Elsberry, MO 63343.

[76] Pyter R, Voigt T, Heaton E, Dohleman F, Long S. Growing Giant Miscanthus in Illinois. Growers Guide, University of Illinois; 2007. Available from: http://miscanthus.illinois.edu/wp-content/uploads/growersguide.pdf

[77] Wood I. Kenaf: The Forgotten Fiber Crop. The Australian New Crops Newsletter 10; 2003. Available from: http://www.newcrops.uq.edu.au/newslett/ncn10212.htm

[78] USDA, FAS. Grain: World Markets and Trade, Jan. 12; 2015. Marketing Year Oct. 1, 2014-Sept. 30, 2015. Available from: http://www.worldofcorn.com/pdf/WOC-2015.pdf

[79] US Weekly Corn Update: Ethanol Demand Weighs on Corn Futures, See It Market; 2015. Available from: http://www.seeitmarket.com/us-weekly-corn-update-ethanol-demand-weighs-on-corn-futures-14780/

[80] UNICA. Statistics of Sugarcane Sector-2009. Available from: http://www.unica.com.br/dadosCotacao/estatistica/

[81] Waclawovsky AJ, Sato PM, Lembke CG, Moore PH, Souza, GM. Sugarcane for bioenergy production: an assessment of yield and regulation of sucrose content. Plant Biotechnol J. 2010;8:263–276. DOI: 10.1111/j.1467-7652.2009.00491.x

[82] Silalertruksa T, Gheewala SH. Environmental sustainability assessment of bio-ethanol production in Thailand. Energy. 2009;34:1933–1946. DOI: 10.1016/j.energy.2009.08.002

[83] Mejean A, Hope C. Modeling the costs of energy crops: a case study of US corn and Brazilian sugarcane. Energy Pol. 2010;38:547–561. DOI: 10.1016/j.enpol.2009.10.006

[84] Laureysens I, Bogaert J, Blust R, Ceulemans R. Biomass production of 17 poplar clones in a short-rotation coppice culture on a waste disposal site and its relation to soil characteristics. Ecol Manage. 2004;187:295–309. DOI: 10.1016/j.foreco.2003.07.005

[85] Kopp RF, Abrahamson LP, White EH, Volk TA, Nowak CA, Fillhart RC. Willow biomass production during ten successive annual harvests. Biomass Bioenergy. 2001;20:1–7. DOI: 10.1016/S0961-9534(00)00063-5

[86] Stanton B, Eaton J, Johnson J, Rice D, Schuette B, Moser B. Hybrid poplar in the pacific northwest: the effects of market-driven management. J Forest. 2002;100:28–33.

[87] Smith GA, Buxton DR. Temperate zone sweet sorghum ethanol production potential. Bioresour Tech. 1993;43:71–75. DOI: 10.1016/0960-8524(93)90086-Q

[88] Rains GC, Cundiff JS, Vaughan DH. Development of a whole-stalk sweet sorghum harvester. Trans ASAE. 1990;33:56–62. DOI: 10.13031/2013.31294

[89] Hunter EL, Anderson IC. Sweet sorghum. In: Janick J, editor. Horticultural reviews 21. New York, NY: John Wiley and Sons; 1997. pp. 73–104. DOI: 10.1002/9780470650660

[90] Stevens G. Sweet Sorghum for Biofuel Production; 2014. Available from: http://articles.extension.org/pages/26634/sweet-sorghum-for-biofuel-production

[91] Kadam KL, McMillan JD. Availability of corn stover as a sustainable feedstock for bioethanol production. Bioresour Technol. 2003;88:17–25. DOI: 10.1016/S0960-8524(02)00269-9

[92] Pordesimo L, Edens WC, Sokhansanj S. Distribution of above ground biomass in corn stover. Biomass Bioenerg. 2004;26:337–343. DOI: 10.1016/S0961-9534(03)00124-7

[93] Bach QV, Chen WH, Chu YS, Skreiberg Ø. Predictions of biochar yield and elemental composition during torrefaction of forest residues. Bioresour Technol. 2016;215:239–246. DOI: 10.1016/j.biortech.2016.04.009

[94] Guo Z, Sun C, Grebner DL. Utilization of forest derived biomass for energy production in the U.S.A.: status, challenges, and public policies. Int Forest Rev. 2007;9:748–758. DOI: 10.1505/ifor.9.3.748

[95] Betancur GJV, Pereira N Jr. Sugar cane bagasse as feedstock for second generation ethanol production. Part I: diluted acid pretreatment optimization. Electron J Biotechnol. 2010;13:1–9. ISSN: 0717-3458

[96] Kingston D, Novelli GF, Cerrutti P, Recupero MN, Blasco M, Galvagno MA. Use of grape pomaces to produce biomass of a *Komagataella pastoris* strain expressing a bovine chymosin activity. Food Sci Nutr. 2014;2:734–743. DOI: 10.1002/fsn3.128

[97] Agger J, Viksø-Nielsen A, Meyer AS. Enzymatic xylose release from pretreated corn bran arabinoxylan: differential effects of deacetylation and deferuloylation on insoluble and soluble substrate fractions. J Agric Food Chem. 2010;58:6141–6148. DOI: 10.1021/jf100633f

6

Developing Tree Biomass Models for Eight Major Tree Species in China

WeiSheng Zeng

Abstract

In the context of climate change, estimating forest biomass for large regions is key to national carbon stocks, but few models have been developed at regional level. Based on mensuration data from large samples (4818 and 1626 trees for above- and belowground biomass, respectively) of eight major tree species in China, the author developed one- and two-variable compatible integrated model systems for aboveground and belowground biomass, biomass conversion factor (*BCF*) and root-to-shoot ratio (*RSR*), using the error-in-variable simultaneous equations. Furthermore, the differences of aboveground and belowground biomass among various species were analyzed using the dummy variable approach. The results indicated that (1) two-variable models were almost better than one-variable models for aboveground biomass estimation, while the two model systems were not significantly different for belowground biomass estimation; (2) the eight species can be ranked in terms of aboveground biomass from *Quercus* (largest), *Betula, Populus, Pinus massoniana, Picea, Larix, Abies* to *Cunninghamia lanceolata* and in terms of belowground biomass from *Quercus* (largest), *Betula, Larix, Picea, Populus, P. massoniana, C. lanceolata* to *Abies*; (3) mean prediction errors (*MPEs*) of aboveground biomass models for the species were less than 5%, whereas *MPEs* of belowground biomass equations were less than 10%, except for *Abies*.

Keywords: aboveground biomass, belowground biomass, biomass conversion factor, root-to-shoot ratio, error-in-variable simultaneous equations

1. Introduction

Increasingly, governments worldwide attach considerable importance to estimating biomass and carbon storage of forest ecosystems in the context of global climate change. To help countries conduct national greenhouse gas inventories, forest biomass estimation and carbon stock assessment, the Intergovernmental Panel on Climate Change (IPCC) provided such carbon-accounting parameters as biomass expansion factors (*BEF*) and root-to-shoot ratios (*RSR*) for estimating different geographic zones in 2003 [1]. However, it probably has great uncertainty to apply these parameters for biomass estimation. Developing individual tree biomass models and parameters for national monitoring and assessment of biomass and carbon storage of forest ecosystems has become fundamentally important.

The earliest research on forest biomass abroad can be traced to the 1870s [2]. In recent years, biomass models for major tree species in America, Canada and some European countries have been developed or improved [3–11]. Their purpose was to assess and monitor forest biomass and carbon storage and to provide a basis for evaluating the contribution of forest ecosystems to the global carbon cycle. Studies on forest biomass in China have only been implemented since the late 1970s when some related articles were published [12, 13], i.e., a century after the earliest study abroad. Due to special historical reasons, China did not participate in the International Biological Program (IBP), initiated by the International Union of Forest Research Organizations (IUFRO), during the period of 1964–1974 and thus missed the golden development stage of forest biomass research [14].

Reviewing the development of forest biomass modeling near 40 years in China, three stages could be classified: the first is estimating biomass and productivity of major forest types toward the end of the twentieth century [13, 15–30]; the second is assessing carbon storage in Chinese forest ecosystems since the beginning of the current century [31–37]; and the third is the new development stage for monitoring and assessing forest biomass and carbon storage at provincial and national levels [14, 38]. To monitor forest biomass and carbon storage in the National Forest Inventory (NFI) system, the National Forest Biomass Modeling Program has been implemented since early 2009. Up to now, many papers on modeling individual tree biomass have been published [39–51], which classified 70 modeling populations for developing individual tree biomass models, determined the sample structure of each population and studied the modeling methods including nonlinear error-in-variable simultaneous equations, mixed-effects modeling, dummy variable modeling and segmented modeling approaches. Also, logarithmic regression and weighted regression were analyzed [52] and goodness evaluation and precision analysis of biomass models were studied [53]. Based on the studying achievements, two ministerial standards on technical regulations and five ministerial standards on biomass models have been approved for application [54–60]. In the near future, more ministerial standards on biomass models for other tree species would be published.

From the published papers and ministerial standards, we could find that the aboveground and belowground biomass models were developed separately owing to the unequal sample sizes and most of the studies were only based on sample trees of one tree species. In this chapter, the author will use the mensuration data of aboveground and belowground biomass from 4818 to 1626 destructive sample trees of eight major tree species, respectively. The main purpose was to develop an integrated individual tree model system for aboveground and belowground biomass, biomass conversion factor (*BCF*) and root-to-shoot ratio (*RSR*), using the approach of nonlinear error-in-variable simultaneous equations with dummy variable. The system could assure aboveground biomass models compatible with stem volume models and *BCF* models and belowground biomass models compatible with aboveground biomass models and *RSR* models. Secondly, the generalized dummy-variable models of aboveground and belowground biomass for eight major tree species were established and compared and the ranks of eight species for aboveground and belowground biomass estimation were provided respectively from the species-specific parameter estimates.

2. Materials and methods

2.1. Data

During the 5 years between 2009 and 2013, a total amount of 4818 sample trees for 31 modeling populations of eight major tree species or species groups, namely, *Picea* spp., *Abies* spp., *Betula* spp., *Quercus* spp., *Populus* spp., *Larix* spp., *Cunninghamia lanceolata* and *Pinus massoniana*, which occupied more than 60% of forest volume in China [39], were felled for aboveground biomass mensuration. The sample trees were evenly distributed in ten diameter classes of 2, 4, 6, 8, 12, 16, 20, 26, 32 and more than 38 cm for each modeling population, and about 15 sample trees in each diameter class were selected by height class as evenly as possible. For example, if three height classes were defined, i.e., low, intermediate and high, then five sample trees should be selected in each height class. For each sample tree, the diameter at breast height of stem was measured in the field. After the tree was felled, total trunk length (tree height, from ground level to the top) and live crown length were also measured. The trunk was divided into 11 sections at points corresponding to 0, 0.05, 0.1, 0.2, 0.3, 0.4, 0.5, 0.6, 0.7, 0.8 and 0.9 of tree height. Base diameters of all sections were measured and the tree volume was computed using Smalian's formula [61], which referred to total volume over bark. Specifically, the formula was written as $V = (A_1 + A_2)/2 \times L$ with V as the volume of a section of tree trunk, A_1 and A_2 as two areas of the small and large ends of the section and L as the section length. The fresh weights of stem, branch and foliage were also measured; subsamples were selected and weighed in the field [54]. Among all sample trees, about one third (1626 trees) were selected for measuring both aboveground and belowground biomass. The whole roots were excavated out, fresh weights of stump, coarse roots (more than 10 mm) and small roots (2–10 mm, not including fine roots less than 2 mm) were measured, respectively and subsamples were selected. After being taken into the laboratory, all subsamples were oven-dried at 85°C until a constant weight was reached. According to the ratio of dry weight to fresh weight, each component biomass was computed and the aboveground biomass of the tree was obtained by summation [54]. **Table 1** shows the general situation for biomass samples of eight major tree species or groups.

2.2 Model construction

The general form of individual tree biomass and stem volume models is as follows [45, 62]:

$$y = \beta_0 {x_1}^{\beta_1} {x_2}^{\beta_2} \cdots {x_j}^{\beta_j} + \varepsilon \tag{1}$$

where y is biomass (kg), x_j are predictive biometric variables, which reflect the dimensions of a tree, such as diameter at breast height D (cm) and tree height H (m), β_j are parameters and ε is the error term. Because the biomass data are significantly heteroscedastic, some measures should be taken to eliminate heteroscedasticity prior to parameter estimation. In this paper, weighted regression was applied and the specific weight functions were derived from the

Species	Samples	Variables	Mean	Min	Max	S.D.	CV (%)
Picea spp.	900/295	Diameter D (cm)	17.0	1.0	65.5	12.8	75.6
		Height H (m)	12.3	1.4	46.9	8.1	66.4
		Stem volume V (dm^3)	343.0	0.6	6770.7	609.9	177.8
		Aboveground biomass M_a (kg)	174.5	0.4	1668.9	251.3	143.9
		Belowground biomass M_b (kg)	41.2	0.1	289.1	61.3	148.8
Abies spp.	751/249	Diameter D (cm)	17.1	1.1	68.0	13.0	76.6
		Height H (m)	11.9	1.5	39.0	7.4	62.7
		Stem volume V (dm^3)	352.4	0.5	4525.0	589.5	167.3
		Aboveground biomass M_a (kg)	168.9	0.3	1817.0	262.7	155.6
		Belowground biomass M_b (kg)	29.0	0.1	393.4	52.4	180.7
Betula spp.	690/236	Diameter D (cm)	15.9	1.0	60.8	11.8	73.7
		Height H (m)	11.3	1.9	33.0	6.2	55.1
		Stem volume V (dm^3)	235.0	0.3	2782.7	345.9	147.2
		Aboveground biomass M_a (kg)	167.4	0.2	1671.0	240.6	143.7
		Belowground biomass M_b (kg)	45.0	0.1	343.6	67.0	148.8
Quercus spp.	670/228	Diameter D (cm)	16.1	1.5	54.0	11.6	72.1
		Height H (m)	10.9	1.4	28.6	6.3	57.6
		Stem volume V (dm^3)	253.2	0.2	2487.1	370.9	146.5
		Aboveground biomass M_a (kg)	208.2	0.3	1664.1	295.2	141.8
		Belowground biomass M_b (kg)	51.4	0.1	385.9	71.6	139.4
Populus spp.	602/207	Diameter D (cm)	16.4	1.2	48.9	11.9	72.3
		Height H (m)	12.9	2.4	31.1	6.9	53.6
		Stem volume V (dm^3)	281.4	0.3	2228.4	385.3	136.9
		Aboveground biomass M_a (kg)	174.1	0.2	1065.1	241.3	138.6
		Belowground biomass M_b (kg)	35.6	0.1	384.5	54.3	152.7
Larix spp.	602/199	Diameter D (cm)	16.7	1.5	54.2	12.3	73.7
		Height H (m)	12.6	1.4	37.5	7.6	60.0
		Stem volume V (dm^3)	316.6	0.6	3016.6	471.7	149.0
		Aboveground biomass M_a (kg)	160.4	0.2	1301.9	231.1	144.1
		Belowground biomass M_b (kg)	41.0	0.1	300.0	61.8	150.9
Cunninghamia lanceolata	302/108	Diameter D (cm)	16.4	1.8	42.0	11.8	71.8
		Height H (m)	11.5	1.9	33.0	7.1	61.7
		Stem volume V (dm^3)	293.7	0.6	1815.2	409.7	139.5
		Aboveground biomass M_a (kg)	75.6	0.3	644.9	105.5	139.5
		Belowground biomass M_b (kg)	25.9	0.1	174.9	37.7	145.8

Species	Samples	Variables	Mean	Min	Max	S.D.	CV (%)
Pinus massoniana	301/104	Diameter D (cm)	16.5	1.2	47.2	11.9	72.4
		Height H (m)	12.1	1.6	30.3	7.2	59.4
		Stem volume V (dm^3)	300.8	0.3	1825.4	405.7	134.9
		Aboveground biomass M_a (kg)	125.1	0.1	1079.3	171.6	137.2
		Belowground biomass M_b (kg)	35.5	0.1	285.0	53.7	151.6

Min—minimum, Max—maximum, S.D.—standard deviation, and CV—coefficient of variation.
The sample sizes are for aboveground and belowground biomass mensuration, respectively.

Table 1. General situation of biomass samples for eight major tree species.

residuals of independently fitted models by ordinary least squares regression [62, 63]. Since models based on one (D) or two variables (D and H) have been commonly used, this paper develops both one- and two-variable models. The aboveground biomass, belowground biomass and stem volume models based on two variables can be expressed respectively as:

$$M_a = a_0 D^{a_1} H^{a_2} + \varepsilon \tag{2}$$

$$M_b = b_0 D^{b_1} H^{b_2} + \varepsilon \tag{3}$$

$$V = c_0 D^{c_1} H^{c_2} + \varepsilon \tag{4}$$

where M_a and M_b are aboveground and belowground biomass (kg), respectively; V is stem volume (dm^3); a_i, b_i and c_i are parameters; and other symbols are the same as above.

2.2.1 Integrated compatible model systems

The aboveground biomass is correlated to stem volume through biomass conversion factor (*BCF*), which is equal to biomass expansion factor (*BEF*) multiplied by basic wood density following the IPCC's approach [64]. Because the *BCF* is an important parameter for forest biomass estimation [65], it is very common to develop both an aboveground biomass model and a *BCF* model that are compatible with stem volume model [45, 51]. Similarly, belowground biomass is connected with aboveground biomass model through root-to-shoot ratio (*RSR*) [66, 67]. Because the *RSR* model is also an important parameter for forest biomass estimation, generally both belowground biomass model and *RSR* model compatible with aboveground biomass model are developed simultaneously [44]. Therefore, we can develop an integrated aboveground and belowground biomass model system through using the nonlinear error-in-variable simultaneous equation approach [51, 68]. Because the belowground biomass observations were only 1/3 of the aboveground biomass observations, a dummy variable (x) was required for those trees for which no belowground biomass observation was available, i.e., 1 for the trees with belowground biomass observation and 0 for the trees with no belowground biomass observation [69]. The system can ensure the compatibility between aboveground biomass, belowground biomass, stem volume, *BCF* and *RSR*. The one- and two-variable integrated systems are as follows, respectively:

$$\begin{cases} M_a = a_0 D^{a_1} + \varepsilon \\ M_b = b_0 D^{b_1} x + \varepsilon \\ V = c_0 D^{c_1} + \varepsilon \\ BCF = a_0 D^{a_1} / c_0 D^{c_1} + \varepsilon \\ RSR = b_0 D^{b_1} x / a_0 D^{a_1} + \varepsilon \end{cases} \tag{5}$$

$$\begin{cases} M_a = a_0 D^{a_1} H^{a_2} + \varepsilon \\ M_b = b_0 D^{b_1} H^{b_2} x + \varepsilon \\ V = c_0 D^{c_1} H^{c_2} + \varepsilon \\ BCF = a_0 D^{a_1} H^{a_2} / c_0 D^{c_1} H^{c_2} + \varepsilon \\ RSR = b_0 D^{b_1} H^{b_2} x / a_0 D^{a_1} H^{a_2} + \varepsilon \end{cases} \tag{6}$$

where, M_a, M_b, V, BCF and RSR are aboveground biomass, belowground biomass, stem volume, biomass conversion factor and root-to-shoot ratio, respectively, which are regarded as error-in-variables; D and H are diameter at breast height and tree height, which are regarded as error-free variables; x is a dummy variable to distinguish if belowground biomass is available; and a_i, b_i and c_i are parameters.

Various methods have been attempted to estimate the parameters of the simultaneous equations. Parresol [63] used the seemingly unrelated regression (SUR) for solving the additivity of simultaneous biomass equations. Tang et al. [70] further developed an error-in-variable modeling approach to estimate the parameters of simultaneous equations, which has been widely used in recent years [40, 45, 49, 51]. In this study, the error-in-variable simultaneous equation approach was used to estimate the parameters of the integrated systems based on maximum likelihood estimation through ForStat software (statistical software with analytical tools for forestry as well as general statistical procedures, developed in the Chinese Academy of Forestry, Beijing, China) [68].

In addition, the weighted regression method was used to eliminate the heteroscedasticity commonly exhibited in biomass and volume data by using specific weight functions, which were derived from the residuals of biomass or volume equations fitted through the ordinary least square (OLS) technique [52, 62]. For biomass conversion factor and root-to-shoot ratio modeling, the OLS regression technique was directly used to estimate the parameters because the *BCF* and *RSR* data mostly exhibited homoscedasticity.

2.2.2 Generalized dummy variable models

The one-variable biomass equation was the most widely used model in estimating individual tree biomass [3, 7]. The power function of one-variable aboveground biomass equation was based on the WBE theory for the origin of allometric scaling laws [71, 72]. According to the results from Zeng and Tang [73], the generalized one-variable aboveground biomass model can be expressed as:

$$M_a = aD^{7/3} + \varepsilon \tag{7}$$

That is, the power parameter of the allometric model is constantly equal to 7/3 (≈2.33), only the parameter a depends on tree species. If a variable vector $\boldsymbol{z}$ was defined as dummy variable to indicate tree species, then the generalized model (7) could be expressed as:

$$M_a = (a + v_a z)D^{7/3} + \varepsilon \tag{8}$$

where a is the global parameter and $\boldsymbol{v_a}$ is tree species-specific parameter vector. The dummy variable vector $\boldsymbol{z}$ includes seven elements, indicating the eight tree species by the following combinations:

$z_1 = 1, z_2 = 0, z_3 = 0, z_4 = 0, z_5 = 0, z_6 = 0$ and $z_7 = 0$ for *Picea* spp.

$z_1 = 0, z_2 = 1, z_3 = 0, z_4 = 0, z_5 = 0, z_6 = 0$ and $z_7 = 0$ for *Abies* spp.

$z_1 = 0, z_2 = 0, z_3 = 1, z_4 = 0, z_5 = 0, z_6 = 0$ and $z_7 = 0$ for *Betula* spp.

$z_1 = 0, z_2 = 0, z_3 = 0, z_4 = 1, z_5 = 0, z_6 = 0$ and $z_7 = 0$ for *Quercus* spp.

$z_1 = 0, z_2 = 0, z_3 = 0, z_4 = 0, z_5 = 1, z_6 = 0$ and $z_7 = 0$ for *Populus* spp.

$z_1 = 0, z_2 = 0, z_3 = 0, z_4 = 0, z_5 = 0, z_6 = 1$ and $z_7 = 0$ for *Larix* spp.

$z_1 = 0, z_2 = 0, z_3 = 0, z_4 = 0, z_5 = 0, z_6 = 0$ and $z_7 = 1$ for *C. lanceolata*

$z_1 = 0, z_2 = 0, z_3 = 0, z_4 = 0, z_5 = 0, z_6 = 0$ and $z_7 = 0$ for *P. massoniana*

Consequently, from comparing the estimated values of species-specific parameter vector $\boldsymbol{v_a}$, the differences among various tree species could be analyzed.

2.3 Model evaluation

Many statistical indices could be used to evaluate individual tree biomass models [63]. According to the study results from Zeng and Tang [53], the following six statistical indices, namely, the coefficient of determination (R^2), standard error of estimate (*SEE*), mean prediction error (*MPE*), total relative error (*TRE*), average systematic error (*ASE*) and mean percent standard error (*MPSE*), were very important for assessing biomass models. In this study, the same six statistical indices were used for model evaluation [50, 51]:

$$R^2 = 1 - \sum (y_i - \hat{y}_i)^2 / \sum (y_i - \bar{y})^2 \tag{9}$$

$$SEE = \sqrt{\sum (y_i - \hat{y}_i)^2 / (n-p)} \tag{10}$$

$$TRE = \sum (y_i - \hat{y}_i) / \sum \hat{y}_i \times 100 \tag{11}$$

$$ASE = \sum (y_i - \hat{y}_i) / \hat{y}_i / n \times 100 \tag{12}$$

$$MPE = t_\alpha \cdot (SEE/\bar{y}) / \sqrt{n} \times 100 \tag{13}$$

$$MPSE = \sum \left| (y_i - \hat{y}_i) / \hat{y}_i \right| / n \times 100 \tag{14}$$

where y_i are observed values, $\hat{y}_i$ are estimated values, $\bar{y}$ is mean value of samples, n is the number of samples, p is the number of parameters and t_α is the t-value at confidence level α with n-p degrees of freedom.

3. Results and analysis

The one- and two-variable integrated systems (Eqs. (5) and (6)) for eight tree species or groups were estimated using the error-in-variable simultaneous equation approach through ForStat (**Tables 2** and **3**). The six fitting statistics, R^2, *SEE, TRE, ASE, MPE* and *MPSE*, were calculated and could be used for evaluating the goodness-of-fit of the three models (**Table 4**). From the

	Aboveground biomass models		**Belowground biomass models**		**Stem volume models**	
Species	a_0	a_1	b_0	b_1	c_0	c_1
Pi	0.17417	2.2270	0.04853	2.1954	0.1528	2.4548
Ab	0.10195	2.3676	0.02873	2.2452	0.1297	2.5106
Be	0.13392	2.3401	0.05767	2.2039	0.1712	2.3653
Qu	0.16592	2.3409	0.10619	2.0373	0.1448	2.4351
Po	0.09198	2.4490	0.02958	2.3200	0.1410	2.4702
La	0.12473	2.3190	0.03154	2.3355	0.1464	2.4737
Cl	0.09782	2.3099	0.02853	2.2500	0.1144	2.5421
Pm	0.13771	2.3243	0.01959	2.4400	0.1514	2.4655

Pi—*Picea* spp., Ab—*Abies* spp., Be—*Betula* spp., Qu—*Quercus* spp., Po—*Populus* spp., La—*Larix* spp., Cl—*Cunninghamia lanceolata,* and Pm—*Pinus massoniana*. Same in **Tables 3–7**.

Table 2. The parameter estimates of the one-variable integrated system (Eq. (5)).

	Aboveground biomass models			**Belowground biomass models**			**Stem volume models**		
Species	a_0	a_1	a_2	b_0	b_1	b_2	c_0	c_1	c_2
Pi	0.11007	2.1369	0.2615	0.03284	2.3516	−0.0527	0.07763	1.7758	1.0122
Ab	0.06720	2.0221	0.5442	0.02412	2.5974	−0.3600	0.07429	1.8135	0.9975
Be	0.08322	2.0749	0.4844	0.04531	2.1630	0.1401	0.08383	1.8246	0.8965
Qu	0.10520	1.9808	0.5939	0.09338	2.1694	−0.1091	0.07796	1.8607	0.9115
Po	0.06304	2.2460	0.3588	0.03216	2.5313	−0.2697	0.07611	1.9503	0.7927
La	0.07437	2.0003	0.5438	0.02195	2.2354	0.2369	0.07610	1.8067	0.9827
Cl	0.06740	1.9253	0.5765	0.02252	2.5080	−0.2072	0.07417	1.7949	1.0121
Pm	0.10462	2.1591	0.2857	0.01744	2.5697	−0.1028	0.09393	1.8696	0.8451

Table 3. The parameter estimates of the two-variable integrated system (Eq. (6)).

Species	Systems	Items	R^2	*SEE*	*MPE* (%)	*TRE* (%)	*ASE* (%)	*MPSE* (%)
Pi	(5)	AB	0.9109	75.05	2.81	1.31	−2.44	24.21
		BB	0.7842	28.55	5.51	−0.25	−3.75	40.29
		SV	0.8380	245.64	4.68	2.63	0.19	27.65
	(6)	AB	0.9061	77.10	2.89	0.37	7.68	25.26
		BB	0.7751	29.19	5.63	0.65	9.28	43.70
		SV	0.9744	97.70	1.86	0.44	6.01	15.87
Ab	(5)	AB	0.9223	73.29	3.10	0.28	−1.24	22.19
		BB	0.5474	35.29	11.97	−1.15	−2.74	49.06
		SV	0.9222	164.52	3.34	−0.60	−0.29	20.79
	(6)	AB	0.9434	62.61	2.65	0.24	4.97	22.50
		BB	0.5547	35.08	11.89	0.37	7.64	52.51
		SV	0.9800	83.41	1.69	0.25	4.34	13.05
Be	(5)	AB	0.9139	70.63	3.15	2.02	0.11	22.52
		BB	0.7734	31.95	6.16	2.47	−1.15	37.64
		SV	0.9118	102.74	3.26	2.87	7.19	26.85
	(6)	AB	0.9332	62.27	2.78	1.05	2.47	21.30
		BB	0.7741	31.97	6.17	0.18	1.16	38.38
		SV	0.9566	72.13	2.29	0.49	6.53	19.60
Qu	(5)	AB	0.9030	91.99	3.35	2.67	0.93	27.82
		BB	0.8168	30.73	5.23	2.21	−1.83	40.37
		SV	0.9262	100.81	3.01	4.11	4.36	27.33
	(6)	AB	0.9285	79.04	2.87	1.67	2.81	24.66
		BB	0.8133	31.09	5.29	1.11	2.35	41.08
		SV	0.9790	53.88	1.61	1.07	3.71	16.72
Po	(5)	AB	0.9379	60.17	2.76	1.51	−0.86	17.55
		BB	0.8440	21.51	5.47	2.21	−2.71	32.15
		SV	0.9539	82.81	2.35	−0.41	1.51	15.87
	(6)	AB	0.9506	53.72	2.47	1.17	1.25	17.11
		BB	0.8618	20.29	5.15	1.77	1.15	32.11
		SV	0.9842	48.49	1.38	0.83	1.04	9.48
La	(5)	AB	0.9123	68.47	3.41	1.26	1.15	23.18
		BB	0.6773	35.22	8.65	−0.52	0.14	37.37
		SV	0.9016	148.10	3.74	0.10	4.26	25.04
	(6)	AB	0.9432	55.14	2.75	0.22	6.91	24.68
		BB	0.6675	35.83	8.81	−0.83	6.78	40.08
		SV	0.9770	71.63	1.81	0.21	5.58	15.05

Species	Systems	Items	R^2	*SEE*	*MPE* (%)	*TRE* (%)	*ASE* (%)	*MPSE* (%)
Cl	(5)	AB	0.9614	30.62	2.98	3.24	1.03	22.99
		BB	0.8414	15.08	7.43	1.66	1.75	38.39
		SV	0.9474	94.14	3.62	1.99	2.28	17.25
	(6)	AB	0.9774	23.47	2.28	1.47	6.10	23.66
		BB	0.8342	15.50	7.64	−0.88	9.87	41.43
		SV	0.9931	34.10	1.31	0.50	5.21	11.30
Pm	(5)	AB	0.9542	48.45	3.21	0.91	0.25	17.29
		BB	0.8509	20.85	7.64	1.55	−2.12	39.21
		SV	0.9503	90.62	3.40	0.17	0.34	20.87
	(6)	AB	0.9572	46.89	3.10	0.41	3.76	16.30
		BB	0.8546	20.69	7.59	−0.63	1.44	39.73
		SV	0.9846	50.59	1.90	0.64	1.13	12.51

AB—aboveground biomass, BB—belowground biomass, SV—stem volume, R^2—coefficient of determination, *SEE*—standard error of estimate, *MPE*—mean prediction error, *TRE*—total relative error, *ASE*—average systematic error, and *MPSE*—mean percent standard error.
Units of *SEE*: dm^3 for volume and kg for biomass.

Table 4. The fitting statistics of two integrated systems (Eqs. (5) and (6)).

fitting results of integrated systems (Eqs. (5) and (6)), the parameter estimates of the *BCF* and *RSR* models could be obtained (**Table 5**).

From comparison of the fitting statistics of two integrated systems (Eqs. (5) and (6)) in **Table 4**, we can found that for aboveground biomass estimation, two-variable models were better than one-variable models except *Picea*. For belowground biomass estimation, one- and two-variable models were not significantly different, even some of one-variable models were slightly better than two-variable models, such as *Picea, Quercus, Larix* and *C. lanceolata*. Considering that tree height measurement is time consuming and two-variable biomass models are not significantly different from one-variable models, especially for belowground biomass estimation, it was commended to apply one-variable models in forestry practice such as National Forest Inventory.

From **Table 2**, it was found that the estimates of parameter a_1 were approximately equal to 7/3, confirming the results of an earlier study [73]. To analyze the difference among various tree species, the dummy model (8) was fitted using the aboveground biomass data of all eight species (**Table 6**).

According to the parameter estimates in **Table 6**, we could rank the eight tree species by aboveground biomass estimates in descending order as *Quercus, Betula, Populus, P. massoniana, Picea, Larix, Abies* and *C. lanceolata*. That is, *Quercus* had the largest aboveground biomass, whereas *C. lanceolata* had the smallest one for the same diameter trees. The aboveground biomass estimates of the dummy model (Eq. (8)) for *Quercus, Betula, Populus, P. massoniana,*

Species	Systems	*BCF* models	*RSR* models
Pi	(5)	$BCF = 1.1401\ D^{-0.2278}$	$RSR = 0.2786\ D^{-0.0316}$
	(6)	$BCF = 1.4183\ D^{0.3611}\ H^{-0.7507}$	$RSR = 0.2983\ D^{0.2147}\ H^{-0.3142}$
Ab	(5)	$BCF = 0.7858\ D^{-0.1430}$	$RSR = 0.2818\ D^{-0.1224}$
	(6)	$BCF = 0.9046\ D^{0.2086}\ H^{-0.4533}$	$RSR = 0.3589\ D^{0.5753}\ H^{-0.9041}$
Be	(5)	$BCF = 0.7821\ D^{-0.0252}$	$RSR = 0.4307\ D^{-0.1362}$
	(6)	$BCF = 0.9928\ D^{0.2504}\ H^{-0.4121}$	$RSR = 0.5445\ D^{0.0880}\ H^{-0.3443}$
Qu	(5)	$BCF = 1.1456\ D^{-0.0942}$	$RSR = 0.6400\ D^{-0.3036}$
	(6)	$BCF = 1.3494\ D^{0.1201}\ H^{-0.3177}$	$RSR = 0.8877\ D^{0.1887}\ H^{-0.7030}$
Po	(5)	$BCF = 0.6522\ D^{-0.0212}$	$RSR = 0.3216\ D^{-0.1290}$
	(6)	$BCF = 0.8283\ D^{0.2958}\ H^{-0.4339}$	$RSR = 0.5102\ D^{0.2853}\ H^{-0.6285}$
La	(5)	$BCF = 0.8522\ D^{-0.1547}$	$RSR = 0.2528\ D^{0.0165}$
	(6)	$BCF = 0.9773\ D^{0.1936}\ H^{-0.4389}$	$RSR = 0.2951\ D^{0.2351}\ H^{-0.3068}$
Cl	(5)	$BCF = 0.8554\ D^{-0.2321}$	$RSR = 0.2917\ D^{-0.0599}$
	(6)	$BCF = 0.9087\ D^{0.1303}\ H^{-0.4356}$	$RSR = 0.3341\ D^{0.5828}\ H^{-0.7836}$
Pm	(5)	$BCF = 0.9096\ D^{-0.1412}$	$RSR = 0.1422\ D^{0.1157}$
	(6)	$BCF = 1.1138\ D^{0.2894}\ H^{-0.5593}$	$RSR = 0.1667\ D^{0.4106}\ H^{-0.3885}$

Table 5. The simultaneously estimated *BCF* and *RSR* models.

Species	Global parameter (a)	Species-specific parameters (v_a)
Pi	0.13485	−0.01084
Ab		−0.01959
Be		0.00443
Qu		0.03908
Po		0.00106
La		−0.01441
Cl		−0.04254
Pm		0.00000

Table 6. The parameter estimates of dummy aboveground biomass model (Eq. (8)).

Picea, Larix and *Abies* were 88%, 51%, 47%, 46%, 34%, 30% and 25% larger, respectively, than that for *C. lanceolata* (see **Figure 1**).

Similarly, for one-variable belowground biomass models, it was found that the estimates of parameter b_1 for eight species were not significantly different. To analyze the difference of belowground biomass estimation among various tree species, we fitted the following dummy model:

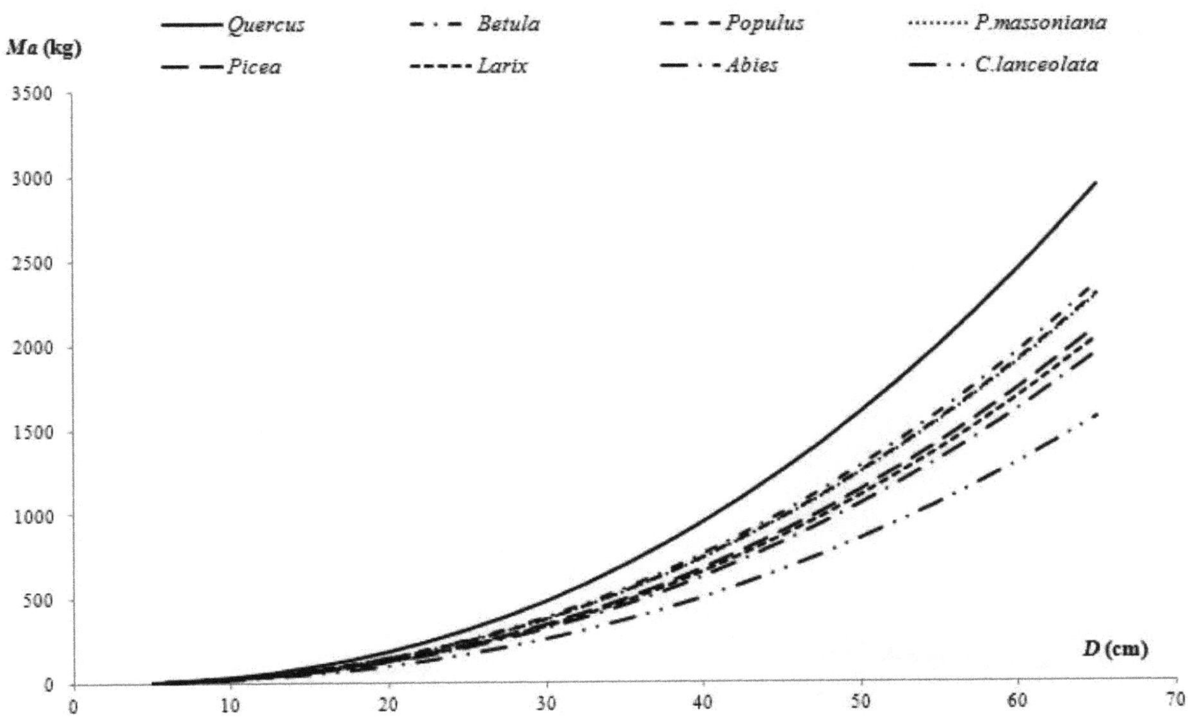

Figure 1. Comparison of aboveground biomass models for eight tree species.

Species	Global parameters b_0	b_1	Species-specific parameters (v_b)
Pi	0.03551	2.2544	0.00424
Ab			−0.00792
Be			0.01437
Qu			0.01835
Po			0.00091
La			0.00583
Cl			−0.00761
Pm			0.00000

Table 7. The parameter estimates of dummy belowground biomass model (Eq. (15)).

$$M_b = (b_0 + v_b z)D^{b_1} + \varepsilon \tag{15}$$

where b_0 and b_1 are global parameters and v_b is species-specific parameter vector. The parameter estimates of dummy model (Eq. (15)) are listed in **Table 7**.

According to the parameter estimates in **Table 7**, we could rank the eight tree species by belowground biomass estimates in descending order as *Quercus, Betula, Larix, Picea, Populus, P.*

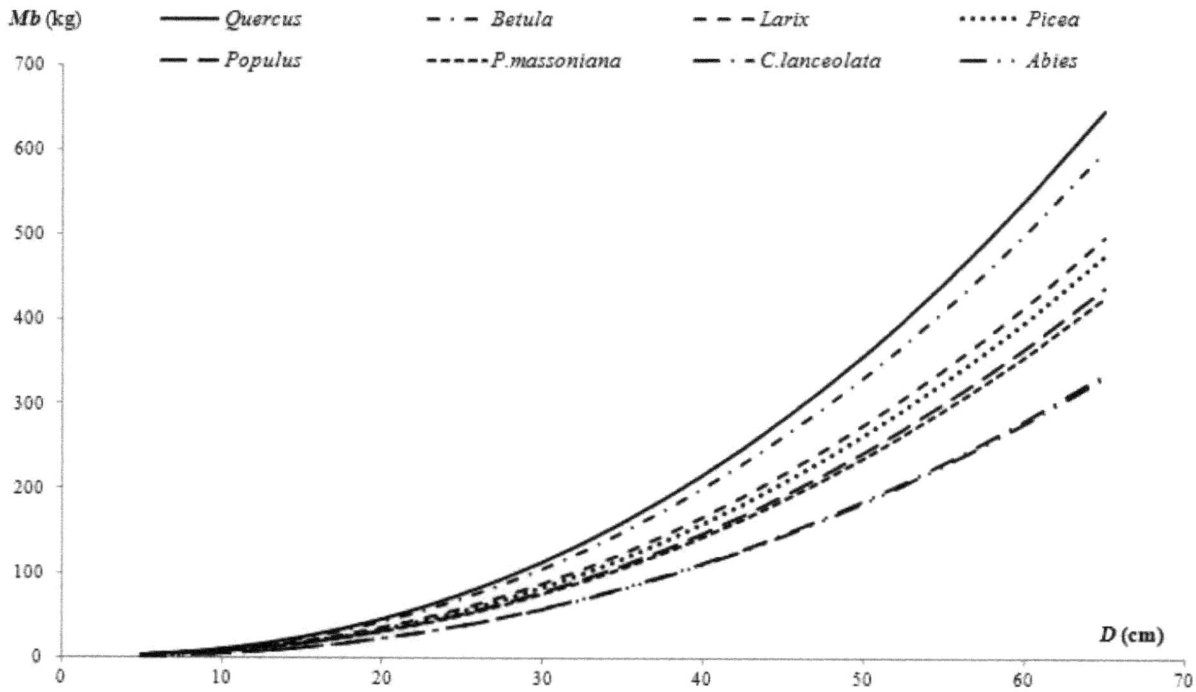

Figure 2. Comparison of belowground biomass models for eight tree species.

massoniana, *C. lanceolata* and *Abies*. That is, *Quercus* had the largest belowground biomass, while *Abies* had the smallest one for the same diameter trees. The belowground biomass estimates of the dummy model (Eq. (15)) for *Quercus, Betula, Larix, Picea, Populus, P. massoniana* and *C. lanceolata* were 95%, 81%, 50%, 44%, 32%, 29% and 1% larger, respectively, than that for *Abies* (see **Figure 2**).

4. Discussion and conclusion

In this study, data on above- and belowground biomass from 4818 to 1626 sample trees, respectively, for eight major tree species in China were used to develop compatible individual tree biomass models. The models included aboveground biomass equations and *BCF* equations compatible with stem volume equations and belowground biomass equations and *RSR* models compatible with aboveground biomass equations. To solve compatibility of the biomass models, the nonlinear error-in-variable simultaneous equations were applied and to solve the issue of unequal sample sizes for above- and belowground biomass, the dummy-variable model approach was used. In the technical regulation on methodology for tree biomass modeling [55], the segmented modeling approach was recommended when the biomass estimate of small trees was obviously biased [43, 46]. Furthermore, for the tree species distributed in various regions, it was generally needed to develop biomass models for different regions. For example, according to the population classification on modeling of single-tree biomass equations [39], it was necessary to establish five sets of biomass models for both *Abies* and *Picea*. But in this study, the segmented modeling approach was not used to develop biomass models for large and small trees, respectively and the differences among various regions were not taken into account, only one set of biomass models, including one- and two-variable models, was developed for each tree species.

The data of three tree species, i.e., *C. lanceolata, P. massoniana* and *Larix* spp., were used or partly used to develop biomass models, which were published as original papers [40–51] or ministerial standards [56, 57]. Comparing with the study results by Zeng et al. [47], the parameter estimates and fitness indices of aboveground biomass and volume models are very close to those for *C. lanceolata* in this study. From the achievements by Zeng and Tang [45], we can find that the parameter estimates of aboveground biomass and volume models are not significantly different from those for *P. massoniana* in this chapter, but this study provided better models considering the statistical indices of goodness-of-fit. Comparing with the biomass models published as ministerial standards [56, 57], the developed models in this study are more generalized and simpler for application in national and regional biomass estimation. There are four sets of biomass models in total for trees (dbh ≥ 5 cm) and saplings (dbh < 5 cm) for two modelling populations of each tree species in the ministerial standards [56, 57] and here we have only one set of biomass models which are suitable for both trees and saplings and for the whole country.

The results indicated that two-variable models were almost better than one-variable models for aboveground biomass estimation, while the two model systems were not significantly different for belowground biomass estimation. The mean prediction errors (*MPEs*) of aboveground biomass models for the eight species were less than 5%, whereas *MPEs* of belowground biomass equations were less than 10%, except for *Abies*. The models developed in this study can provide a basis for estimating biomass for the eight major tree species in China and will fill in the lack for China on the web platform GlobAllomeTree [74]. Also, they will have the potential to support the implementation of policies and mechanisms designed to mitigate climate change (e.g., CDM and REDD+) and to calculate costs and benefits associated with forest carbon projects. In addition, the overall modeling methodology presented in this study can be taken into consideration in any case that involves individual tree biomass modeling.

Acknowledgements

The author acknowledges the National Biomass Modeling Program in Continuous Forest Inventory (NBMP-CFI), which was funded by the State Forestry Administration of China, for providing the mensuration biomass data of eight tree species. This study was financially supported by the National Natural Science Foundation of China (Grant No. 31370634).

Author details

WeiSheng Zeng

Address all correspondence to: zengweisheng@forestry.gov.cn

Academy of Forest Inventory and Planning, State Forestry Administration, Beijing, China

References

[1] IPCC. Good practice guidance for land use, land use change and forestry. The Institute for Global Environmental Strategies for the IPCC, Japan; 2003

[2] Ebermeyr E. A comprehensive study of forest litter with regard to silvicultural effects on soil chemistry. Springer, Berlin; 1876

[3] Jenkins JC, Chojnacky DC, Heath LS, Birdsey RA. National-scale biomass estimators for United States tree species. For Sci. 2003; 49(1):12–35

[4] Bi H, Turner J, Lambert MJ. Additive biomass equations for native eucalypt forest trees of temperate Australia. Trees. 2004; 18:467–479

[5] Lambert MC, Ung CH, Raulier F. Canadian national tree aboveground biomass equations. Can J For Res. 2005; 35:1996–2018

[6] Snorrason A, Einarsson SF. Single-tree biomass and stem volume functions for eleven tree species used in Icelandic forestry. Icelandic Agric Sci. 2006; 19:15–24

[7] Muukkonen P. Generalized allometric volume and biomass equations for some tree species in Europe. Eur J For Res. 2007; 126:157–166

[8] Návar J. Allometric equations for tree species and carbon stocks for forests of northwestern Mexico. For Ecol Manage. 2009; 257: 427–434

[9] Blujdea VNB, Pilli R, Dutca I, Ciuvat L, Abrudan IV. Allometric biomass equations for young broadleaved trees in plantations in Romania. For Ecol Manage. 2012; 264:172–184

[10] Fayolle A, Doucet JL, Gillet JF, Bourland N, Lejeune P. Tree allometry in central Africa: testing the validity of pantropical multi-species allometric equations for estimating biomass and carbon stocks. For Ecol Manage. 2013; 305:29–37

[11] Sileshi GW. A critical review of forest biomass estimation models, common mistakes and corrective measures. For Ecol Manage. 2014; 3: 237–254

[12] Li WH. Concept of forest biomass productivity and its basic studying approach. Nat Res. 1978; 1:71–92

[13] Pan WC, Li LC, Gao ZH. Biomass and nutrient elements distribution of two different forest types of Chinese fir. Hunan For Sci Technol. 1980; 4:1–14

[14] Zeng WS. Development of monitoring and assessment of forest biomass and carbon storage in China. For Ecosyst. 2014; 1:20. DOI: 10.1186/s40663-014-0020-5

[15] Li WH, Deng KM, Li F. Research on biomass productivity of major ecosystems in the Changbai Mountain. For Ecosyst Res. 1981; (Test issue):34–50

[16] Feng ZW, Chen CY, Zhang JW, Wang KP, Zhao JL, Gao H. Determination of biomass of *Pinus massoniana* stand in Huitong county, Hunan province. Sci Silv Sin. 1982; 18(2):127–134

[17] Ye JZ. Annual dynamic of the biomass of Chinese fir forests on Yangkou forestry farm, J Nanjing For Univ. 1984; 4:1–9

[18] Chen LZ, Chen QL, Bao XC, Ren JK, Miu YG, Hu YH. Studies on Chinese arborvitae (*Platycladus orientalis*) forest and its biomass in Beijing. J Plant Ecol. 1986; 10(1):17–25

[19] Xu ZB. Biomass productivity of major forest types in Daxinganling. Chin J Ecol. 1988; 7 (Sp):49–54

[20] Ma QY. A study on the biomass of Chinese pine forests. J Beijing For Univ. 1989; 11(4):1–10

[21] Chen CG, Zhu JF. Manual of tree biomass for main species in northern China. Chinese Forestry Press, Beijing; 1989.

[22] Liu SR, Chai YX, Cai TJ, Peng CH. Research on biomass and net primary productivity of *Larix gmelini* plantations. J Northeast For Univ. 1990; 18(2): 40–46

[23] Liu ZG. Research on biomass and productivity of *Larix principis-rupprechtii* plantations. J Beijing For Univ. 1992; 14(Sp.1):114–123

[24] Liu XZ. Study on biomass of Masson pine forests at different ages. For Res Manage. 1993; 2: 77–80

[25] Fang JY, Liu GH, Xu SL. Biomass and net production of forest vegetation in China. Acta Ecol Sin. 1996; 16(4): 497–508

[26] Luo TX. Patterns of net primary productivity for Chinese major forest types and their mathematical models. Dissertation, Commission for Integrated Survey of Natural Resources, the Chinese Academy of Sciences and State Planning Commission, Beijing; 1996.

[27] Tian DL, Pan HH, Kang WX, Fang HB Study on biomass of second generation of Chinese fir plantations. J Cent S Univ For Technol. 1998; 18(3):14–19

[28] Feng ZW, Wang XK, Wu G. The biomass and productivity of forest ecosystem in China. Science Press, Beijing; 1999

[29] Zeng WS, Luo QB, He DB. Study on compatible nonlinear tree biomass models. Chin J Ecol. 1999; 18(4):19–24

[30] Tang SZ, Zhang HR, Xu H. Study on establishing and estimating method of compatible biomass model. Sci Silv Sin. 2000; 36(Sp.1):19–27

[31] Liu GH, Fu BJ, Fang JY. Carbon dynamics of Chinese forests and its contribution to global carbon balance. Acta Ecol Sin. 2000; 20(5):733–740

[32] Zhou YR, Yu ZL, Zhao SD. Carbon storage and budget of major Chinese forest types. Acta Phytoecologica Sinica. 2000; 24(5):518–522

[33] Fang JY, Chen AP, Peng CH, Zhao SQ, Ci LJ. Changes in forest biomass carbon storage in China between 1949 and 1998. Science. 2001; 292:2320–2322

[34] Xu B, Guo ZD, Piao SL, Fang JY. Biomass carbon stocks in China's forests between 2000 and 2050: a prediction based on forest biomass–age relationships. Sci China Life Sci. 2010; 53(7):776–783. DOI: 10.1007/s11427-010-4030-4

[35] Li HK, Lei YC. Estimation and evaluation of forest biomass carbon storage in China. Chinese Forestry Press, Beijing; 2010, 60 p.

[36] Li HK, Lei YC, Zeng WS. Forest carbon storage in China estimated using forest inventory data. Sci Silv Sin. 2011; 47(7):7–12

[37] Li HK, Zhao PX, Lei YC, Zeng WS. Comparison on estimation of wood biomass using forest inventory data. Sci Silv Sin. 2012; 48(5):44–52

[38] Zheng DX, Liao XL, Li CW, Ye QL, Chen PL. Estimation and dynamic change analysis of forest carbon storage in Fujian province. Acta Agric Univ Jiangxiensis. 2013; 35(1):112–116

[39] Zeng WS, Tang SZ, Huang GS, Zhang M. Population classification and sample structure on modeling of single-tree biomass equations for national biomass estimation in China. For Res Manage. 2010; 3:16–23

[40] Zeng WS, Tang SZ. Using measurement error modeling method to establish compatible single-tree biomass equations system. For Res. 2010; 23(6):797–803

[41] Zeng WS, Zhang HR, Tang SZ. Using the dummy variable model approach to construct compatible single-tree biomass equations at different scales—a case study for Masson pine (*Pinus massoniana*) in southern China. Can J For Res. 2011; 41(7): 1547–1554. DOI: 10.1139/X11-068

[42] Zeng WS. Methodology on modeling of single-tree biomass equations for national biomass estimation in China [thesis]. Chinese Academy of Forestry, Beijing; 2011.

[43] Zhang LJ, Zeng WS, Tang SZ. Comparison of nonlinear regression equation with intercept and segmented modeling approach for estimation of single tree biomass. For Res. 2011; 24(4): 453–457

[44] Zeng WS, Tang SZ. Establishment of belowground biomass equations for larch in northeastern and Masson pine in southern China. J Beijing For Univ. 2011; 33(2):1–6

[45] Zeng WS, Tang SZ. Modeling compatible single-tree biomass equations of Masson pine (*Pinus massoniana*) in southern China. J For Res. 2012; 23(4): 593–598. DOI: 10.1007/s11676-012-0299-4

[46] Dang YF, Wang XJ, Zeng WS. Using segmented modeling approach to construct tree volume and biomass equations for larch in northeastern China. For Res. 2012; 25(5): 558–563

[47] Zeng M, Nie XY, Zeng WS. Compatible tree volume and aboveground biomass equations of Chinese fir in China. Sci Silv Sin. 2013; 49(10):74–79. DOI: 10.11707/j.1001-7488.20131012

[48] Fu LY, Zeng WS, Zhang HR, Wang GX, Lei YC, Tang SZ. Generic linear mixed-effects individual-tree biomass models for *Pinus massoniana* Lamb. in southern China. South For. 2014; 76(1):47–56. DOI: 10.2989/20702620.2013.870389

[49] Zou WT, Zeng WS, Zhang LJ, Zeng M. Modeling crown biomass for four pine species in China. Forests. 2015; 6(2):433–449. DOI: 10.3390/f6020433

[50] Zeng WS. Using nonlinear mixed model and dummy variable model approaches to construct origin-based single tree biomass equations. Trees Struct Funct. 2015; 29(1): 275–283. DOI: 10.1007/s00468-014-1112-0

[51] Zeng WS. Integrated individual tree biomass simultaneous equations for two larch species in northeastern and northern China. Scand J For Res. 2015; 30(7): 594–604. DOI: 10.1080/02827581.2015.1046481

[52] Zeng WS, Tang SZ. Bias correction in logarithmic regression and comparison with weighted regression for non-linear models. Nat Prec. 2011; Available from http://dx.doi.org/10.1038/npre.2011.6708.1

[53] Zeng WS, Tang SZ. Goodness evaluation and precision analysis of tree biomass equations. Sci Silvae Sin. 2011; 47(11): 106–113

[54] State Forestry Administration of China. Technical regulation on sample collections for tree biomass modeling. China Standard Press, Beijing; 2015; 11 p.

[55] State Forestry Administration of China. Technical regulation on methodology for tree biomass modeling. China Standard Press, Beijing; 2015; 7 p.

[56] State Forestry Administration of China. Tree biomass models and related parameters to carbon accounting for *Cunninghamia lanceolata*. China Standard Press, Beijing; 2015; 14 p.

[57] State Forestry Administration of China. Tree biomass models and related parameters to carbon accounting for *Pinus massoniana*. China Standard Press, Beijing; 2015; 14 p.

[58] State Forestry Administration of China. Tree biomass models and related parameters to carbon accounting for *Pinus yunnanensis*. China Standard Press, Beijing; 2015; 9 p.

[59] State Forestry Administration of China. Tree biomass models and related parameters to carbon accounting for *Pinus tabulaeformis*. China Standard Press, Beijing; 2015; 9 p.

[60] State Forestry Administration of China. Tree biomass models and related parameters to carbon accounting for *Pinus elliottii*. China Standard Press, Beijing; 2015; 9 p.

[61] Jayaraman K. A statistical manual for forestry research. FAO Regional Office for Asia and the Pacific, Bangkok, 1999; 231 p.

[62] Parresol BR. Assessing tree and stand biomass: a review with examples and critical comparisons. For Sci. 1999; 45: 573–593

[63] Parresol BR. Additivity of nonlinear biomass equations. Can J For Res. 2001; 31: 865–878

[64] IPCC. IPCC guidelines for national greenhouse gas inventories—agriculture, forestry and other land use (Volume 4). Published: IGES, Japan, 2006.

[65] Pajtík J, Konôpka B, Lukac M. Biomass functions and expansion factors in young Norway spruce (*Picea abies* L. Karst) trees. For Ecol Manage. 2008; 256: 1096–1103

[66] Wang XP, Fang JY, Zhu BA. Forest biomass and root–shoot allocation in northeast China. For Ecol Manage. 2008; 255: 4007–4020

[67] Mugasha WA, Eid T, Bollandsas OM, Malimbwi RE, Chamshama SAO, Zahabu E, Katani JZ. Allometric models for prediction of above- and belowground biomass of trees in the miombo woodlands of Tanzania. For Ecol Manage. 2013; 310: 87–101

[68] Tang SZ, Lang KJ, Li HK. Statistics and computation of biomathematical models. Science Press, Beijing; 2008. 584 p.

[69] Crecente-Campo F, Soares P, Tomé M, Diéguez-Aranda U. Modelling annual individual-tree growth and mortality of Scots pine with data obtained at irregular measurement intervals and containing missing observations. For Ecol Manage. 2010; 260: 1965–1974

[70] Tang SZ, Li Y, Wang YH. Simultaneous equations, error-in-variable models and model integration in systems ecology. Ecol Model. 2001; 142: 285–294

[71] West GB, Brown JH, Enquist BJ. A general model for the origin of allometric scaling laws in biology. Science. 1997; 276(5309): 122–126

[72] West GB, Brown JH, Enquist BJ. A general model for the structure and allometry of plant vascular systems. Nature. 1999; 400: 664–667

[73] Zeng WS, Tang SZ. A new general allometric biomass model. Nat Prec. 2011; Available from http://dx.doi.org/10.1038/npre.2011.6704.2

[74] Henry M, Bombelli A, Trotta C, et al. GlobAllomeTree: international platform for tree allometric equations to support volume, biomass and carbon assessment. iForest (early view): e1–e5 [Internet]. 2013. Available from: http://www.sisef.it/iforest/contents/?id=ifor0901-006 [Assessed 2016-09-04]

7

Valorisation of Lignocellulosic Biomass Wastes for the Removal of Metal Ions from Aqueous Streams

Carlos Escudero-Oñate, Núria Fiol, Jordi Poch and Isabel Villaescusa

Abstract

Heavy metal pollution derived from anthropogenic activities is a relevant environmental threat nowadays due to their toxic nature, persistence and accumulation potential in the food chain. A wide variety of lignocellulosic-based biomaterials have been thoroughly assessed by the scientific community as sorbents for the removal of metals from aqueous streams. This kind of biomaterials, mainly constituted by lignin and cellulose, bear functional groups such as alcohol, ketone and carboxylates that provide active sorption points for the effective removal of heavy metals. The role of lignin in the sorption process is especially relevant, since this substance provides polyhydroxy and polyphenol functional groups—especially effective in the coordination of metals—and that provide ion exchange functionality to the material. Depending on their nature, these materials can be used either in their raw form or chemically modified form so as to enhance their sorption capacity and/or to achieve improved mechanical and mass transfer properties.

Keywords: lignocellulosic wastes, immobilisation, heavy metals, sorption, desorption, kinetics, equilibrium, modelling, recovery

1. Introduction

1.1. Heavy metal pollution

Heavy metals occur as natural constituents of the earth crust and some of them considered as persistent environmental pollutants. They may vary in oxidation state and hydrochemical speciation, but cannot be degraded or destroyed. The problems related to heavy metal pollution are transversal; water, air and soil components are susceptible of being severely polluted [1].

To a large degree, industrial activities are mainly responsible for environmental discharges and pollution. There are many industrial sources of pollution including manufacturing processes such as smelting and refining, electricity generation and nuclear power, agricultural fertilisation, wastewater treatment, fuel combustion, production of batteries, alloys manufacturing, electroplating, waste incineration, ceramics production and glass colouring [2].

Human exposure and intake of hazardous levels of heavy metals may occur through food, air and water [1]. It should be highlighted as well that some metals such as barium, cadmium, chromium, mercury, lead and hazardous metalloids such as arsenic can be bioaccumulated in different organs and may exert their toxic effects after long time of periodic ingestion of even low levels.

1.2. Structure of lignocellulosic biomass

Lignocellulosic biomass has been pointed out as a valuable source of chemicals and materials for different applications and a very relevant actor in the design of strategies to reduce the social and economic reliance on fossil resources. These materials are naturally produced from incorporation of CO_2 and water (driven by solar power) through the photosynthesis process. In this context, lignocellulosic biomass is the most abundant and biorenewable biomass on earth [3].

The major components of these materials are cellulose, hemicellulose and lignin, and cellulose and hemicellulose are considered to be polysaccharides. The structure of cellulose is based on the molecular formula $(C_6H_{10}O_5)_n$ and is one of the most relevant polysaccharides occurring in the plant cell wall. Hundreds of glucose units are linked through glucosidic linkage (**Figure 1a**) and individual chains usually interact with one another through hydrogen bonds. This material has found an extensive use in the paper industry. Hemicellulose is another important polysaccharide usually contained in lignocellulosic biomass with a more intricate structure and linkages than cellulose. Hemicellulose is composed of heteropolymers and may contain xyloglucan, xylan, glucomannans and galactoglucomannans in variable ratios depending on the type of biomass.

When it comes to heavy metal sorption by raw, unmodified lignocellulosic biomass, lignin has been found to play a key role. Lignin is a three-dimensional structure made of phenolic polymers that consists of three types of phenylpropanoid units: *p*-coumaryl alcohol, coniferyl alcohol and sinapyl alcohol (**Figure 1b**). A short section of a lignin polymer is presented in **Figure 1(c)**. This substance acts as cellular glue, joining the individual fibres and conferring strength to the plant tissue. The richness in electron-donor active sites that are provided in the polyhydroxy and polyphenol functional groups of lignin offers a unique frame for the interaction and binding of cationic heavy metals [6].

(a)

(b)

(c)

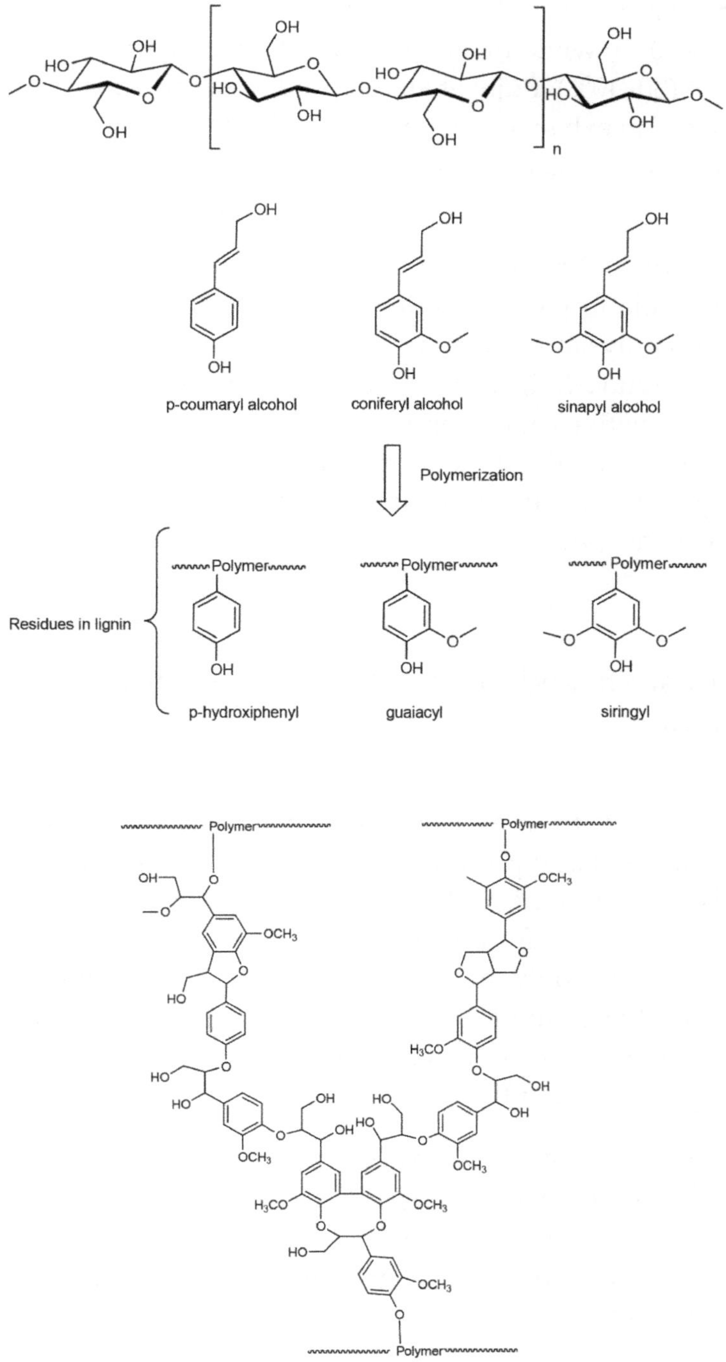

Figure 1. Structure of (a) cellulose, (b) main phenylpropane units and their related residues found in lignin and (c) structure of a lignin template [4, 5].

Abdolali et al. reported the chemical composition of some common lignocellulosic materials studied in metal sorption processes [6]. In general terms, lignocellulosic sorbents are composed of cellulose (30–35%), hemicellulose (20–40%), lignin (15–25%) and small amounts of water, ash, cyclic hydrocarbons and extractives. The chemical compositions in lignocellulosic materials are in different percentages depending on plant and also on the part of plant: leaves contain lower percentage of cellulose (15–25%) and lignin (5–10%) but higher percentage of hemicellulose (70–80%), while in stones and nuts, the lignin content is higher than cellulose and hemicellulose (30–40%). Pujol et al. reported a composition of about 25% total lignin and 23% polysaccharides in exhausted coffee wastes [7]. They also reported a relative composition depending on the particle size when characterising the different components of grape stalks (GS) [8].

The main constituents of lignocellulosic materials contain a variety of functional groups that play an important role in metal sorption. It was reported that acetamido groups, carbonyl, phenolic, structural polysaccharides, amino, amido, sulphydryl, carboxyl groups, alcohol and esters, present in lignocellulosic materials have affinity for metal complexation [9]. Ion exchange between positive cations and sodium, potassium, calcium and magnesium present in the adsorbent was identified as an important mechanism in several studies. These two mechanisms, together with chelation, are the main mechanisms known for metal sorption in lignocellulosic sorbents [9].

2. Lignocellulosic-based materials for the removal of heavy metals from aqueous effluents

Lignocellulosic materials from agricultural or plant wastes have been widely studied due to their renewable nature, large production and great local availability. A review of the recent literature reveals that hundreds of lignocellulosic materials around the world were tested to be used as low-cost sorbents for heavy metals and most of them are considered efficient and promising sorbents. The only negative aspects related to the use of raw lignocellulosic materials are connected to the lower sorption capacity compared to activated carbons and commercial ion exchange resins, as well as the potential release of organic matter that might cause a secondary pollution in the water treated.

Recently, Malik et al. reviewed low-cost adsorbents from different plant parts: husks, shells, straws, stems and woods, leaves, barks, grasses, stalks, seeds and hulls, fibres, fruit peels and pulps, bagasse and other lignocellulosic-based materials such as corn cob, oaks or fruit stones to be used as biosorbents [10]. The particular physicochemical properties of each material determine its properties as a sorbent.

Metal sorption by these heterogeneous materials is a complex process affected not only by the main mechanism, but also by other secondary mechanisms and combinations of different phenomena including chemisorption, adsorption on surface pores, adsorption by physical forces, entrapment in inter- and intra-fibrillar capillaries and spaces of the structural polysaccharides network, diffusion through cell walls and membranes, surface precipitation and metal reduction [9, 11]. The main mechanisms involved in the interaction between a specific sorbent material and metals could be predicted and verified by combination of different spectroscopic techniques and conventional techniques such as titration, chemical blocking of functional groups and related release of cations from sorbent [6].

The efficiency of each sorbent extremely depends on the operating conditions such as temperature, sorbent particle size, pH, contact time or initial metal concentration. For this reason, a thorough study to find out the optimal sorption conditions is required for each couple sorbent metal. Hundreds of these studies can be found in the literature, where the optimal conditions for a specific pair of lignocellulosic sorbent metal were reported. The main results in this field can be found in reviews published in prestigious journals, helping the researchers to update the information about potential biosorbents suitable for a given application [6, 9, 11–14]. In these reviews, the performance of different types of lignocellulosic sorbents is usually expressed as maximum sorption capacity (q_{max}) and the efficiency comparison between materials used to be based in these values. Nevertheless, additional information is needed to compare sorption efficiency between sorbents for a specific metal due to differences in particle sorbent sizes, sorbent dosages and other experimental conditions used in each work.

A large share of current research on metal sorption by lignocellulosic materials has focused on the removal of heavy metal cations, such as Pb(II), Cd(II), Zn(II) and Cu(II) [4], but other toxic metals such as Ni(II), Al(III), Hg(II), Fe(II), chromium and metalloids (arsenic, selenium, molybdenum, and vanadium) have been also studied as target pollutant in biosorption processes [15].

Nowadays however, there is a growing interest in biosorption processes that use low-cost materials as sorbents for the recovery of rare earths or precious metals [16, 17]. The recovery of valuable metals by biosorption opens a challenging and exciting new scenario in sorption studies far beyond the regular water treatment, since the recovered material is expected to have an intrinsic economic value.

Despite metal sorption using biomass has been regarded as an environmental-friendly technique and provides a set of potential advantages against traditional technologies; most of the studies have been focused on synthetic solutions. To the best of our knowledge, studies reporting the use of biosorption-based technologies for metal removal from real scenarios of contaminated wastewaters are scarce. Few studies have so far explored the use of lignocellulosic materials in the treatment of real polluted wastewater. Fruit shell of gulmohar and olive stones were investigated for the removal of Cr(VI) from an electroplating wastewater [18, 19], and rice agro-wastes, coconut shell, neem leaves and hyacinth roots were used for the removal of Pb(II) from wastewater in a battery industry [20], and recently Liu et al. [21] published a complete method for chromium electroplating wastewater treatment based on biosorption using exhausted coffee.

3. Structural characterisation of lignocellulosic wastes and their interactions with heavy metals

3.1. Infrared spectroscopy

Infrared spectroscopy (IR) is a technique based on the vibrations of the atoms on a given molecule and is a result of the molecular vibration mechanism, which refers to energy-matter interaction [22]. Usually, an infrared spectrum is obtained by passing infrared radiation through a sample and quantifying the fraction of incident radiation of each frequency that has been absorbed. The energy at which a peak appears in an absorption (or transmission) spectra corresponds to the characteristic energy of the vibration in a part of the molecule. The selection rule for IR spectroscopy is that an electric dipole moment in the molecule has to change during the vibration.

Techniques based on IR spectroscopy have become a powerful tool for determining the functional groups and the mechanisms involved in the removal of heavy metals by different type of biomass. A magnification in the region 800–1800 cm^{-1} of a raw lignocellulosic material (grape stalk wastes) and exposed to either Cr(III) or Cr(VI) solutions are presented in **Figure 2**. In the spectra, the positions of the main bands that have suffered modifications in their frequencies have been indicated with arrows.

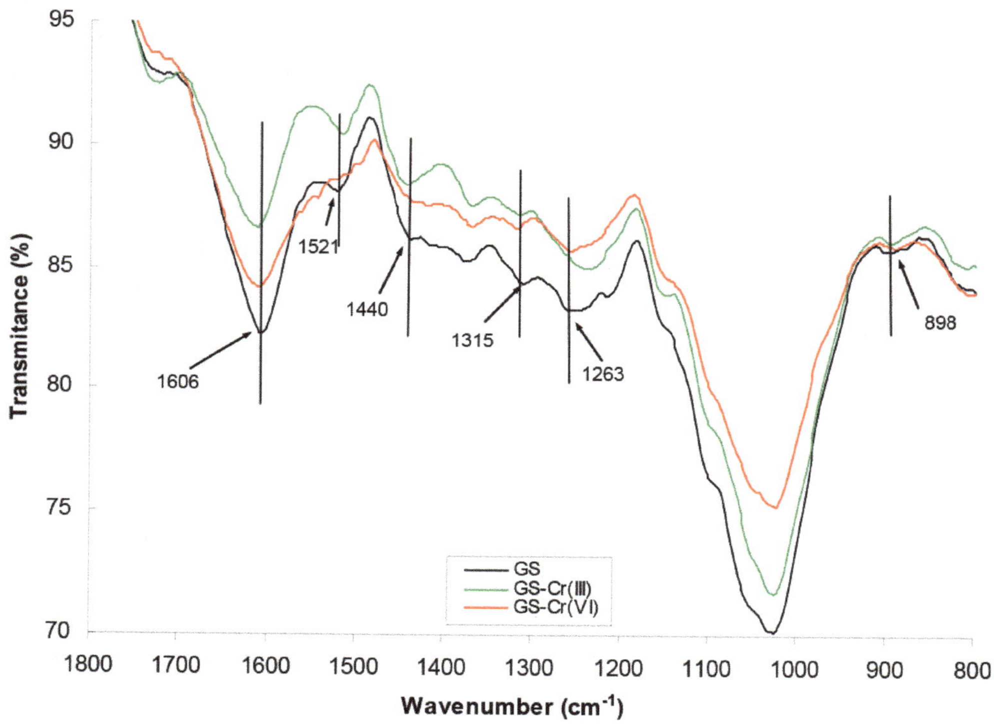

Figure 2. Grape stalks (GS) FTIR-ATR spectra before and after treatment with Cr(III) and Cr(VI) solutions. Initial metal concentration: 800 $mg \cdot L^{-1}$. Agitation time: 24 h, $pH_0 = 3$.

The initial and final positions of the modified bands have been summarised in **Table 1**, including the assignment of the band to the functional group/s involved in Cr(III) and Cr(VI) sorption.

Grape stalks treated with Cr(III) and Cr(VI) solutions show modifications in the characteristic bands of syringyl and guaiacyl moieties (1263 and 1315 cm^{-1}, respectively), the methoxy deformation (1440 cm^{-1}), the aromatic skeleton vibration (1521 cm^{-1}) and the aromatic ring vibration (1606 cm^{-1}). All these bands, characteristic of the lignin macromolecule, indicate that both Cr(III) and Cr(VI) are adsorbed onto this component of the sorbent, making the cellulose almost unaltered during the sorption process. The important role of lignin in sorption of copper and nickel onto the cork had been previously noticed in results of ^{13}CP-MAS-NMR on the solid phase of the sorbent [25].

Frequency (cm^{-1})				
GS	**GS-Cr(III)**	**GS-Cr(VI)**	**Assignment**	**Ref.**
898	897	898	Carbohydrates (unaltered)	[23]
1263	1238	1247	Guayacyl/C–O phenolic	[24]
1315	1319	1321	Syringyl	[23]
1440	1444	1421	Methoxy deformation	[23]
1521	1515	1511	Aromatic skeleton vibration	[23]
1606	1614	1612	C=C Aromatic vibration	[23]

Table 1. Observed frequencies and assignment.

3.2. Scanning electron microscopy-energy dispersive X-ray analysis

Scanning electron microscopy (SEM) coupled to energy dispersive X-ray analysis (EDX) is a powerful instrumental combination to assess the mechanisms governing heavy metal removal by lignocellulosic biomass. SEM provides a great magnification of the surfaces and allows gathering high-quality images, from which the morphology and topography of the materials can be assessed. The additional use of detection of backscattered electrons (BSE) helps finding target regions where metals may have been selectively accumulated in lignocellulosic biomaterials. EDX is widely used coupled to SEM and allows obtaining local elemental analysis of the surfaces under the observation. The technique becomes especially relevant in the assessment of sorption processes where local microprecipitation is likely to occur.

Metal microprecipitation takes place when the solubility of the sorbate reaches its limit. Microprecipitation in metal-removal processes—despite not being directly related to sorption —positively contributes to the overall detoxification of the effluent, since the metal microprecipitate remains immobilised in the surface and remains separated from the solution.

This may occur even due to local conditions, e.g. on or inside the sorbent, and not necessarily in the bulk of the solution. These favourable conditions for microprecipitation may be created by local deviations in physical conditions such as pH or by the presence of materials released from the sorbent itself. Escudero et al. [26] using SEM-BSE-EDX techniques observed microprecipitation phenomena when exploring the removal of hexavalent chromium by grape stalk wastes (GS) entrapped into calcium alginate gel beads. Micrographs in the SEM mode (a, d), BSE mode (b) and characteristic EDX spectra (c) are presented in **Figure 3**.

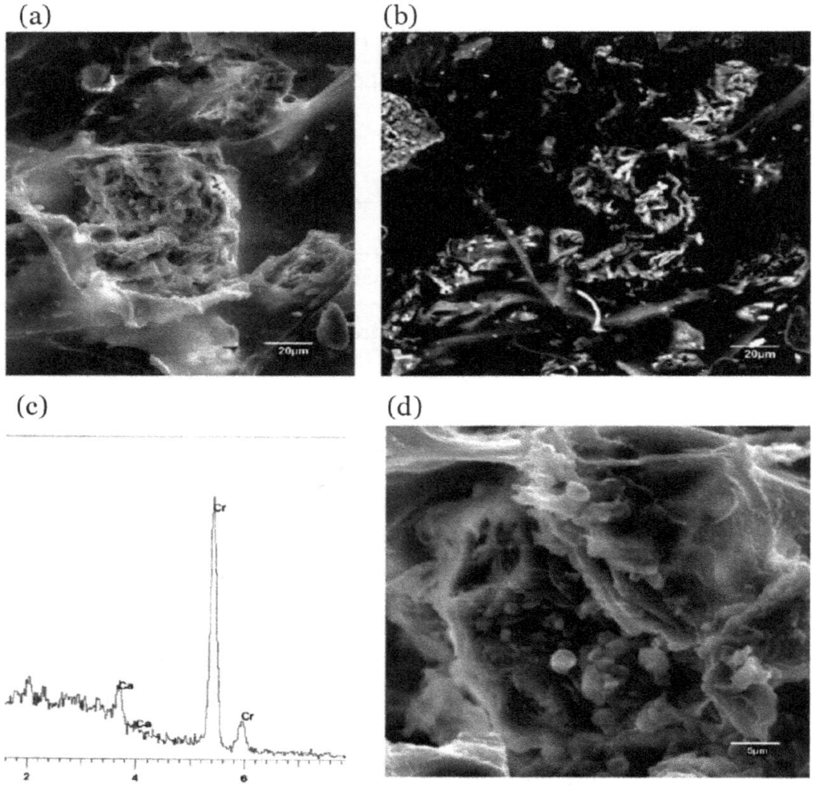

Figure 3. Electronic micrograph of grape stalks entrapped into a calcium alginate gel matrix exposed to a Cr(VI) solution. (a) Scanning mode, (b) backscattered electrons mode, (c) EDX local analysis and (d) focus on the grape stalk region in the SEM mode [26].

The observations indicated that chromium is mostly accumulated in the grape stalk surface, as it may be evidenced by the bright colour observed in the BSE picture (b). While magnifying the region, small nodules appear onto the surface of the grape stalks, indicating the formation of microprecipitates (d). The evidence of chromium accumulation in these nodules was further demonstrated through the EDX local analysis (c). Using further electron spin resonance analysis of the solids, the authors evidenced the formation of Cr(III). It was hypothesised then that Cr(VI) undergoes a reduction process that partially converts hexavalent into trivalent chromium with an associated large proton consumption (**Scheme 1a**). In this local alkalinised environment, Cr(III) precipitates onto the surface of the lignocellulosic material as hydroxide through reaction 1 (b) and remains isolated by the bulk of the solution by the surrounding calcium alginate gel.

$$\text{(a)} \quad HCrO_4^- + 7H^+ + 3e^- \xrightarrow{\text{Raw GS} \rightarrow \text{Oxidized GS}} Cr^{3+} + 4H_2O$$

$$\text{(b)} \quad Cr^{3+} + 3OH^- \longrightarrow Cr(OH)_3 \ (\downarrow)$$

Scheme 1. Hexavalent chromium removal through a combined reduction-precipitation process promoted by grape stalks [26].

4. Immobilisation techniques: the use of hydrogels as high-permeable immobilising materials

Native biomass shows low density and poor mechanical strength and rigidity [27]. These properties make its application difficult in biosorption processes either in batch or in column. In the batch mode, mechanical stirring and shear forces can break fragile particles of the biosorbent resulting in particle attrition (breakage, fragmentation, fines formation) and then, a continuous change on the original particle size distribution. Particle size is a key parameter in the performance of sorption processes; it is desirable that size distribution remains homogeneous throughout the entire process. It is known that, to maximise mass transfer and sorption efficiency, the sorbent particle size of the sorbent must be as small as possible (maximising surface-to-mass ratio). Limitations in the use of small particle sizes are (i) sorbent should be easily handled and (ii) clogging in filters, columns, valves and pipes must be prevented.

In the case of continuous bed up-flow sorption processes (columns), additional problems due to the native biomass physical properties must be considered. Biomass needs to be wetted to allow free swelling of the particles prior to column filling and to remove the finest particles by natural flotation. In addition to this, the non-uniform shape of the native biomass hinders the estimation of a form factor (sphere, cylinder, ellipsoid) of the particles. This parameter is needed when formulating a model to describe sorption processes. Most of the aforementioned drawbacks and limitations can be overcome, however, by immobilisation of the biomass in a water-permeable polymeric matrix. Immobilisation of biomass provides several advantages over native biomass: (i) increase of mechanical resistance, (ii) increase of density, (iii) possibility of enhancement of effective surface area due to the use of biomass powder, (iv) easy handling and (v) achievement of precision in form factor, since immobilisation can yield quasi-spherical beads or granules.

Biomass immobilisation consists of the attachment or entrapment of biomass on a support. Immobilisation can be carried out by four different techniques: adsorption, covalent binding, entrapment and membrane confinement. Entrapment in a gel matrix or gel encapsulation is among the most widely studied methods for immobilisation of enzymes, microbial biomass and animal and plant cells, being calcium alginate one of the regularly used matrixes [28]. The support selection is of crucial importance and must be chosen according to the target application of the immobilised material. Applications of immobilisation techniques are found in different scopes of several fields such as biotechnology and pharmaceutical, environment, food and biosensor industries [29].

For the treatment of wastewater, support materials need to meet (among others) the following criteria: insoluble, non-biodegradable, non-toxic, high mechanical and chemical stability, high diffusivity and ease of immobilisation procedure [30]. Natural polymer derivatives of algal polysaccharides (alginate, carrageenan, agar and agarose) and chitosan (an amino polysaccharide derived from chitin) have been experimentally used. Crini [31] has recently reviewed the most important features of polysaccharide-based materials used as adsorbents.

One of the most used immobilisation supports is alginate [32]. Alginate is a common term used for natural polymers composed of linear, unbranched chains of varying lengths, proportions and sequences of 1,4 linked residues of β-D-mannuronic acid and α-L-guluronic acid residues [33]. As the composition of alginates depends on the source from which they are extracted, they can exhibit different physical and chemical properties. The main source of alginates are different species of brown seaweeds *Laminaria digitata* [34] and *Macrocystis pyrifera* [35]. One of the properties of alginates is that they can cross-link with divalent metal ions such as calcium, barium, copper, zinc and lead [36, 37] to form M^{2+}-alginate hydrogels. Alginate gelation takes place when divalent metals interact with blocks of acid residues resulting in the formation of a 3D network, which is usually described by an 'egg-box' model [38]. Calcium alginate is one of the most utilised entrapment matrixes in environmental studies. Heavy metals adsorption takes place via ion exchange between Ca^{2+} ions from the hydrogel and metal ions. Pandey et al. [39] studied Cr(VI, Pb(II) and Cu(II) by using calcium alginate beads and Jodra and Mijangos [40] determined the ion exchange selectivity of calcium alginate gels for heavy metals and found the selectivity order: Pb > Cu > Cd > Ni > Zn > Co. During the last years different types of biomass used for heavy metals sorption have been entrapped in calcium alginate with the aim of enhancing their sorption performance. Studies on fungi [41, 42], algae [43, 44], bacteria [45] and yeast [46] encapsulated in calcium alginate have been recently carried out. Nevertheless, studies on metal sorption by using vegetable wastes and specifically lignocellulosic materials encapsulated in calcium alginate are scarce. The most probable reason is the high capacity of calcium alginate itself to sorb heavy metal ions. The only example of vegetable waste can be found in the work of Ansari et al. [47]. The authors encapsulated rose waste biomass in calcium alginate beads to sorb Pb(II) and found that the immobilised biomass yielded to higher metal ion sorption. Unfortunately, the authors did not report the amount of lead sorbed by pure calcium alginate beads.

Grape stalks, a lignocellulosic waste produced in wine production, was entrapped in calcium alginate and used for Cr(VI) removal at pH 3 in batch mode [48–50] and in column [51] from single metal solutions and from binary mixtures containing Cr(VI) and Cr(III) [52]. A scheme of a typical arrangement to obtain gel beads by the dropping technique is presented in **Figure 4**.

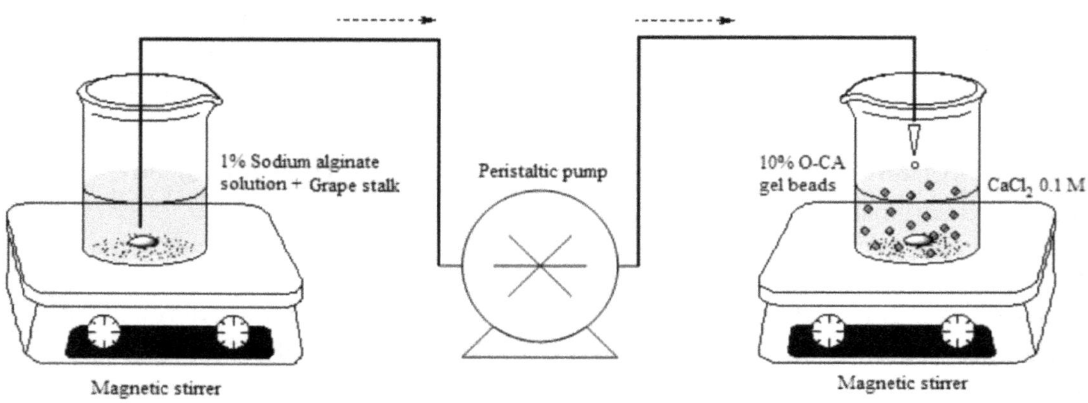

Figure 4. Entrapment procedure to obtain gel beads of grape stalks using calcium alginate [48].

Grape stalks were reported to be a very effective reducing agent to reduce Cr(VI) to Cr(III) [53]. Results of the above-mentioned studies with grape stalk demonstrated that both sorption and reduction processes were enhanced when the sorbent was immobilised into calcium alginate. Furthermore, the Cr(III) formed as a consequence of the reduction reaction was sorbed onto calcium alginate beads via ion exchange with calcium ions. It must be remarked that at pH 3 major chemical species of Cr(VI) is $HCrO_4^-$ that is hardly sorbed by calcium alginate beads. Therefore, the joint action of both sorbents under appropriate conditions can lead to the total elimination of Cr(VI) [26].

Treatment phase			
Electrochemical treatment		**Biosorption-based treatment**	
1. Electrochemical reduction	**Subtotal**	**1. Batch reactor**	**Subtotal**
Electricity	1.560	Stirring	780
2. Reagent addition	**Subtotal**	**2. Biosorbent addition**	**Subtotal**
Sodium hydroxide (NaOH)	500	Sodium hydroxide (NaOH)	500
Hydrochloric acid (HCl)	100	Hydrochloric acid (HCl)	100
Coagulant	2000	Grape stalks	21
3. Control sensors and separation unit	**Subtotal**	**3. Control sensors and separation unit**	**Subtotal**
Electrodes (pH and redox)	1000	Electrodes (pH and redox)	1000
Filter clothes	100	Filter clothes	100
Energy	200	Energy	200
4. Waste management	**Subtotal**	**4. Waste management**	**Subtotal**
Waste collection	240	Waste collection	240
post-treatment phase			
Electrochemical treatment		**Biosorption-based treatment**	
5. Treatment with activated carbon	**Subtotal**	**5. Treatment with biosorbent**	**Subtotal**
Activated carbon	134.4	Grape stalks	1.2
		Sodium alginate	339.4
Other expenses			
Electrochemical treatment		**Biosorption-based treatment**	
Internal labour costs	3000	Internal labour costs	1500
External labour costs	1000	External labour costs	750
Pumps and valves	1000	Pumps and valves	1000
Total	**10,834€**	**Total**	**6532€**

Table 2. Comparison between electrolytic and biosorption-based schemes for the treatment of 300 m^3/year of wastewater from an electroplating water industry. Costs are expressed in €.

The authors of the present chapter carried out a thorough techno-economical study concerning implementation of biosorption-based technology for the treatment of wastewater from an electroplating industry (unpublished data). A brief economic summary showing the different technological processes involved and their related expenses is presented in **Table 2**. The table compares the costs related to the use of regular technology (based on electroreduction) with the ones associated with a water treatment based on biosorption/bioreduction by grape stalks (GS) and precipitation/refining using GS entrapped into calcium alginate (CA).

In a regular wastewater treatment, Cr(VI) is reduced to Cr(III) using electricity. In acidic media, 3 mols of electrons are consumed in this process per mol of Cr(VI) reduced. In the case of the GS sorption/reduction process, there is no electric current applied to the solution and the only energetic requirements are related to the mixing of the material with the Cr(VI) solution. Avoiding electrochemical reduction in the treatment phase, decreases the costs related to energy consumption to half (from 1560 to 780€). In terms of reagents, NaOH and HCl are reagents used in both treatment schemes. In the secondary phase (post-treatment or refining), the costs related to the use of activated carbon are about 134€. Using GS and CA for the removal of residual Cr(VI) and the formed Cr(III) involve about 341€. In terms of other expenses, there is both an internal and external labour costs reduction when biosorption-based technologies are used. This saving is due to the frequent cleaning and, in the last term, replacement of the electrodes in the electrochemical cells. In the case of a biosorption unit, repairs are less frequent and maintenance less expensive.

Overall, the use of GS and CA in the detoxification of Cr(VI) polluted effluents involves a saving close to 40% if compared to the electrochemical reduction process. These data are just an example that clearly shows the viability of lignocellulosic materials for real wastewater treatment. There is a clear opportunity to go beyond the laboratory studies and implement properly scaled biosorption processes to the treatment of real industrial effluents.

5. Equilibrium and kinetics modelling

5.1. Equilibrium models in single component solutions

A precise mathematical description of the equilibrium isotherms is of paramount importance for the effective design of sorption systems. The most widely used adsorption isotherms found in the literature to describe the amount of solute adsorbed as a function of the equilibrium concentration in solution are summarised in **Table 3**.

From these models, by far the most widely employed are Langmuir and Freundlich. While Langmuir isotherm model relies on the adsorption theory and assumes the formation of a sorbate monolayer, Freundlich isotherm is an empirical model to correlate the concentration of a sorbate on the solid phase and the concentration of the sorbate in the fluid.

Isotherm model	Equation	Linear expression	Ref.
Langmuir	$q_e = \frac{Q_m b C_e}{1+bC_e}$	$\frac{C_e}{q_e} = \frac{1}{bQ_m} + \frac{C_e}{Q_m}$	[54–56]
Freundlich	$q_e = k_F C_e^{\frac{1}{n_F}}$	$\ln q_e = \ln k_F + \frac{1}{n}\ln C_e$	[54–56]
Redlich-Peterson	$q_e = \frac{K_{RP} C_e}{1+\alpha_{RP} C_e^{\beta}}$	$\ln\left(\frac{K_{RP} C_e}{q_e} - 1\right) = \ln\alpha_{RP} + \beta \ln C_e$	[55, 56]
Sips	$q_e = \frac{Q_m b C_e^{n_S}}{1+bC_e^{n_S}}$		[54–56]
Temkin	$q_e = \frac{RT}{b_T}\ln\left(K_T C_e\right)$	$q_e = \frac{RT}{b_T}\ln K_T + \frac{RT}{b_T}\ln C_e$	[54–56]
Dubinin-Radushkevich	$q_e = Q_m \exp\left(-B_D \varepsilon^2\right)$ $\varepsilon = RT ln\left(1+\frac{1}{C_e}\right)$	$\ln q_e = \ln Q_m - 2B_D R^2 T^2 \ln\left(1+\frac{1}{C_e}\right)$	[55, 56]
Toth	$q_e = \frac{Q_m b C_e}{\left(1+\left(bC_e\right)^t\right)^{\frac{1}{t}}}$		[54, 56]

Table 3. Mathematical equations of single component isotherm models.

Obtaining isotherm parameters from experimental data is a key aspect to consider. Using linear regression to determine the sorption isotherm parameters and decide which model provides the best fit was not only the easiest way at the time this approach was proposed, but also has become a 'custom' extended to our days. Obviously, the linearisation of the equations involves a transformation of the data in a process that alters the structure of errors, violates the error variance assumptions and the normality of the least squares method [57–59]. This set of error sources explains that the linear parameters obtained through Freundlich model produces isotherms that best fit data at low concentrations, while the linear parameters of the Langmuir tend to better match isotherm data at higher concentrations [57, 60]. Taking into account the aforementioned drawbacks of the linear procedures, direct calculation of the isotherm parameters through non-linear optimisation methods is strongly recommended. The non-linear optimisation is a more complex method than the linearisation approach and requires the proper choice of an error function to accurately evaluate the fit of the experimental results to the chosen isotherm, since the choice of the error function may affect the derived parameters. The most widely used error functions to fit sorption isotherm data are summarised in **Table 4**.

Error function	Expression	Ref.
The sum of the squares of the errors (SSR)	$\sum_{i=1}^{N}\left(q_{i,\exp}-q_{i,cal}\right)^2$	[54, 57, 61]
Hybrid fractional error function (HYBRD)	$\frac{100}{N-p}\sum_{i=1}^{N}\frac{\left(q_{i,\exp}-q_{i,cal}\right)^2}{q_{i,\exp}}$	[54, 57, 61]
Marquardt's percent standard deviation (MPSD)	$100\sqrt{\frac{1}{N-p}\sum_{i=1}^{N}\left(\frac{q_{i,\exp}-q_{i,cal}}{q_{i,\exp}}\right)^2}$	[54, 57, 61]
Average relative error (ARE)	$\frac{100}{N}\sum_{i=1}^{N}\left\|\frac{q_{i,\exp}-q_{i,cal}}{q_{i,\exp}}\right\|$	[54, 57, 61]
Sum of the absolute errors (EABS)	$\sum_{i=1}^{N}\left\|q_{i,\exp}-q_{i,cal}\right\|$	[54, 57, 61]

Table 4. Error functions.

The sum of the squares of the errors (SSR) is the most widely used error function, but it shows an important drawback; the isotherm parameters obtained provides a better fit in the case of high concentrations due to the fact that errors and therefore their squares increase, when it does the concentration.

Hybrid fractional error function (HYBRD) is an error function developed with the attempt of improving the adjustment of the SSR at low concentrations divided by the measured value and includes the degrees of freedom of the system.

Marquardt's percent standard deviation (MPSD) is similar in some aspect to a geometric mean error distribution modified according to the number of degrees of freedom of the system.

Average relative error function (ARE) attempts to minimise the fractional error distribution across the entire concentration range.

Sum of the absolute errors (EABS) uses a similar approach to that used in the SSR. Isotherm parameters determined using this error function provide a better fit as the magnitude of the error increases, biasing the fit towards the high concentration data [61].

In terms of least-squares regression analysis, one of the central premises is that the independent variable x remains fixed, i.e. there is no measured error in x. Therefore, there should be no error in the measurement of the equilibrium concentration in solution (C_e). However, the determination of the remaining concentration in the solution is not free from error, despite careful calibration and repetition of sample analysis is performed. Hence, the application of orthogonal distance regression (ODR) may be appropriate [59, 62–65].

The ODR takes into account the errors in both variables (C_e and q). The most widely used error functions are summarised in **Table 5**.

Error function	Expression	Ref.
Theoretical orthogonal distance regression	$\sum_{i=1}^{N}\left[\left(\frac{q_{i,\exp}-q_{i,cal}}{\sigma_{\varepsilon i}}\right)^2+\left(\frac{C_{ei,\exp}-C_{ei,cal}}{\sigma_{\delta i}^{e}}\right)^2\right]$	[59, 63]
Orthogonal distance regression	$\sum_{i=1}^{N}\left[\left(\frac{q_{i,\exp}-q_{i,cal}}{q_{i,\exp}}\right)^2+\left(\frac{C_{ei,\exp}-C_{ei,cal}}{C_{ei,\exp}}\right)^2\right]$	[62, 63, 65, 66]
Experimental weighted orthogonal distance regression	$\sum_{i=1}^{N}\left[\left(\frac{q_{i,\exp}-q_{i,cal}}{sdy_i}\right)^2+\left(\frac{C_{ei,\exp}-C_{ei,cal}}{sdx_i}\right)^2\right]$	[63]
Triplicate orthogonal distance regression	$\sum_{i=1}^{N}\left[\left(\frac{q_{i,\exp}-q_{i,cal}}{q_{i,\exp}}\right)^2+\left(\frac{C_{ei,\exp}-C_{ei,cal}}{C_{ei,\exp}}\right)^2\right]$	[63]

Table 5. Error functions.

5.2. Equilibrium models in multicomponent solutions

Since the interaction of a component with other components in a mixture can be synergistic, antagonistic or non-interactive, the biosorption results cannot be predicted on the basis of studies derived from single component solutions. The behaviour of each species in a multi-component system depends largely on the physical and chemical properties of both, sorbent and sorbate. However, most of the isotherms of a single component can be extended to describe a multi-component adsorption system. The most frequently used extensions are summarised in **Table 6**.

In the scientific literature, there are few models to describe synergistic effects in sorption processes. A straightforward and simple way to develop a synergistic model is based on the use of monocomponent models and introduction of correction factors, for example, by combining a single isotherm (Langmuir, Freundlich, etc.) with a form factor $1+F_i(C)$ [72, 73].

When it comes to obtaining the parameters of these models, the same comments stated previously for monocomponent sorption isotherms applies. The error function most widely used is based on the sum of the SSR for each component although the most suitable method would be the use of ODR.

Isotherm model	Equation	Ref.
Extended Langmuir:	$q_{e,i} = \frac{q_{0,i} b_i C_{e,i}}{1+\sum_j b_j C_{e,j}}$	[67–69]
Modified competitive Langmuir:	$q_{e,i} = \frac{q_{0,i} b_i \frac{C_{e,i}}{\eta_i}}{1+\sum_j b_j \frac{C_{e,j}}{\eta_j}}$	[67, 70, 71]
Extended Freundlich:	$q_{e,i} = \frac{K_{F,i} C_{e,i}^{\frac{1}{n_i}+x_i}}{C_{e,i}^{x_i} + y_i C_{e,j}^{z_j}}$	[67, 68]
Competitive non modified Redlich Peterson model:	$q_{e,i} = \frac{K_{RP,i} C_{e,i}}{1+\sum_j a_{RP,j} C_{e,j}^{\beta_j}}$	[67, 68]
Competitive modified Redlich Peterson model:	$q_{e,i} = \frac{K_{RP,i} \frac{C_{e,i}}{\eta_i}}{1+\sum_j a_{RP,j} \left(\frac{C_{e,j}}{\eta_j}\right)^{\beta_j}}$	[68]

Table 6. Multicomponent equilibrium models.

5.3. Kinetics models

A summary on the most widely used models to describe the time-course profile of metal removal in sorption processes is presented in **Table 7**.

The scientific literature to date shows that, in most cases, the obtention of the characteristic parameters of the aforementioned models is performed through linearisation. As previously discussed, this process involves a transformation of the data set and the subsequent alteration of the error structure and therefore a given bias in the so obtained parameters. To avoid this, non-linear regression should be used using the SSR as target error function. In this case, the use of ODR is not strictly needed since it can be considered that there is no error in the accurate determination of the time variable.

As a final remark, it should be highlighted that a good fitting of an experimental data set to a given kinetic or equilibrium model does not necessarily imply that the sorption process is governed by the mechanisms on which the model relies [74].

Kinetic model	Equation	Linear expression	Ref.
Pseudo-first-order or Lagergren equation	$\frac{dq_t}{dt} = K_1(q_e - q_t)$ $q_t = q_e(1-\exp(-K_1 t))$	$\ln(q_e - q_t) = \ln q_e - K_1 t$	[56, 74]
Pseudo-second-order	$\frac{dq_t}{dt} = K_2(q_e - q_t)^2$ $q_t = \frac{K_2 q_e^2 t}{1+K_2 q_e t}$	$\frac{t}{q_t} = \frac{1}{K_2 q_e^2} + \frac{t}{q_e}$	[56, 74]
Weber-Morris or intra-particle diffusion	$q_t = K_p t^{\frac{1}{2}}$		[56, 74]
Elovich equation	$\frac{dq_t}{dt} = a_E \exp(-\alpha_E q_t)$ $q_t = \frac{1}{\alpha_E}\ln(t + t_0) - \frac{1}{\alpha_E}\ln(t_0)$ $t_0 = \frac{1}{a_E \alpha_E}$		[74]

Table 7. Kinetic models.

6. Metal desorption and regeneration of lignocellulosic wastes

Metal laden biomass can either be directly disposed or incinerated, delivering the ashes to a regular hazardous waste landfill. Incineration reduces waste volume, but enhances the metal content per mass unit and may cause environmental issues due to potential toxic metal leaching. Disposal or incineration of metal-laden biomass is recommended when biomass is abundant and cheap and metal is not worth recovering. Another alternative is regeneration of the biomass by elution of the loaded metal using desorbing agents. The selection criteria that guides to choose the appropriate desorbing agent are: (i) small volume of eluent should yield high metal concentration in the resulting solution, (ii) the structural integrity of the biomass must not be severely affected, and (iii) the eluent should be economic and environmentally friendly.

Most of the desorbing agents used for sorbents regeneration and metal recovery are based on strong mineral acids (HCl, HNO_3, H_2SO_4), short-chain organic acids (HCOOH, CH_3COOH), bases (NaOH, $NaHCO_3$, Na_2CO_3, KOH, K_2CO_3), salts (NaCl, KCl, $CaCl_2$, KNO_3), chelating agents (ethylenediaminetetraacetic (EDTA), diethylenediaminepentacetic (DTPA), nitrilotri-acetic (NTA)) or buffer solutions (phosphate, bicarbonate). Recently, the most widely used desorbing agents have been reviewed [75].

The most used desorption agents to remove metal-laden lignocellulosic biomass are acids. An acidic solution (pH 2.0) was found to be the best desorbing solution for Cu(II) and Cr(VI) loaded onto commercial coffee wastes [76]. A 0.1 M HCl solution resulted to be effective to desorb Cd(II) (83.9%) from rice husk in both batch and column modes [77]; Cu(II) (99.4%), Cd(II) (98.5%) and Zn(II) (99.3%) from papaya wood [78]; As(V) from coconut coir pith [79]; Cu(II) from cork and yohimbe bark [80]; and 0.2 and 0.5 M HCl for chromium recovery from avena by-products and Ga(III) desorption from coir [71]. A 0.1 M HNO_3 solution was used by Gupta and Nayak [81] to desorb 98.2% Cd(II) loaded onto orange peel powder with Fe_3O_4. Higher concentrations than 0.01 M of HNO_3 and HCl were needed to effectively desorb Cd(II) loaded onto coffee beans [82]. *Ficus religiosa* lead(II) laden was regenerated by using 0.05 M HNO_3 [83]. HCl and EDTA solutions were used to desorb U(VI) from citrus waste material. The best desorption yields were obtained by 0.1 EDTA (94.7%) followed by 0.1 HCl (89.71%) [84]. HCl and EDTA were also tested by Martinez et al. to recover Pb, Cu, Cd and Ni from grape stalk and olive stones wastes, the former being the most effective [85, 86].

Alkaline solutions were also successfully tested as desorbing agents for metal ions from biomass. Note that 0.5 M sodium citrate was able to remove lead(II) laden on hop by-products [87]. Elution of arsenic-laden rice polish and rice husk packed columns was achieved by passing through the column of 10% NaOH [88] and 1 M KOH [89] solutions, respectively.

The main conclusion retrieved from the literature survey is that recovery of metals from exhausted lignocellulosic materials and other low-cost sorbent materials and the regeneration of the sorbent is not nowadays the focus of researchers. It may be foreseen however that in the future, when some metals become scarce, further research will be conducted to find out selective metal desorbing agents that meet the most of the aforementioned criteria.

7. Conclusion

Lignocellulosic biomass has a huge potential as a low cost, renewable source and environmentally friendly alternative to conventional methods for the removal of heavy metals from aqueous polluted streams. This technology is of special interest to treat large volumes of effluent containing low metal concentration to produce a final effluent that does not pose environmental hazards. A thorough screening and selection of the most effective low-cost sorbents with sufficiently high metal-binding capacity and selectivity for heavy metal ions are prerequisites for full-scale implementation in industrial processes. Despite further research, efforts towards full understanding of sorption mechanisms and development of more accurate mathematical models might be required, the technology can be considered mature enough as to face scale-up scenarios to large scale. Industrial stakeholders, policymakers and regulators have nowadays a challenging and exciting opportunity to take a step forward towards environmental sustainability considering sorption onto biomaterials on their wastewater treatment schemes.

Annex 1: Notation

q_e	Amount of sorbate adsorbed per unit mass of sorbent at equilibrium
Q_m	Maximum uptake of sorbate per unit mass of sorbent
b	Langmuir constant related to sorbent affinity
C_e	Concentration of sorbate in solution at equilibrium
k_F	Freundlich constant related to the amount of sorbate adsorbed
n_F	Freundlich constant related to sorption intensity
K_{RP}	Redlich-Peterson constant related to the amount of sorbate adsorbed
α_{RP}	Redlich-Peterson constant
β	Redlich-Peterson exponent $0 \leq \beta \leq 1$
n_S	Sips constant related to sorption intensity
R	Universal gas constant
T	Temperature
b_T	Temkin constant
K_T	Temkin constant related to binding
B_D	Dubinin-Radushkevich constant related to sorption energy
ε	Polanyi potential
τ	Toth model exponent
$q_{i,exp}$	Experimental data values
$q_{i,cal}$	Data values calculated from the model
N	Number of data
p	Number of model parameters
$\sigma_{\varepsilon i}$	Population standard deviation of measurement error in dependent variable
$\sigma_{\delta i}^{e}$	Population standard deviation of measurement error in equilibrium concentration
sdy_i	Estimates of population standard deviation for dependent variable
sdx_i	Estimates of population standard deviation for independent variable
η	Dimensionless interaction factor
q_t	Amount of sorbate adsorbed per unit mass of sorbent at time t
K_1	First-order reaction rate equilibrium constant
K_2	Second-order reaction rate equilibrium constant
t	Time
K_p	Intra-particle diffusion rate constant
a_E	Elovich equation constant
α_E	Elovich equation constant

Author details

Carlos Escudero-Oñate[1*], Núria Fiol[2], Jordi Poch[3] and Isabel Villaescusa[2]

*Address all correspondence to: carlos.escudero@niva.no

1 Norwegian Institute for Water Research (NIVA), Gaustadalléen, Oslo, Norway

2 Chemical Engineering Department, EPS, University of Girona, Girona, Spain

3 Applied Mathematics Department, EPS, University of Girona, Girona, Spain

References

[1] Järup L. Hazards of heavy metals contamination. British Medical Bulletin, 2003. 68; 167–182. DOI: 10.1093/bmb/ldg032.

[2] Wang J, Chen C. Biosorbents for heavy metals removal and their future. Biotechnology Advances. 2009. 27; 195–226. DOI: 10.1016/j.biotechadv.2008.11.002.

[3] Zhou CH, Xia X, Lin CX, Tong DS, Beltramini J. Catalytic conversion of lignocellulosic biomass to fine chemicals and fuels. Chemical Society Reviews. 2011. 40(11); 5588–5617. DOI: 10.1039/C1CS15124J.

[4] Vanholme R, Demedts B, Morreel K, Ralph J, Boerjan W. Plant Physiology, 2010. 153(3); 895–905. DOI: 10.1104/pp.110.155119.

[5] Walker GM. Bioethanol: Science and technology of fuel alcohol. Ventus Publishing ApS. Frederiksberg, Denmark, 2010. ISBN 978-87-7681-681-0.

[6] Abdolali A, Guo WS, Ngo HH, Chen SS, Nguyen NC, Tung KL. Typical lignocellulosic wastes and by-products for biosorption process in water and wastewater treatment: A critical review. Bioresource Technology. 2014. 160; 57–66. DOI: 10.1016/j.biortech. 2013.12.037.

[7] Pujol D, Liu C, Gominho J, Olivella MA, Fiol N, Villaescusa I, Pereira H. The chemical composition of exhausted coffee waste. Industrial Crops and Products. 2013. 50; 423–429. DOI: 10.1016/j.indcrop.2013.07.056.

[8] Pujol D, Liu C, Fiol N, Olivella MA, Gominho J, Villaescusa I, Pereira H. Chemical characterization of different granulometric fractions of grape stalks waste. Industrial Crops and Products. 2013. 50; 494–400. DOI: 10.1016/j.indcrop.2013.07.051.

[9] Sud D, Mahajan G, Kaur MP. Agricultural waste material as potential adsorbent for sequestering heavy metal ions from aqueous solutions – A review. Bioresource Technology. 2008. 14(99); 6017–6027. DOI: 10.1016/j.biortech.2007.11.064.

[10] Malik DS, Jain CK, Yadav AK. Removal of heavy metals from emerging cellulosic low-cost adsorbents: A review. Applied Water Science. 2016. 1–24. DOI: 10.1007/s13201-016-0401-8.

[11] Miretzky P, Fernandez Cirelli A. Cr(VI) and Cr(III) removal from aqueous solution by raw and modified lignocellulosic materials: A review. Journal of Hazardous Materials. 2010. 180(1–3); 1–19. DOI: 10.1016/j.jhazmat.2010.04.060.

[12] Demirbas A. Heavy metal adsorption onto agro-based waste materials: A review. Journal of Hazardous Materials. 2008. 2–3; 220–229. DOI: 10.1016/j.jhazmat.2008.01.024.

[13] Hubbe MA, Hasan SH, Ducoste JJ. Cellulosic substrates for removal of pollutants from aqueous systems: A review. Metals BioResources. 2011. 1. 2161–2287 ISSN: 1930–2126.

[14] Michalak I, Chojnacka K, Witek-Krowiak A. State of the art for the biosorption process —A review. Applied Biochemistry and Biotechnology. 2013. 170; 1389. DOI: 10.1007/s12010-013-0269-0.

[15] Volesky B. Sorption and biosorption. BV Sorbex, Inc, Montreal, Canada. 2004.

[16] Das N. Recovery of precious metals through biosorption—A review. Hydrometallurgy. 2010. 103(1–4); 180–189. DOI: 10.1016/j.hydromet.2010.03.016.

[17] Das N, Das D. Recovery of rare earth metals through biosorption: An overview. Journal of Rare Earths. 2013. 31(10); 933–943. DOI: 10.1016/S1002-0721(13)60009-5.

[18] Prasad AGD, Abdullah MA. Biosorption of Cr (VI) from synthetic wastewater using the fruit shell of gulmohar (*Delonix regia*): Application to electroplating wastewater. BioResources. 2010. 5(2); 838–853. ISSN: 1930–2126.

[19] Martín-Lara MA, Blázquez G, Trujillo MC, Pérez A, Calero M. New treatment of real electroplating wastewater containing heavy metal ions by adsorption onto olive stone. Journal of Cleaner Production. 2014. 81; 120–129. DOI: 10.1016/j.jclepro.2014.06.036.

[20] Singha B, Das SK. Removal of Pb(II) ions from aqueous solution and industrial effluent using natural biosorbents. Environmental Science and Pollution Research. 2012. 19; 2212. DOI: 10.1007/s11356-011-0725-8.

[21] Liu C, Fiol N, Poch J, Villaescusa I. A new technology for the treatment of chromium electroplating wastewater based on biosorption. Journal of Water Process Engineering. 2016. 11. 143–151 DOI: 10.1016/j.jwpe.2016.05.002.

[22] Chipara M, Hamilton J, Chipara AC, George T, Chipara DM, Ibrahim EE, Lozano K, Sellmyer DJ. Fourier transform infrared spectroscopy and wide-angle X-ray scattering: Investigations on polypropylene–vapor-grown carbon nanofiber composites. Journal of Applied Polymer Science. 2012. 125(1); 353–360. DOI: 10.1002/app.35576.

[23] Dupont L, Guillon E. Removal of hexavalent chromium with a lignocellulosic substrate extracted from wheat bran. Environmental Science and Technology. 2003. 15(18); 4235–4241. DOI: 10.1021/es0342345.

[24] Twardowska I, Kyziol J. Sorption of metals onto natural organic matter as a function of complexation and adsorbent-adsorbate contact mode. Environment International. 2003. 8; 783–791. DOI: 10.1016/S0160-4120(02)00106-X.

[25] Villaescusa I, Fiol N, Cristiani F, Floris C, Lai S, Nurchi VM. Copper(II) and nickel(II) uptake from aqueous solutions by cork wastes: A NMR and potentiometric study. Polyhedron. 2002. 21(14); 1363–1367. DOI: 10.1016/S0277-5387(02) 00957-9.

[26] Escudero C, Fiol N, Villaescusa I, Bollinger JC. Effect of chromium speciation on its sorption mechanism onto grape stalks entrapped into alginate beads. Arabian Journal of Chemistry. 2013. DOI: 10.1016/j.arabjc.2013.03.011.

[27] Misra SP. Adsorption-desorption of heavy metal ions. Current Science. 2014. 107(4); 601–612.

[28] Chaplin MF, Bucke C. The preparation and kinetics of immobilised enzymes in enzyme technology. Cambridge University Press. Cambridge, United Kingdom. 1990. Chapter 3. pp. 81–89.

[29] Peinado RA, Moreno JJ, Maestre O, Mauricio JC. Use of a novel immobilization yeast system for winemaking. Biotechnology Letters. 2005. 27; 1421–1424. DOI: 10.1007/ s10529-005-0939-2.

[30] Martins SCS, Martins CM, Fiúza LMCG, Santaella ST. Immobilization of microbial cells: A promising tool for treatment of toxic pollutants in industrial wastewater. African Journal of Biotechnology. 2013. 12(28); 4412–4418. DOI: 10.5897/AJB12.2677.

[31] Crini G. Recent developments in polysaccharide-based materials used as adsorbents in wastewater treatment. Progress in Polymer Science. 2005. 30; 38–70. DOI: 10.1016/ j.progpolymsci.2004.11.002.

[32] Eroglu E, Smith SM, Raston CL. Application of various immobilization techniques for algal bioprocesses. Biofuel and Biorefinery Technologies. 2015. 2; 19–44. DOI: 10.1007/978-3-319-16640-7_2.

[33] Feng L, Cao Y, Su D, You S, Han F. Influence of sodium alginate pretreated by ultrasound on papain properties: Activity, structure, conformation and molecular weight and distribution. Ultrasonic Sonochemistry. 2016. 32; 224–230. DOI: 10.1016/j.ultsonch. 2016.03.015.

[34] Vauchel P, Kaas R, Arhaliass A, Baron R, Legrand J. A new process for extracting alginates from Laminaria digitata: Reactive extrusion. Food and Bioprocess Technology. 2008. 1(3); 297–300. DOI: 10.1007/s11947-008-0082-x.

[35] Hernández-Carmona G, McHugh DJ, Arvizu-Higuera DL, Rodríguez-Montesinos E. Pilot plant scale extraction of alginate from *Macrocystis pyrifera*. 1. Effect of pre-extraction treatments on yield and quality of alginate. Journal of Applied Phycology. 1999. 10; 507–513. DOI: 10.1023/A:1008004311876.

[36] Bajpai SK, Sharma S. Investigation of swelling degradation behaviour of alginate beads crosslinked with Ca^{2+} and Ba^{2+} ions. Reactive & Functional Polymers. 2004. 59; 129–140. DOI: 10.1016/j.reactfunctpolym.2004.01.002.

[37] Hassan R, Tirkistani F, Zaafarany I, Fawzy A, Khairy M, Iqbal S. Polymeric biomaterial hydrogels. I. Behavior of some ionotropic cross-linked metal-alginate hydrogels especially copper-alginate membranes in some organic solvents and buffer solutions. Advances in Bioscience and Biotechnology. 2012. 3; 845–854. DOI: 10.4236/abb.2012.37105.

[38] Grant GT, Morris ER, Rees DA, Smith PJC, Thom D. Biological interactions between polysaccharides and divalent cations: The egg-box model. FEBS Letters. 1973. 32(1); 195–198. DOI: 10.1016/0014-5793(73)80770-7.

[39] Pandey A, Bera D, Shukla A, Ray L. Studies on Cr(VI), Pb(II) and Cu(II) adsorption-desorption using calcium alginate as biopolymer. Chemical Speciation & Bioavailability. 2007. 19(1); 17–24. DOI: 10.3184/095422907X198031.

[40] Jodra Y, Mijangos F. Ion exchange selectivities of calcium alginate gels for heavy metals. Waters Science & Technology. 2001. 43(2); 237–244.

[41] Bishnoi NR, Kumar R, Bishnoi K. Biosorption of Cr(VI) with *Trichoderma viride* immobilised fungal biomass and cell free Ca-alginate beads. Indian Journal of Experimental Biology. 2007. 45(7); 657–664.

[42] Wuyep PA, Chuma AG, Awodi S, Nok AJ. Biosorption of Cr, Mn, Fe, Ni, Cu and Pb metals from petroleum refinery effluent by calcium alginate immobilized mycelia of *Polyporus squamosus*. Scientific Research & Essay. 2007. 2(7); 217–221.

[43] Mata YN, Blázquez ML, Ballester A, González F, Muños JA. Biosorption of cadmium, lead and copper with calcium algiante xerogels and immobilized *Fucus vesiculosus*. Journal of Hazardous Materials. 2009. 163(2–3); 555–562. DOI: 10.1016/j.jhazmat.2008.07.015.

[44] Olusola B, Aransiola MN. Biosorption of zinc (Zn^{2+}) and Iron (Fe^{2+}) from wastewater using *Botrydium granulatum* and *Euglena texta*. Bioscience and Bioengineering. 2015. 1(2); 17–21.

[45] Samuel J, Pulimi M, Paul ML, Maurya A, Chandrasekaran N, Mukherjee A. Batch and continuous flow studies of adsorptive removal of Cr(VI) by adapted bacterial consortia immobilized in alginate beads. Bioresource Technology. 2013. 128; 423–430. DOI: 10.1016/j.biortech.2012.10.116.

[46] Gad A, Attia M, Ahmed HA. Heavy metals bio-remediation by immobilizaed *Saccharomyces cerevisiae* and *Opuntia ficus indica*. Journal of American Science. 2010. 6(8); 79–87.

[47] Ansari TM, Hanif MA, Mahmood A, Ijaz U, Khan MA, Nadeem R, Ali M. Immobilization of rose waste biomass for uptake of Pb(II) from aqueous

solutions. Biotechnology Research International. 2011. 2011; 1–9. DOI: 10.4061/2011/685023.

[48] Fiol N, Poch J, Villaescusa I. Chromium(VI) uptake by grape stalks wastes encapsulated in calcium alginate beads: Equilibrium and kinetics studies. Chemical Speciation & Bioavailability. 2004. 16(1/2). 25–33 DOI: 10.3184/095422904782775153.

[49] Fiol N, Poch J, Villaescusa I. Grape stalks wastes encapsulated in calcium alginate beads for Cr(VI) removal from aqueous solutions. Separation Science and Technology. 2005. 40; 1013–1028. DOI: 10.1081/SS-200048041.

[50] Sillerová H, Komárek M, Liu C, Poch J, Villaescusa I. Biosorbent encapsulation in calcium alginate: Effects of process variables on Cr(VI) removal from solutions. International Journal of Biological Macromolecules. 2015. 80; 260–270. DOI: 10.1016/j.ijbiomac.2015.06.032.

[51] Fiol N, Escudero C, Poch J, Villaescusa I. Preliminary studies on Cr(VI) removal from aqueous solution using grape stalk wastes encapsulated in calcium alginate beas in a packed bed up-flow column. Reactive & Functional Polymers. 2006. 66; 795–807. DOI: 10.1016/j.reactfunctpolym.2005.11.006.

[52] Escudero C, Fiol N, Villaescusa I. Chromium sorption on grape stalks encapsulated in calcium alginate beads. Environmental Chemistry Letters. 2006. 4; 239–242. DOI: 10.1007/s10311-006-0055-0.

[53] Fiol N, Escudero C, Villaescusa I. Chromium sorption and Cr(VI) reduction to Cr(III) by grape stalks and yohimbe bark. Bioresource Technology. 2008. 99(11); 5030–5036. DOI: 10.1016/j.biortech.2007.09.007.

[54] Anirudhan TS, Radhakrishnan PG. Kinetic and equilibrium modelling of cadmium(II) ions sorption onto polymerized tamarind fruit shell. Desalination. 2009. 249(3); 1298–1307. DOI: 10.1016/j.desal.2009.06.028.

[55] Rangabhashiyam S, Anu S, Giri Nandagopal MS, Selvaraju N. Relevance of isotherm models in biosorption of pollutants by agricultural by-products. Journal of Environmental Chemical Engineering. 2014. 2(1); 398–414. DOI: 10.1016/j.jece.2014.01.014.

[56] Abdolali A, Ngo HH, Guo WS, Lee DJ, Tung KL, Wang XC. Development and evaluation of a new multi-metal binding biosorbent. Bioresource Technology. 2014. 160; 98–106. DOI: 10.1016/j.biortech.2013.12.038.

[57] Allen SJ, Mckay G, Porter JF. Adsorption isotherm models for basic dye adsorption by peat in single and binary component systems. Journal of Colloid and Interface Science. 2004. 280(2); 322–333. DOI: 10.1016/j.jcis.2004.08.078.

[58] Badertscher M, Pretsch E. Bad results from good data. Trends in Analytical Chemistry. 2006. 25(11); 1131–1138. DOI: 10.1016/j.trac.2006.09.003.

[59] El-Khaiary MI. Least-squares regression of adsorption equilibrium data: Comparing the options. Journal of Hazardous Materials. 2008. 158(1); 73–87. DOI: 10.1016/j.jhazmat.2008.01.052.

[60] Ayoob S, Gupta AK. Insights into isotherm making in the sorptive removal of fluoride from drinking water. Journal of Hazardous Materials. 2008. 152; 976–985. DOI: 10.1016/j.jhazmat.2007.07.072.

[61] Foo KY, Hameed BH. Insights into the modeling of adsorption isotherm systems. Chemical Engineering Journal. 2010. 156; 2–10. DOI: 10.1016/j.cej.2009.09.013.

[62] Poch J, Villaescusa I. Orthogonal distance regression: A good alternative to least squares for modeling sorption data. Journal of Chemical & Engineering Data. 2012. 57(2); 490–499. DOI: 10.1021/je201070u.

[63] Marković DD, Lekić BM, Rajaković-Ognjanović VN, Onjia AE, Rajaković LV. A new approach in regression analysis for modeling adsorption isotherms. The Scientific World Journal. 2014. 1–17 DOI: 10.1155/2014/930879.

[64] Yan F, Chu Y, Zhang K, Zhang F, Bhandari N, Ruan G, Dai Z, Liu Y, Zhang Z, Kan AT, Tomson MB. Determination of adsorption isotherm parameters with correlated errors by measurement error models. Chemical Engineering Journal. 2015. 281(1); 921–930. DOI: 10.1016/j.cej.2015.07.021.

[65] Millar GJ, Miller GL, Couperthwaite SJ, Papworth S. Factors influencing kinetic and equilibrium behaviour of sodium ion exchange with strong acid cation resin. Separation and Purification Technology. 2016. 163; 79–91. DOI: 10.1016/j.seppur.2016.02.045.

[66] Beltrán JL, Pignatello JJ, Teixidó M. ISOT_Calc: A versatile tool for parameter estimation in sorption isotherms. Computers & Geosciences. 2016. 94; 11–17. DOI: 10.1016/j.cageo.2016.04.008.

[67] Gupta A, Balomajumder C. Simultaneous adsorption of Cr(VI) and phenol onto tea waste biomass from binary mixture: Multicomponent adsorption, thermodynamic and kinetic study. Journal of Environmental Chemical Engineering. 2015. 3; 785–796. DOI: 10.1016/j.jece.2015.03.003.

[68] Chan OS, Cheung WH, McKay G. Single and multicomponent acid dye adsorption equilibrium studies on tyre demineralised activated carbon. Chemical Engineering Journal. 2012. 191; 162–170. DOI: 10.1016/j.cej.2012.02.089.

[69] Villaescusa I, Fiol N, Martínez M, Miralles N, Poch J, Serarols J. Removal of copper and nickel ions from aqueous solutions by grape stalks wastes. Water Research. 2004. 38(4); 992–1002. DOI: 10.1016/j.watres.2003.10.040.

[70] Escudero C, Poch J, Villaescusa I. Modelling of breakthrough curves of single and binary mixtures of Cu(II), Cd(II), Ni(II) and Pb(II) sorption onto grape stalks waste. Chemical Engineering Journal. 2013. 217(1); 129–138. DOI: 10.1016/j.cej.2012.11.096.

[71] Suryavanshi US, Shukla SR. Adsorption of Ga(III) on oxidized coir. Industrial and Engineering Chemical Research. 2009. 48; 870–876. DOI: 10.1021/ie801259c.

[72] Aksu Z, Alper Isoglu I. Use of dried sugar beet pulp for binary biosorption of Gemazol Turquoise Blue-G reactive dye and copper(II) ions: Equilibrium modeling. Chemical Engineering Journal. 2007. 127; 177–188. DOI: 10.1016/j.cej.2006.09.014.

[73] Pujol D, Bartrolí M, Fiol N, de la Torre F, Villaescusa I, Poch J. Modelling synergistic sorption of Cr(VI), Cu(II) and Ni(II) onto exhausted coffee wastes from binary mixtures Cr(VI)–Cu(II) and Cr(VI)–Ni(II). Chemical Engineering Journal. 2013. 230; 396–405. DOI: 10.1016/j.cej.2013.06.033.

[74] Liu Y, Liu YJ. Biosorption isotherms, kinetics and thermodynamics. Separation and Purification Technology. 2008. 61(3); 229–242. ISSN 1383-5866. DOI: 10.1016/j.seppur.2008.07.017.

[75] Lata S, Singh PK, Samadder SR. Regeneration of adsorbents and recovery of heavy metals. International Journal of Environmental Science and Technology. 2015. 12; 1461–1478. DOI: 10.1007/s13762-014-0714-9.

[76] Kyzas GZ. Commercial coffee wastes as material for adsorption of heavy metals from aqueous solutions. Materials. 2012. 5; 1826–1840. DOI: 10.3390/ma5101826.

[77] Ajmal M, Rao RAK, Anwar S, Ahmad J, Ahmad R. Adsorption studies on rice husk: Removal and recovery of Cd(II) from wastewater. Bioresource Technology. 2003. 86(2); 147–149. DOI: 10.1016/S0960-8524(02)00159-1.

[78] Saeed A, Akhter MW, Iqbal M. Removal and recovery of heavy metals from aqueous solution using papaya wood as a new biosorbent. Separation & Purification Technology. 2005. 45; 25–31. DOI: 10.1016/j.seppur.2005.02.004.

[79] Anirudhan TS, Unnithan MR. Arsenic (V) removal from aqueous solutions using an anion exchanger derived from coconut coir pith and its recovery. Chemosphere. 2007. 66(1); 60–66. DOI: 10.1016/j.chemosphere.2006.05.031.

[80] Villaescusa I, Martínez M, Miralles N. Heavy metal uptake from aqueous solution by cork and yohimbe bark. Journal of Chemical Technology & Biotechnology. 2000. 75; 812–816. DOI: 10.1002/1097-4660(200009)75:9.

[81] Gupta VK, Nayak A. Cadmium removal and recovery from aqueous solutions by novel adsorbents prepared from orange peel and Fe_2O_3 nanoparticles. Chemical Engineering. 2012. 180; 81–90. DOI: 10.1016/j.cej.2011.11.006.

[82] Kaikake K, Hoaki K, Sunada H, Dhakal RP, Baba Y. Removal characteristics of metal ions using degreased coffee beans: Adsorption equilibrium of cadmium(II). Bioresource Technology. 2007. 98(15); 2787–2791. DOI: 10.1016/j.biortech.2006.02.040.

[83] Qaiser S, Saleemi AR, Umar M. Biosorption of lead from aqueous solution by *Ficus*

religiosa leaves: Batch and column study. Journal of Hazardous Materials. 2009. 166(2–3); 998–1005. DOI: 10.1016/j.jhazmat.2008.12.003.

[84] Saleem N, Bhatti HM. Adsorptive removal and recovery of U(VI) by citrus waste biomass. BioResources. 2011. 6(3); 2522–2538. ISSN: 1930–2126.

[85] Martínez M, Miralles N, Hidalgo S, Fiol N, Villaescusa I, Poch J. Removal of lead(II) and cadmium(II) from aqueous solutions using grape stalk waste. Journal of Hazardous Materials. 2006. B133; 203–211. DOI: 10.1016/j.jhazmat.2005.10.030.

[86] Fiol N, Villaescusa I, Martínez M, Miralles N, Poch J, Serarols J. Sorption of Pb(II), Ni(II), Cu(II) and Cd(II) from aqueous solution by olive stone waste. Separation & Purification Technology. 2006. 50(1); 132–140. DOI: 10.1016/j.seppur.2005.11.016.

[87] Gardea-Torresdey J, Hejazi M, Tiemann K, Parsons JG, Duarte-Gardea M, Henning J. Use of hop (*Humulus lupulus*) agricultural by-products for the reduction of aqueous lead(II) environmental health hazards. Journal of Hazardous Materials. 2002. B91; 95–112. DOI: 10.1016/S0304-3894(01)00363-6.

[88] Ranjan D, Talat M, Hasan SH. Rice polish: An alternative to conventional adsorbents for treating arsenic. Industrial and Engineering Chemical Research. 2009. 48; 10180–10185. DOI: 10.1021/ie900877p.

[89] Amin MN, Kaneco S, Kitagawa T, Begum A, Katsumata H, Suzuki T, Ohta K. Removal of arsenic in aqueous solutions by adsorption onto waste rice husk. Industrial and Engineering Chemical Research. 2006. 45(24); 8105–8110. DOI: 10.1021/ie060344j.

8

Biomass Compositional Analysis for Conversion to Renewable Fuels and Chemicals

C. Luke Williams, Rachel M. Emerson and
Jaya Shankar Tumuluru

Abstract

As the world continues to deplete its nonrenewable resources, there has begun a shift toward using renewable materials for the production of fuels and chemicals. Terrestrial biomass, as well as municipal solid wastes, provides renewable feedstocks for fuel and chemical production. However, one of the major challenges to using biomass as a feedstock for fuel and chemical production is the great amount of innate variability between different biomass types and within individual biomass species. This inconsistency arises from varied growth and harvesting conditions and presents challenges for conversion processes, which frequently require physically and chemically uniform materials. This chapter will examine intrinsic biomass compositional characteristics including cellulose, hemicellulose, lignin, extractives/volatiles, and ash for a wide array of biomass types. Additionally, extrinsic properties, such as moisture content and particle grind size, will be examined for their effect on biomass conversion to fuels using four major conversion processes: direct combustion, pyrolysis, hydrothermal liquefaction, and fermentation. A brief discussion on recent research for the production of building block chemicals from biomass will also be presented.

Keywords: biomass, composition, variability, renewable, fuels, chemicals

1. Introduction

CO_2 and other greenhouse gases are contributing to increased concerns about global warming. Biomass growth and utilization provides one solution to reduce greenhouse gas emissions and balance the ecosystem. For instance, carbon sequestered by the world's forests accounts for about 77% of terrestrial ecosystem [1]. In concert with fears about CO_2 production and global warming, concerns over dwindling and limited petroleum resources have given momentum to the production of renewable fuels, chemicals, and other materials from biomass [2]. In addition to providing carbon sequestration benefits for the world, biomass is also positioned to have a large impact on the domestic production of fuels and chemicals in the United States. According to the "Billion-Ton" Study [3] and its update [4] by the US Department of Energy (USDOE) and US Department of Agriculture (USDA), biomass has the potential to sustainably supply one third of the nation's petroleum consumption. This large available supply makes biomass one of the most abundant, inexpensive, and currently underutilized products from the agricultural and biorefinery industries [5]. While the first generation of biofuels is being produced using sugarcane in places such as Brazil, the second generation of biofuels will likely be derived from the over 1 billion tons of lignocellulosic biomass that is produced annually in the United States or from the estimated 10 to 50 billion tons of waste lignocellulose that is produced worldwide [3, 6]. The beginning of the second-generation enzymatic conversion of lignocellulosic material to fermentable sugars for fuel ethanol is becoming more economically viable in the United States [7] with companies such as DuPont [8] and POET [9] making gains in the production of second-generation cellulosic ethanol using agricultural wastes such as corn stover. In fact, to meet the US congressional mandate to manufacture 36 billion gallons of biofuels per year by 2022, several new types of biomass including energy crops, forest residues, and municipal wastes will have to be processed through modern conversion systems.

Besides just the production of fuel petroleum also produces a significant portion (basically all) of our plastics and other materials. The use of petroleum to produce chemical building blocks and materials has resulted in a global interest in using renewable bio-based polymers and composites derived from biomass to reduce our environmental impact [10, 11]. Additionally, from a broad energy standpoint, biomass has a significant advantage over renewable energy sources such as hydropower, wind, geothermal, and solar in that biomass is the only renewable energy source that can be turned directly into fuels and chemicals, as opposed to just generate electricity.

The output of these conversion systems, whether it is fuels or chemicals, is highly dependent on the quality of biomass input to the system. The quality of the biomass is dependent on inherent species variability, production conditions, and differing harvest, collection, and storage practices. Some of the most important parameters related to the biomass composition, in regard to the impact on biofuels production, are moisture content, ash content and speciation, carbohydrate distribution, and higher heating value. For example, moisture content impacts the storage, supply, and transportation of feedstock to biorefineries, and ash content often reduces oil yields in thermochemical conversion processes and, to a lesser extent, reduces the effectiveness of dilute alkali pretreatment in biochemical processes. Commercialization of biorefineries in the United States has resulted in an understanding of the importance of biomass quality (moisture, ash, and sugar content) and physical properties (particle size and shape), especially in regard to feed and handling issues. The overall objective of the present research is to understand the impact of the chemical and physical composition of various biomass feedstocks on the production of fuels and chemicals through a variety of conversion pathways. This work will focus on everything from the feeding and handling of biomass all the way through the effect of feedstock variability on the final oil product. Specific objectives seek to understand chemical composition of woody and herbaceous crops, agricultural residues, and municipal solid wastes and suggest their suitability for different biofuel production conversion pathways. This research will also briefly touch on platform chemical production from biomass and consider methods for mitigating the problems associated with feedstock variability while converting biomass to fuels and chemicals.

2. Biomass compositional analysis

To effectively produce fuels and chemicals from biomass, it is critical to understand the composition of the feedstock material. The chemical composition of biomass, whether it is lignocellulosic or herbaceous, can be characterized by five primary components: cellulose, hemicellulose, lignin, extractives/volatiles, and ash. The most abundant biopolymer on earth, cellulose, is a polysaccharide of glucose monomers held together by $\beta(1\rightarrow4)$ linkages. These $\beta(1\rightarrow4)$ linkages are what make cellulose resistant to hydrolysis. The second major component of biomass, hemicellulose, is an amorphous heteropolymer comprised of several different carbohydrates including xylose, mannose, and glucose, among others. Due to its amorphous structure, hemicellulose is significantly more susceptible to hydrolysis than crystalline cellulose. Cellulose and hemicellulose, combined with the third major component of biomass, lignin, make up over 90% of lignocellulosic biomass and 80% of herbaceous biomass. Lignin is an intricate array of aromatic alcohols and is intertwined with the cellulose and hemicellulose fraction of the biomass structure. This interwoven nature of the lignin helps provide rigidity to lignocellulosic materials, such as trees.

The other minor components of biomass are extractives/volatiles and ash. While these components make up a smaller portion of the biomass composition, they can still have a major influence on what ends up being the optimal conversion process. The components comprising the extractives/volatiles include both water and ethanol solubles. Water-soluble compounds include nonstructural sugars and proteins, and ethanol-soluble components are typically represented by chlorophyll and waxes. Ash, which comprises the inorganic content in biomass, can be intrinsic to the biomass or added anthropogenically. Intrinsic ash includes material-like calcium and potassium ions, while anthropogenic ash is mostly silica (dirt) collected during harvest.

There is obviously significant compositional variation between different biomass types, but there is also a lot of variation within a single feedstock. This variation, while substantial across terrestrial feedstocks, varies even more widely when municipal solid wastes are included as renewable energy feedstocks. **Table 1** illustrates the large difference in composition across three broad categories of renewable feedstocks including lignocellulosic, herbaceous biomass, as well as municipal solid wastes. Algal biomass was not included in this study due to a lack of available data and the difficulty in obtaining consistent analysis methods across institutions [12].

As seen in **Table 1**, there exists a significant amount of variability in overall composition (i.e., cellulose, hemicellulose, and lignin) between different types of feedstocks. These

Feedstock composition	Woody	Herbaceous	Wastes
Proximate			
Volatiles (%)	84.0 (2.1)[193]	79.1 (5.8)[284]	76.7 (5.5)[21]
Ash (%)	1.3 (0.9)[193]	5.5 (3.2)[284]	6.6 (6.7)[21]
Fixed carbon (%)	14.7 (1.6)[193]	15.4 (4.0)[284]	14.8 (5.0)[21]
Ultimate			
Hydrogen (%)	6.0 (0.1)[192]	5.8 (0.3)[276]	5.9 (0.4)[21]
Carbon (%)	50.7 (4.71)[192]	47.4 (1.9)[276]	46.0 (4.0)[21]
Nitrogen (%)	0.32 (0.01)[192]	0.75 (0.49)[276]	1.3 (1.6)[21]
Oxygen (%)	41.9 (1.4)[134]	41.0 (2.4)[107]	38.3 (4.2)[7]
Sulfur (%)	0.03 (0.01)[135]	0.10 (0.32)[107]	0.15 (0.16)[7]
Structural			
Cellulose (%)	51.2 (8.7)[241]	32.1 (4.5)[2425]	28.4 (13.2)[27]
Hemicellulose (%)	21.0 (8.7)[241]	18.6 (3.4)[2425]	16.4 (5.5)[27]
Lignin (%)	26.1 (5.3)[241]	16.3 (3.3)[2425]	12.5 (2.7)[15]

Table 1. Feedstock compositions for woody, herbaceous, and waste materials; average (standard deviation)$^{\text{number of samples}}$.

differences are large enough that conversion reactors have to be operated under different conditions based on the type of material supplied to the conversion facility (such as lower pyrolysis temperatures for herbaceous feedstocks). Also, herbaceous feedstocks, in addition to having higher ash content, exhibit more variability in their composition of volatiles (and ash) than woody biomass.

While a high degree of variability is expected across broad categories such as lignocellulosic material and municipal solid waste (MSW), there also exists significant variability within individual feedstock categories. **Tables 2–4** highlight the differences within an individual feedstock category for lignocellulosic material, herbaceous material, and municipal solid waste, respectively.

While it is obvious that compositional differences can be stark between different biomass types, there is also a substantial compositional variability between different anatomical fractions of the same type of biomass. **Table 5** compiles information on the chemical composition of different plant fractions for woody biomass, corn, and wheat.

It can be seen that lignocellulosic biomass contains a large fraction of cellulose in the heartwood (shown by *whole tree*), while the *bark* contains a high percent lignin. In woody biomass, the extractives are fairly evenly distributed. Conversely, corn stover contains a majority of the extractives in the leaves and internodes (the links between different stalk segments). Taking advantage of processing a specific anatomical fraction could allow for greater control over product output by tailoring the composition of the reactor feed. Additionally, utilizing anatomical fractionation separation could increase the economic viability of a process by

Feedstock composition	Shrub willow	Hybrid poplar	Pine	Other softwoods	Other hardwoods
Proximate					
Volatiles (%)	84.7 (0.8)^{76}	84.0 (1.3)^{41}	83.5 (2.5)^{46}	81.3 (2.9)^{18}	85.1 (3.0)^{11}
Ash (%)	1.5 (0.4)^{76}	1.3 (0.5)^{41}	0.7 (0.6)^{46}	2.1 (2.0)^{18}	1.8 (1.2)^{11}
Fixed carbon (%)	13.8 (0.7)^{76}	14.6 (0.1)^{41}	15.7 (1.9)^{46}	16.5 (1.6)^{18}	13.1 (1.8)^{11}
Ultimate					
Hydrogen (%)	6.0 (0.2)^{76}	6.0 (0.1)^{41}	6.1 (0.1)^{45}	6.1 (0.1)^{18}	6.1 (0.1)^{11}
Carbon (%)	50.3 (0.9)^{76}	50.0 (1.1)^{41}	51.5 (1.0)^{45}	51.8 (0.9)^{18}	50.2 (0.5)^{11}
Nitrogen (%)	0.36 (0.10)^{76}	0.35 (0.17)^{41}	0.17 (0.12)^{45}	0.27 (0.21)^{18}	0.55 (0.49)^{11}
Oxygen (%)	42.6 (0.4)^{44}	42.8 (1.2)^{28}	41.4 (1.0)^{38}	39.7 (1.8)^{14}	41.1 (1.6)^{10}
Sulfur (%)	0.04(0.01)^{44}	0.03 (0.01)^{28}	0.02 (0.01)^{39}	0.03 (0.01)^{14}	0.05 (0.05)^{10}
Structural					
Cellulose (%)	–	43.8 (1.2)^{43}	47.4 (2.2)^{55}	42.1 (7.1)^{26}	50.8 (6.9)^{24}
Hemicellulose (%)	–	14.7 (0.1)^{43}	21.9 (4.9)^{55}	25.1 (5.2)^{26}	29.7 (4.3)^{24}
Lignin (%)	–	25.7 (0.3)^{43}	28.6 (0.7)^{55}	29.1 (1.7)^{26}	19.5 (4.1)^{24}

Table 2. Feedstock compositions for specific woody feedstocks; average (standard deviation)$^{\text{number of samples}}$.

Feedstock composition	Corn stover	Switchgrass	*Sorghum*	Energy cane (bagasse)	Mixed grasses	*Miscanthus*
Proximate						
Volatiles (%)	78.1 (5.0)^{50}	82.4 (4.1)^{43}	77.0 (3.7)^{44}	82.2 (1.9)^{48}	78.6 (2.8)^{47}	82.5 (3.5)^{35}
Ash (%)	6.3 (3.5)^{50}	4.0 (2.0)^{43}	7.2 (2.6)^{44}	3.4 (1.6)^{48}	6.6 (1.7)^{47}	2.6 (1.3)^{35}
Fixed carbon (%)	15.6 (4.4)^{50}	13.6 (3.0)^{43}	15.7 (2.3)^{44}	14.4 (1.0)^{48}	14.8 (2.4)^{47}	14.8 (2.9)^{35}
Ultimate						
Hydrogen (%)	5.7 (0.3)^{40}	5.9 (0.2)^{43}	5.7 (0.2)^{44}	6.1 (0.1)^{48}	5.8 (0.3)^{47}	5.8 (0.1)^{35}
Carbon (%)	47.1 (2.3)^{40}	47.1 (1.1)^{43}	46.4 (1.3)^{44}	48.8 (0.9)^{48}	47.6 (1.1)^{47}	48.9 (1.5)^{35}
Nitrogen (%)	0.63 (0.32)^{40}	0.60 (0.26)^{43}	1.04 (0.38)^{44}	0.43 (0.20)^{48}	1.38 (0.54)^{47}	0.35 (0.17)^{35}
Oxygen (%)	40.3 (2.2)^{39}	42.4 (2.3)^{42}	40.3 (0.6)^{3}	-	39.5 (0.7)^{2}	42.3 (1.1)^{4}
Sulfur (%)	0.14 (0.53)^{39}	0.06 (0.03)^{42}	0.11 (0.01)^{3}	-	0.12 (0.02)^{2}	0.04 (0.02)^{4}
Structural						20.1 (1.4)^{274}
Cellulose (%)	34.3 (2.5)^{251}	34.2 (2.7)^{348}	28.6 (2.6)^{488}	32.1 (3.2)^{479}	28.9 (2.9)^{465}	38.9 (3.2)^{274}
Hemicellulose (%)	20.7 (2.0)^{251}	21.9 (2.6)^{348}	15.4 (1.6)^{488}	19.5 (1.9)^{479}	16.7 (3.9)^{465}	20.1 (1.4)^{274}
Lignin (%)	15.2 (1.6)^{251}	19.2 (1.4)^{348}	12.2 (1.9)^{488}	16.3 (1.8)^{479}	15.7 (1.7)^{465}	21.1 (1.6)^{274}

Table 3. Feedstock compositions for specific herbaceous feedstocks; average (standard deviation)$^{\text{number of samples}}$.

Feedstock composition	MSW	C&D waste	Woody residues
Proximate			
Volatiles (%)	76.5 (1.1)^{11}	76.5 (3.7)^{9}	81.1 (2.4)^{2}
Ash (%)	11.8 (5.2)^{11}	0.8 (0.4)^{9}	1.2 (0.3)^{2}
Fixed carbon (%)	11.2 (5.2)^{11}	18.9 (2.1)^{9}	17.8 (2.0)^{2}
Ultimate			
Hydrogen (%)	5.6 (0.4)^{11}	6.2 (0.2)^{9}	6.0 (0.0)^{2}
Carbon (%)	43.3 (3.3)^{11}	48.3 (1.2)^{9}	52.5 (0.2)^{2}
Nitrogen (%)	1.52 (1.72)^{11}	1.09 (1.47)^{9}	0.22 (0.06)^{2}
Oxygen (%)	36.3 (4.8)^{4}	42.4 (0.1)^{2}	40.1 (0.6)^{2}
Sulfur (%)	0.25 (0.14)^{4}	0.02 (0.01)^{2}	0.01 (0.01)^{2}
Structural			
Cellulose (%)	28.4 (13.2)^{15}	–	–
Hemicellulose (%)	16.4 (5.5)^{15}	–	–
Lignin (%)	12.5 (2.7)^{15}	–	–

Table 4. Feedstock compositions for specific waste feedstocks; average (standard deviation)$^{\text{number of samples}}$.

Structural component	Cellulose	Hemicellulose	Lignin	Extractives
Woody biomass (wt%–daf)[a]				
Whole tree	51.2	23.4	25.4	3.0
Bark	22.0	47.0	31.0	3.3
Twigs	15.4	62.3	22.3	1.6
Leaves	26.5	47.2	26.3	3.7
Corn (wt%–db)[b]				
Corn cobs	35.92	30.7	16.44	5.89
Corn leaves	34.33	22.77	13.99	10.54
Corn husk	37.73	31.18	10.52	5.80
Corn internodes	40.21	20.03	17.24	12.29
Wheat (wt%–db)[b]				
Internode 1	34.34	21.30	16.36	16.24
Internode 2	39.04	21.07	18.58	10.98
Internodes 3/4/5	38.92	21.56	19.50	9.67

Source: [a]Vassilev et al. [13]; [b]INL Library [14]; daf—dry ash-free, db—dry basis.

Table 5. Compositional variation with anatomical fraction for woody biomass, corn, and wheat.

extricating high value components. Profitable use of a coproduct is exemplified by the use of distiller's dried grains with solubles (DDGS) in ethanol production for high-protein animal feed [15, 16].

The majority of the data included in **Tables 1–5** can be found in Idaho National Laboratory's (INL) Bioenergy Feedstock Library. The woody materials in **Table 1** include a wide variety of softwoods, hardwoods, and other wood varieties making up around 23 different woody species. The herbaceous materials include those listed in **Table 3** along with sugarcane, sugarcane bagasse, and wheat. The waste materials from **Table 1** are all represented in **Table 4**. In **Table 4** municipal solid waste (MSW) includes fractions of paper, cardboard, and grass clippings. No food-based waste is currently accounted for by this data. The construction and demolition (C&D) waste included oriented strand board, particle board, and a variety of lumber conditions. Woody residues included forest thinning and logging residues.

All of the values reported for proximate and ultimate (reported on a dry basis) for **Tables 1–5** were collected at INL [17] and stored in the Bioenergy Feedstock Library [14]. The reported cellulose, hemicellulose, and lignin values were a combination of glucose (representing the cellulose fraction) and xylose, galactose, and arabinose (representing the hemicellulose) values measured using NREL's LAP determination of structural carbohydrates and lignin in biomass [18], glucose, and xylose values predicted using an NIR-based predictive models developed at NREL [19], and cellulose and hemicellulose values reported in literature are all reported on a dry basis. It should be noted that the value for volatiles in the previous tables is determined by heating samples to 950°C in an inert atmosphere. This value for volatile will therefore include all thermal decomposition products, in addition to molecules that could be removed without thermal decomposition.

Understanding the degree of biomass compositional variability is crucial to developing a robust conversion process. However, in addition to understanding compositional variability, it is useful to know where this variability originates. Kenney et al. have produced a thorough review discussing several sources of biomass variability [20]. Briefly, some of the major sources of biomass compositional variation derive from local agronomic conditions [21], drought [22], harvest season and year [23], and harvest method [24]. A further analysis of the sources of biomass variability and its impact on conversion processes has been compiled by Williams et al. [25].

3. Biomass conversion to renewable fuels and chemicals

As can be seen in **Tables 1–5**, biomass has a broad range of compositional variability, even within an individual feedstock. This variation has a substantial impact on biomass conversion to fuels and value-added chemicals that varies depending on the chosen conversion process. The following section investigates how feedstock quality impacts four common conversion processes: biochemical fermentation, direct combustion, pyrolysis, and hydrothermal liquefaction (HTL). General impacts of feedstock physical and chemical properties will be discussed before a more in-depth look at each conversion process.

The physical properties of biomass have a myriad of effects on its conversion to fuels and chemicals. Arguably, the two most important physical properties of biomass, regardless of conversion process, are particle size and moisture content. Practically all conversion methods require some degree of size reduction. Biochemical conversion processes can accept a greater range of particle sizes, and the final size needed tends to be dependent on the processing system utilized [26, 27]. On the thermochemical side, hydrothermal liquefaction is much more insensitive to particle size due to high heating rates in the liquid media [28], but a significant amount of size reduction is needed to pump biomass sludges in a continuous system [29]. Pyrolysis uses particles smaller than 0.5 mm because small particles decrease char yields and have higher heating rates [30]. Optimal combustion particle size is often larger and varies for different biomass types at approximately 6 mm for straw, 4 mm for *Miscanthus*, and 2–4 mm for wood [31]. While particle size is obviously important, others have argued that moisture content is likely the single most problematic property affecting feedstock supply and biorefining operations [20]. Moisture increases heating rates during steam pretreatment for biological conversion [32], reduces bio-oil quality and thermochemical conversion [33], and causes low thermal efficiency in combustion processes [34]. Aside from particle size and moisture content, other physical properties of interest include bulk density, elastic properties, and microstructure. Bulk density has a strong effect on transportation and handling costs (lower densities greatly increase transportation costs), and the elastic properties/microstructure can increase compressibility and interparticle interactions at constricted flow points such as hopper openings.

Biomass chemical properties also have a large influence on best conversion process and the quality of the final product. The three primary chemical components of interest in biomass conversion are ash content, volatiles, and lignin. High ash content generally has a negative effect on biomass conversion across the board by reducing the effectiveness of dilute acid pretreatment for biological processes [35] and increasing char yields and fouling in thermochemical processes such as HTL [36], pyrolysis [37], and combustion [38]. However, there exist several strategies for ash removal including leaching and air classification [39]. Volatiles are generally represented by light organic acids (such as acetic acid) and furans. The furan fraction of the volatiles can reduce fermentation efficiency in biological processes [40] and lower energy content and stability in bio-oils produced by thermochemical processes [41]. Lignin, on the other hand, can have a variety of effects on biomass conversion depending on the process chosen. Lignin generally has a negative effect on ethanol production by blocking enzyme access to cellulose [42] but can increase oil yields for pyrolysis [43] and heating values for combustion [34] during thermochemical conversion.

3.1. Biochemical conversion—fermentation

Ethanol production from biomass occurs via two primary steps: depolymerization of the cellulose and hemicellulose to fermentable sugars and fermentation of these sugars to ethanol. Biomass conversion to ethanol has been evaluated in many reviews [42–45] which vary focuses from pretreatment and enzymatic hydrolysis [42] to optimization of the cellulase enzyme for improving sugar conversion to ethanol [44] and evaluation of current and future economic aspects of fuel ethanol production [46]. This work will build upon these previous reviews to explain how biomass compositional variability can influence fermentation processes for fuel production.

Mixed rangeland grasses are a prime example of a feedstock with high compositional variability. These grasses are an emerging alternative to traditional energy crops. Mixed rangeland grasses also preserve natural habitat and typically require less maintenance than traditional energy crops. However, the naturally high variability of these grasses can lead to reduced product yields in biochemical conversion processes. Adler et al. have shown that ethanol yield per unit area decreases as plant species diversity increases. Ethanol yields are maximized when there is increased targeted coverage of C_4 prairie grass energy crops, such as switchgrass, which sequester more carbon than typical C_3 conservation grassland varieties [47]. This preference for C_4 grasses illustrates how the production of ethanol using fermentation is typically much more dependent on biomass carbohydrate content. In fact, technoeconomic analysis has indicated that adjusting total carbohydrate content by 1% of total dry matter can change the minimum ethanol selling price (MESP) by $0.018/gal [46]. Given the compositional data above, fermentation is better matched to herbaceous crops than lignocellulosic material due to the higher carbohydrate content of grasses. Additionally, fermentation processes are typically more tolerant of the higher ash contents of herbaceous feedstocks [48]. However, it should be repeated that high alkali metal content from excess soil collected during harvest can increase acid neutralization during pretreatment and lower the xylan digestibility for corn stover, consequently lowering ethanol yields [35].

Despite compositional variability generally being a disadvantage in feedstock processing, there exists at least one aspect to variability that could be advantageous. Changes in structural carbohydrate content with anatomical fraction in corn stover significantly affect glucose yield. After hydrolysis, glucose concentration can be three times greater in the cobs, leaves, and husks than stalks [49]. Additionally, the corn cobs, leaves, and husks respond better than stocks to simultaneous saccharification and fermentation (SSF) despite having similar glucan levels [50]. Therefore, selective fermentation of specific anatomical fractions could increase process efficiency if a cost-effective separation process could be devised and there is a value-added coproduct that could be produced from the stalks. The advantage of separating biomass by anatomical fraction extends to other biomass types as well. For example, different fractions within wheat stover exhibit an almost 10% difference in glucan content, and some parts are much more susceptible to chemical saccharification [51].

3.2. Thermochemical conversion—general

While biomass as a feedstock exhibits a significant amount of compositional variability, illustrated in the tables above, the different options for thermochemical conversion are almost diverse. Thermochemical conversion operations utilize reactions using both solids (pyrolysis and combustion) and liquids (hydrothermal liquefaction). Products from thermochemical processes also span a wide range of states from solid (biochar), through liquid (bio-oil), all the way to gas (syngas). The wide variety of processing options and product outputs, along with short reaction times (on the order of seconds), allows thermochemical conversion operations to utilize a wide array of diverse process inputs.

3.3. Thermochemical conversion—combustion

The combustion of biomass, which is still common in developing countries, has been used for thousands of year to do everything from managing agricultural lands to producing heat and energy for industrial processes [52]. Currently, developed countries use nonrenewable fossil fuels such as oil, coal, and natural gas as a primary source of energy; however, these energy supplies could be depleted in the next 40–50 years [53]. In an effort to reduce the rate at which these nonrenewable resources are being depleted and reduce environmental impact, there is a shift toward the combustion of renewable biomass and other waste products (such as paper and plastics). Literature reviews focus on the combustion of biomass as an energy source both with [54–56] and without [30, 53, 57, 58] torrefaction as a pretreatment to improve combustion efficiencies and material grinding and storage properties. One of the major problems with combusting biomass in a traditional coal plant is slagging, a mineral buildup due to the higher ash content in biomass than in coal. This problem means that low-ash content biomass, such as woody feedstocks, is better to use than herbaceous materials (which have intrinsic ash contents about five times greater than woody materials) in combustion applications. While biomass combustion does present problems with slagging, it does have the benefit of reducing harmful greenhouse gas emissions as compared to coal [59], and the energy produced can be incorporated directly into the current energy grid without infrastructure changes.

3.4. Thermochemical conversion—pyrolysis

Pyrolysis is a thermochemical process that starts with a solid and can be tuned to produce either a solid (biochar) or a liquid (bio-oil). However, this chapter will focus on the production of bio-oil and the effects of biomass composition on the resulting oil yields and quality. Pyrolysis is performed at temperatures from 400 to 600°C [60] and often includes a catalyst with the aim of increasing the energy density of the product by removing oxygen (as water and volatiles) [61]. Pyrolysis of biomass to produce fuels has been thoroughly reviewed in the academic literature [33, 61–64].

The pyrolysis process is well suited for low-moisture-content material with low ash and high lignin content, meaning that pyrolysis processes favor lignocellulosic feedstocks. For example, lignin content increases the average molecular weight of resulting pyrolysis oil by 100 Da as lignin content rises from 5 to 15% [43]. The high ash content of herbaceous feedstock can decrease oil yields by 1–5% for every 1% increase in ash over an ash range of 1.5–7.5% [43]. In addition to decreasing oil yields, the alkali metals common in herbaceous crops can also have damaging effects on reactors and reduce catalyst lifetimes [65]. However, more recent studies have taken into account not only the production of pyrolysis oil but also the upgrading of that oil to the final fuel for a range of feedstocks including pines, poplars, switchgrass, and corn stover. In these integrated fast pyrolysis/hydrotreating studies, the effects of ash content on oil and upgraded fuel yields were relatively insignificant over a narrow ash range of 0.7–1.6% [66] (an ash range common for woody materials but low for grasses).

3.5. Thermochemical conversion—hydrothermal liquefaction (HTL)

Hydrothermal liquefaction (HTL) is a unique thermal conversion process that utilizes biomass and water slurries. This makes HTL particularly well suited to turning high water content material, such as algae, municipal solid wastes, or grasses into bio-based oils. Additionally, HTL bio-oils tend to be higher quality than pyrolysis oils because they have less oxygen. However, the oil yields for HTL are lower than pyrolysis and the oxygen content is still higher than crude oil [67]. Performing the dissolution of biomass in a water media also saves energy on drying the feedstock, and the high heat transfer rates in a liquid media reduce particle size reduction requirements [36]. HTL can operate over a wide range of temperatures (200–600°C) to create products that range from solid biochars to gases. Reaction temperatures from 200 to 275°C are suitable for solid production [68], while temperatures from 275 to 350°C produce liquid products, and temperatures above 400°C are suitable for gas production [36]. Due to the liquid nature of the reaction media and the high temperatures, these reactors often operate at high pressures (5–40 MPa) to keep the reaction media as a liquid or supercritical fluid. Since the operating conditions and products of hydrothermal reactors are so diverse, the reviews of this material span a wide range. Some reviews cover both sub- and supercritical temperature regimes, with an array of model compounds and biomass feedstocks, and product arrays including liquid bio-oils and gases [67]. Other reviews focus on narrower operating regimes and liquid products from a variety of feedstocks with both high and low ash content [28, 36] or simply the processing of lignin (which is usually considered a waste product) [69]. While the hydrothermal processing of biomass offers advantages in being more feedstock agnostic, it has drawbacks in high capital equipment cost due to the extreme operating conditions, high energy input to heat the water, and lower yields (even though the oil quality is generally high).

3.6. Chemical production

Aside from the production of biochar and bio-oil, hot liquid water can also be used to convert biomass into value-added chemicals. Luterbacher et al. have achieved a 65% yield of sugars from woody biomass and a 55% yield from switchgrass using a biphasic CO_2/H_2O system. This biphasic system improves process separations and can use larger particles (~1 cm) at a high solids loading (40 wt%) [70]. The targeted production of sugars from biomass, instead of a bio-oil destined for fuel blending, could facilitate the production of high-value chemicals and materials. For instance, biomass-derived sugars can be used to make renewable plastics by producing *p*-xylene [71–73]. The conversion of biomass-derived cellulose to *p*-xylene could take place using a scheme such as the one in **Figure 1**. In this scheme cellulose is converted to *p*-xylene in a four-step process: step one uses a biphasic CO_2/H_2O system to convert biomass into sugars [70], step two isomerizes glucose to fructose [74], step three converts fructose 2,5-dimethylfuran (DMF) [75], and step four converts DMF to *p*-xylene [71]. The final step of converting *p*-xylene to polyethylene terephthalate (PET) would take place in a typical refinery because this renewable *p*-xylene would act as a standard drop in feedstock.

The production of chemicals from biomass has the potential to produce a wide array of drop in building blocks. The top twelve most promising drop in chemical building blocks can be

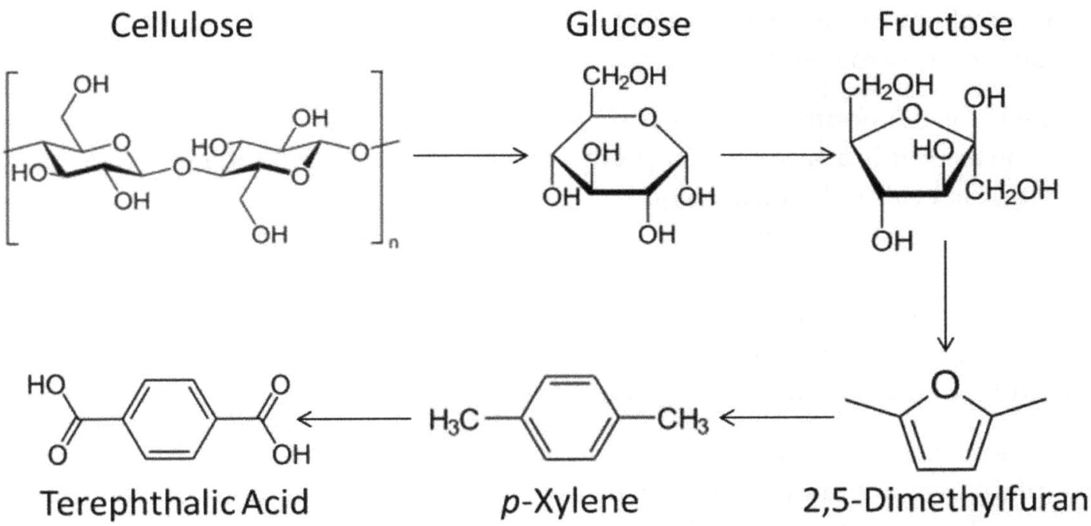

Figure 1. A representative pathway for conversion of cellulose to terephthalic acid through the transformation of cellulose-derived sugars to furans.

found in the Department of Energy's report on Top Value-Added Chemicals from Biomass [76]. This report lists several chemicals that could be made from biomass with an emphasis on the conversion of sugars to building block chemicals and the conversion of these building block chemicals to intermediates. After examining both biochemical and thermochemical pathways, it was noted that biochemical pathways focused on the conversion of sugars to building block chemicals, and thermochemical pathways dominated the conversion of building block chemicals to final products.

4. Pretreatments to improve biomass feedstock chemical composition and suggestions for optimal biomass conversion pathways

Raw herbaceous biomass has a chemical composition which is low in carbon content and high in oxygen, volatiles, and ash; is high in moisture; and has low energy content. This combination of properties does not make herbaceous crops suitable for thermochemical applications such as gasification, pyrolysis, and co-firing [77]. The shortcoming of many types of raw biomass, in terms of chemical and physical properties, can be overcome by pretreatment to produce a conversion-ready feedstock. Currently, there exist a variety of pretreatment methods including pelletization, air classification, dry torrefaction, hydrothermal carbonization, steam explosion, ionic liquid dissolution, acid and alkali leaching, and ammonia fiber expansion (AFEX). These pretreatment techniques are being looked at to improve biomass quality to produce a conversion-ready feedstock for both thermochemical and biochemical applications [25, 78]. Pretreatment can reduce biomass chemical and physical heterogeneity and lessen problems in (a) conversion applications (removing using air classification to remove ash prior to co-firing biomass could reduce slagging), (b) supply chain logistics (pelletizing biomass reduces transportation costs), (c) operational constraints (certain forms of pretreatment allow for utilization of coal infrastructure for feeding, milling, etc. of biomass, without costly modifications or installation of separate processing lines), and (d) technical constraints (e.g., reduction of corrosion due to biomass washing).

Pretreatment for the optimization of chemical production from biomass is very much in its infancy. However, it is a safe bet that pretreatment will be required to get a consistent product, given that specialty chemicals require a much higher purity than the fuels currently being produced. Current research is ongoing for the production of many different value-added chemicals such as *p*-xylene [71], dimethylfuran [75], and levulinic acid [79] to name just a few, but at this point, all of these studies start with pure feedstocks, such as cellulose, and not biomass. To move the industry, past fuels to value-added chemicals will require a greater understanding of how biomass composition effects its conversion to fuels and chemicals.

As the previous pages have illustrated, the transformation of biomass to fuels and chemicals can take place over a wide variety of pathways with numerous influences from the biomass composition. These conversion pathways can be generally grouped as either biochemical or thermochemical. A greatly simplified process diagram for the production of renewable liquid fuels and chemicals from biomass can be seen in **Figure 2**. This figure groups feedstocks with their most likely conversion pathway based on the previous discussion regarding biomass composition.

Given the current variability in biomass resources, it is apparent that conversion technology will have to be tailored to regional renewable supply, be it lignocellulosic, herbaceous, a municipal solid waste stream, or algae. Given the high ash content of herbaceous biomass and the high water content of some municipal solid wastes, it is likely that these streams will be destined for use in biochemical pathways to produce sugars through enzymatic hydrolysis or methane using anaerobic digestion. However, there is also a chance that these materials could be passed through the thermochemical process of hydrothermal liquefaction to produce oils or undergo a more mild hydrothermal treatment to produce a platform chemical stream based on biomass-derived sugars. The abundant lignocellulosic biomass will likely be converted to bio-oil or energy using a thermochemical process such as pyrolysis or combustion, respectively. Thermochemical processes make use of lignocellulosic feedstocks in part due to their low ash content and because a high lignin content is unsuitable for enzymatic digestion in biochemical fermentation.

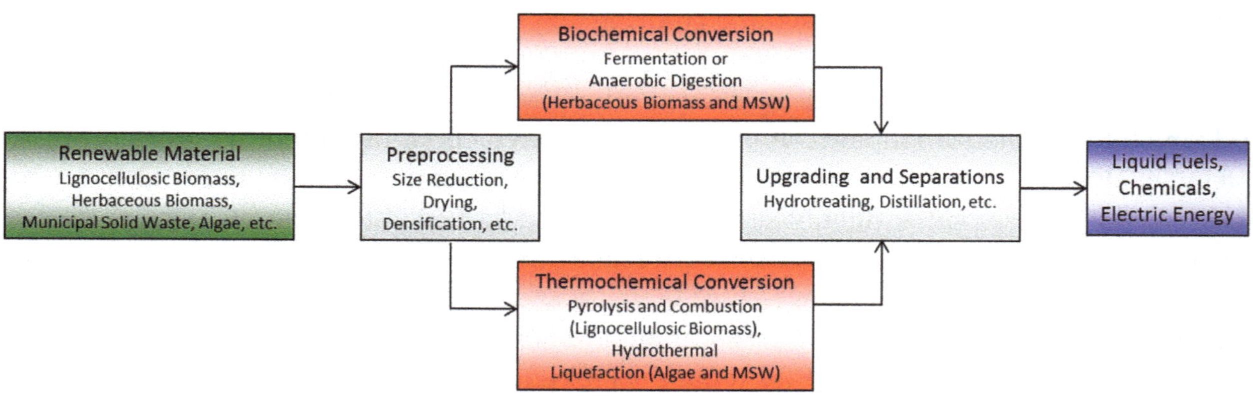

Figure 2. Broad scheme for conversion of renewable material to fuels, chemicals, and energy.

5. Conclusions

The large degree of variability between biomass resources, both currently available and emerging, is a significant barrier to the utilization of biomass as a feedstock for fuel and chemical production. The impacts of physical characteristics such as moisture content and particle size, as well as chemical characteristics such as ash content, extractives/volatiles, and lignin, all play varying, and intricate, roles during biomass conversion. Adding to the complexity of this system is the fact that, in addition to a myriad of compositionally diverse feedstocks, there also exist numerous conversion pathways to the final fuel or chemical products. To alleviate this problem, it will be necessary to develop techniques to reduce biomass variability and develop a consistent, conversion-ready feedstock for biorenewable fuel and chemical production.

Financial and competing interests disclosure

The US government retains, and the publisher, by accepting the article for publication, acknowledges that the US government has a nonexclusive, paid-up, irrevocable, worldwide license to publish or reproduce the published form of this manuscript, or allow others to do so, for US government purposes. The authors have no other relevant affiliations or financial involvement with any organization or entity with a financial interest in or financial conflict with the subject matter or materials discussed in the manuscript apart from those disclosed. No writing assistance was utilized in the production of this manuscript.

Acknowledgements

This research was supported by the US Department of Energy under the Department of Energy Idaho Operations Office Contract No. DE-AC07-05ID14517.

Author details

C. Luke Williams*, Rachel M. Emerson and Jaya Shankar Tumuluru

* Address all correspondence to: luke.williams@inl.gov

Idaho National Laboratory, Idaho Falls, ID, USA

References

[1] Noble IR, Bolin B, Ravindranath NH, Verardo DJ, Dokken DJ. Land Use, Land Use Change, and Forestry. Cambridge University Press: Cambridge England; 2000. p. 375

[2] Chu S, Majumdar A (2012) Opportunities and challenges for a sustainable energy future. Nature 488 (7411):294–303

[3] Perlack RD, Wright LL, Turhollow AF, Graham RL, Stokes BJ, Erbach DC. Biomass as Feedstock for a Bioenergy and Bioproducts Industry: The Technical Feasibility of a Billion- Ton Annual Supply. DTIC Document. April 2005. 78p. DOE/GO-102005-2135. ORNL/TM-2005/66

[4] Perlack RD, Eaton LM, Turhollow Jr AF, Langholtz MH, Brandt CC, Downing ME, Graham RL, Wright LL, Kavkewitz JM, Shamey AM (2011) U.S. Billion-Ton Update: Biomass Supply for a Bioenergy and Bioproducts Industry. August 2011. 195 p. DOE/EE-0363

[5] Brennan L, Owende P (2010) Biofuels from microalgae—a review of technologies for production, processing, and extractions of biofuels and co-products. Renewable and Sustainable Energy Reviews 14 (2):557–577

[6] Lynd LR, Cushman JH, Nichols RJ, Wyman CE (1991) Fuel ethanol from cellulosic biomass. Science (Washington) 251 (4999):1318–1323

[7] Humbird D, Davis R, Tao L, Kinchin C, Hsu D, Aden A, Schoen P, Lukas J, Olthof B, Worley M (2011) Process design and economics for biochemical conversion of lignocellulosic biomass to ethanol. National Renewable Energy Laboratory; March 2011. 147 p. TP-5100-47764

[8] DuPont (2011) DuPont industrial biosciences announces new leader and name for cellulosic ethanol business. http://www2.dupont.com/BioFuel/en_US/whats_new/DuPont_Cellulosic_Ethanol_111511.html. Accessed 03-11-2015

[9] Poet-DSM (2013) POET and DSM to make advanced biofuels a reality by 2013. http://poet-dsm.com/news/poet-and-dsm.aspx. Accessed 03/30/2015

[10] Nagarajan V, Mohanty AK, Misra M (2013) Sustainable green composites: value addition to agricultural residues and perennial grasses. ACS Sustainable Chemistry & Engineering 1 (3):325–333

[11] Ragauskas AJ, Williams CK, Davison BH, Britovsek G, Cairney J, Eckert CA, Frederick WJ, Jr., Hallett JP, Leak DJ, Liotta CL, Mielenz JR, Murphy R, Templer R, Tschaplinski T (2006) The path forward for biofuels and biomaterials. Science 311 (5760):484–489. doi:10.1126/science.1114736

[12] Laurens LML, Dempster TA, Jones HDT, Wolfrum EJ, Van Wychen S, McAllister JSP, Rencenberger M, Parchert KJ, Gloe LM (2012) Algal biomass constituent analysis: method uncertainties and investigation of the underlying measuring chemistries. Analytical Chemistry 84 (4):1879–1887. doi:10.1021/ac202668c

[13] Vassilev SV, Baxter D, Andersen LK, Vassileva CG, Morgan TJ (2012) An overview of the organic and inorganic phase composition of biomass. Fuel 94:1–33. doi:10.1016/j.fuel.2011.09.030

[14] INL (2015) DOE biomass feedstock library. bioenergy.inl.gov. Accessed 3-10-2015

[15] Wyman CE, Goodman BJ (1993) Biotechnology for production of fuels, chemicals, and materials from biomass. Applied Biochemistry and Biotechnology 39 (1):41–59

[16] Dale BE, Allen MS, Laser M, Lynd LR (2009) Protein feeds coproduction in biomass conversion to fuels and chemicals. Biofuels, Bioproducts and Biorefining 3 (2):219–230

[17] ASTM D3172-13, standard practice for proximate analysis of coal and coke, ASTM International, West Conshohocken, PA, 2013, www.astm.org

[18] Sluiter A, Hames B, Ruiz R, Scarlata C, Sluiter J, Templeton D, Crocker D (2008) Determination of structural carbohydrates and lignin in biomass. Laboratory analytical procedure. April 2008. 15 p. NREL/TP-510-42618

[19] Wolfrum EJ, Sluiter AD (2009) Improved multivariate calibration models for corn stover feedstock and dilute-acid pretreated corn stover. Cellulose 16 (4):567–576

[20] Kenney KL, Smith WA, Gresham GL, Westover TL (2013) Understanding biomass feedstock variability. Biofuels 4 (1):111–127

[21] Templeton DW, Sluiter AD, Hayward TK, Hames BR, Thomas SR (2009) Assessing corn stover composition and sources of variability via NIRS. Cellulose 16 (4):621–639. doi:10.1007/s10570-009-9325-x

[22] Emerson R, Hoover A, Ray A, Lacey J, Cortez M, Payne C, Karlen D, Birrell S, Laird D, Kallenbach R, Egenolf J, Sousek M, Voigt T (2014) Drought effects on composition and yield for corn stover, mixed grasses, and Miscanthusas bioenergy feedstocks. Biofuels 5 (3):275–291. doi:10.1080/17597269.2014.913904

[23] Lemus R, Brummer EC, Moore KJ, Molstad NE, Burras CL, Barker MF (2002) Biomass yield and quality of 20 switchgrass populations in southern Iowa, USA. Biomass and Bioenergy 23 (6):433–442

[24] Eisenbies MH, Volk TA, Posselius J, Shi S, Patel A (2014) Quality and variability of commercial- scale short rotation willow biomass harvested using a single pass cut-and-chip forage harvester. BioEnergy Research.(8):546. doi:10.1007/s12155-014-9540-7

[25] Williams CL, Westover TL, Emerson RM, Tumuluru JS, Li C (2015) Sources of biomass feedstock variability and the potential impact on biofuels production. BioEnergy Research (9):1–14

[26] Dibble CJ, Shatova TA, Jorgenson JL, Stickel JJ (2011) Particle morphology characterization and manipulation in biomass slurries and the effect on rheological properties and enzymatic conversion. Biotechnology Progress 27 (6):1751–1759. doi:10.1002/btpr.669

[27] Van Walsum GP, Allen S, Spencer M, Laser M, Antal M, Jr., Lynd L (1996) Conversion of lignocellulosics pretreated with liquid hot water to ethanol. In: Wyman C, Davison B (eds) Seventeenth symposium on biotechnology for fuels and chemicals, vol 57/58. ABAB Symposium. Humana Press: New York USA pp 157–170. doi:10.1007/978-1-4612-0223-3_14

[28] Akhtar J, Amin NAS (2011) A review on process conditions for optimum bio-oil yield in hydrothermal liquefaction of biomass. Renewable and Sustainable Energy Reviews 15 (3):1615–1624. doi:10.1016/j.rser.2010.11.054

[29] Jazrawi C, Biller P, Ross AB, Montoya A, Maschmeyer T, Haynes BS (2013) Pilot plant testing of continuous hydrothermal liquefaction of microalgae. Algal Research 2 (3):268–277. doi:10.1016/j.algal.2013.04.006

[30] Demirbas A (2004) Effects of temperature and particle size on bio-char yield from pyrolysis of agricultural residues. Journal of Analytical and Applied Pyrolysis 72 (2):243–248. doi:10.1016/j.jaap.2004.07.003

[31] Spliethoff H, Hein K (1998) Effect of co-combustion of biomass on emissions in pulverized fuel furnaces. Fuel Processing Technology 54 (1):189–205

[32] Brownell HH, Yu EKC, Saddler JN (1986) Steam-explosion pretreatment of wood: effect of chip size, acid, moisture content and pressure drop. Biotechnology and Bioengineering 28 (6):792–801. doi:10.1002/bit.260280604

[33] Bridgwater A, Meier D, Radlein D (1999) An overview of fast pyrolysis of biomass. Organic Geochemistry 30 (12):1479–1493

[34] Jenkins B, Baxter L, Miles Jr T, Miles T (1998) Combustion properties of biomass. Fuel Processing Technology 54 (1):17–46

[35] Weiss ND, Farmer JD, Schell DJ (2010) Impact of corn stover composition on hemicellulose conversion during dilute acid pretreatment and enzymatic cellulose digestibility of the pretreated solids. Bioresource Technology 101 (2):674–678. doi:http://dx.doi.org/10.1016/j.biortech.2009.08.082

[36] Toor SS, Rosendahl L, Rudolf A (2011) Hydrothermal liquefaction of biomass: a review of subcritical water technologies. Energy 36 (5):2328–2342. doi:10.1016/j.energy.2011.03.013

[37] Tumuluru JS, Hess JR, Boardman RD, Wright CT, Westover TL (2012) Formulation, pretreatment, and densification options to improve biomass specifications for co-firing high percentages with coal. Industrial Biotechnology. 8 (3):113–132

[38] Jenkins BM, Bakker RR, Wei JB (1996) On the properties of washed straw. Biomass and Bioenergy 10 (4):177–200. doi:http://dx.doi.org/10.1016/0961-9534(95)00058-5

[39] Lacey JA, Aston JE, Westover TL, Cherry RS, Thompson DN (2015) Removal of introduced inorganic content from chipped forest residues via air classification. Fuel 160:265–273

[40] Palmqvist E, Hahn-Hägerdal B (2000) Fermentation of lignocellulosic hydrolysates. I: inhibition and detoxification. Bioresource Technology 74 (1):17–24. doi:http://dx.doi.org/10.1016/S0960-8524(99)00160-1

[41] Carpenter D, Westover TL, Czernik S, Jablonski W (2014) Biomass feedstocks for renewable fuel production: a review of the impacts of feedstock and pretreatment on the yield and product distribution of fast pyrolysis bio-oils and vapors. Green Chemistry 16 (2):384. doi:10.1039/c3gc41631c

[42] Sun Y, Cheng J (2002) Hydrolysis of lignocellulosic materials for ethanol production: a review. Bioresource Technology 83 (1):1–11

[43] Fahmi R, Bridgwater AV, Donnison I, Yates N, Jones JM (2008) The effect of lignin and inorganic species in biomass on pyrolysis oil yields, quality and stability. Fuel 87 (7):1230–1240. doi:10.1016/j.fuel.2007.07.026

[44] Lin Y, Tanaka S (2006) Ethanol fermentation from biomass resources: current state and prospects. Applied Microbiology and Biotechnology 69 (6):627–642. doi:10.1007/s00253-005-0229-x

[45] Bothast RJ, Schlicher MA (2005) Biotechnological processes for conversion of corn into ethanol. Applied Microbiology and Biotechnology 67 (1):19–25. doi:10.1007/s00253-004-1819-8

[46] Ruth MF, Thomas SR The effect of corn stover composition on ethanol process economics. In: 25th Symposium on Biotechnology for Fuels and Chemicals, 2003. Humana Press Breckenridge, CO.

[47] Adler PR, Sanderson MA, Weimer PJ, Vogel KP (2009) Plant species composition and biofuel yields of conservation grasslands. Ecological Applications 19 (8):2202–2209. doi:10.1890/07-2094.1

[48] McKendry P (2002) Energy production from biomass (part 1): overview of biomass. Bioresource Technology 83 (1):37–46. doi:http://dx.doi.org/10.1016/S0960-8524(01)00118-3

[49] Montross MD, Crofcheck CL (2004) Effect of stover fraction and storage method on glucose production during enzymatic hydrolysis. Bioresource Technology 92 (3):269–274. doi:http://dx.doi.org/10.1016/j.biortech.2003.09.007

[50] Duguid KB, Montross MD, Radtke CW, Crofcheck CL, Wendt LM, Shearer SA (2009) Effect of anatomical fractionation on the enzymatic hydrolysis of acid and alkaline pretreated corn stover. Bioresource Technology 100 (21):5189–5195. doi:http://dx.doi.org/10.1016/j.biortech.2009.03.082

[51] Duguid KB, Montross MD, Radtke CW, Crofcheck CL, Shearer SA, Hoskinson RL (2007) Screening for sugar and ethanol processing characteristics from anatomical fractions of wheat stover. Biomass and Bioenergy 31 (8):585–592. doi:10.1016/j.biombioe.2007.03.002

[52] Andreae MO (1991) Biomass burning: its history, use, and distribution and its impact on environmental quality and global climate, in Global Biomass Burning: Atmospheric, Climatic and Biospheric Implications,Cambridge MA: MIT Press; 1991 p[3–21]

[53] Saidur R, Abdelaziz EA, Demirbas A, Hossain MS, Mekhilef S (2011) A review on biomass as a fuel for boilers. Renewable and Sustainable Energy Reviews 15 (5):2262–2289. doi:10.1016/j.rser.2011.02.015

[54] Shankar Tumuluru J, Sokhansanj S, Hess JR, Wright CT, Boardman RD (2011) REVIEW: a review on biomass torrefaction process and product properties for energy applications. Industrial Biotechnology 7 (5):384–401

[55] Tumuluru JS, Wright CT, Hess JR, Kenney KL (2011) A review of biomass densification

systems to develop uniform feedstock commodities for bioenergy application. Biofuels, Bioproducts and Biorefining 5 (6):683–707. doi:10.1002/bbb.324

[56] Tumuluru J, Tabil L, Song Y, Iroba K, Meda V (2014) Impact of process conditions on the density and durability of wheat, oat, canola, and barley straw briquettes. BioEnergy Research (8):1–14

[57] Demirbas A (2004) Combustion characteristics of different biomass fuels. Progress in Energy and Combustion Science 30 (2):219–230. doi:10.1016/j.pecs.2003.10.004

[58] Demirbas A (2005) Potential applications of renewable energy sources, biomass combustion problems in boiler power systems and combustion related environmental issues. Progress in Energy and Combustion Science 31 (2):171–192. doi:10.1016/j.pecs.2005.02.002

[59] Veijonen K, Vainikka P, Järvinen T, Alakangas E (2003) Biomass co-firing–an efficient way to reduce greenhouse gas emissions. A report from European Bioenergy Networks (EUBIONET),VTT Processes, Altener Industries, Espoo, Finland (March 2003)

[60] Carlson TR, Vispute TP, Huber GW (2008) Green gasoline by catalytic fast pyrolysis of solid biomass derived compounds. ChemSusChem 1 (5):397–400

[61] Czernik S, Bridgwater A (2004) Overview of applications of biomass fast pyrolysis oil. Energy & Fuels 18 (2):590–598

[62] Bridgwater A (1999) Principles and practice of biomass fast pyrolysis processes for liquids. Journal of Analytical and Applied Pyrolysis 51 (1):3–22

[63] Bridgwater A, Peacocke G (2000) Fast pyrolysis processes for biomass. Renewable and Sustainable Energy Reviews 4 (1):1–73

[64] Bridgwater AV (2012) Review of fast pyrolysis of biomass and product upgrading. Biomass and Bioenergy 38:68–94. doi:10.1016/j.biombioe.2011.01.048

[65] Davidsson KO, Pettersson JBC, Nilsson R (2002) Fertiliser influence on alkali release during straw pyrolysis. Fuel 81 (3):259–262

[66] Howe DT, Westover T, Carpenter DL, Santosa D, Emerson R, Deutch S, Starace A, Kutnyakov I, Lukins C (2015) Field-to-fuel performance testing of lignocellulosic feedstocks: an integrated study of the fast pyrolysis/hydrotreating pathway. Energy & Fuels 29 (5):3188–3197

[67] Peterson AA, Vogel F, Lachance RP, Fröling M, Antal Jr MJ, Tester JW (2008) Thermochemical biofuel production in hydrothermal media: a review of sub-and supercritical water technologies. Energy & Environmental Science 1 (1):32–65

[68] Reza MT, Lynam JG, Uddin MH, Coronella CJ (2013) Hydrothermal carbonization: fate of inorganics. Biomass and Bioenergy 49:86–94. doi:http://dx.doi.org/10.1016/j.biombioe.2012.12.004

[69] Kang S, Li X, Fan J, Chang J (2013) Hydrothermal conversion of lignin: a review. Renewable and Sustainable Energy Reviews 27:546–558. doi:10.1016/j.rser.2013.07.013

[70] Luterbacher JS, Tester JW, Walker LP (2012) Two-temperature stage biphasic CO_2–H_2O pretreatment of lignocellulosic biomass at high solid loadings. Biotechnology and Bioengineering 109 (6):1499–1507

[71] Williams CL, Chang C-C, Do P, Nikbin N, Caratzoulas S, Vlachos DG, Lobo RF, Fan W, Dauenhauer PJ (2012) Cycloaddition of biomass-derived furans for catalytic production of renewable *p*-xylene. ACS Catalysis 2 (6):935–939. doi:10.1021/cs300011a

[72] Chang C-C, Green SK, Williams CL, Dauenhauer PJ, Fan W (2014) Ultra-selective cycloaddition of dimethylfuran for renewable *p*-xylene with H-BEA. Green Chemistry 16 (2):585–588. doi:10.1039/c3gc40740c

[73] Williams CL, Vinter KP, Patet RE, Chang C-C, Nikbin N, Feng S, Wiatrowski M, Caratzoulas S, Fan W, Vlachos DG, Dauenhauer PJ (2016) Inhibition of xylene isomerization in the production of renewable aromatic chemicals from biomass-derived furans. ACS Catalysis. doi:10.1021/acscatal.5b02329

[74] Moliner M, Román-Leshkov Y, Davis ME (2010) Tin-containing zeolites are highly active catalysts for the isomerization of glucose in water. Proceedings of the National Academy of Sciences 107 (14):6164–6168

[75] Roman-Leshkov Y, Barrett CJ, Liu ZY, Dumesic JA (2007) Production of dimethylfuran for liquid fuels from biomass-derived carbohydrates. Nature 447 (7147):982–985. doi:http://www.nature.com/nature/journal/v447/n7147/suppinfo/nature05923_S1.html

[76] Werpy T, Petersen G, Aden A, Bozell J, Holladay J, White J, Manheim A, Eliot D, Lasure L, Jones S (2004) Top value added chemicals from biomass. Volume 1—Results of screening for potential candidates from sugars and synthesis gas. Aug. 2004. 76 p. ADA436528

[77] Tumuluru JS (2015) Comparison of chemical composition and energy property of torrefied switchgrass and corn stover. Frontiers in Energy Research 3:46

[78] Bonner IJ, Thompson DN, Plummer M, Dee M, Tumuluru JS, Pace D, et al. Impact of ammonia fiber expansion (AFEX) pretreatment on energy consumption during drying, grinding, and pelletization of corn stover. Drying Technol. 2016 Jan;34(11):1319-1329

[79] Weingarten R, Conner WC, Huber GW (2012) Production of levulinic acid from cellulose by hydrothermal decomposition combined with aqueous phase dehydration with a solid acid catalyst. Energy & Environmental Science 5 (6):7559–7574

9

Biomass Blending and Densification: Impacts on Feedstock Supply and Biochemical Conversion Performance

Allison E. Ray, Chenlin Li, Vicki S. Thompson, Dayna L. Daubaras, Nicholas J. Nagle and Damon S. Hartley

Abstract

The success of lignocellulosic biofuels and biochemical industries depends on an economic and reliable supply of high-quality biomass. However, research and development efforts have been historically focused on the utilization of agriculturally derived cellulosic feedstocks, without considerations of their low energy density, high variations in compositions and potential supply risks in terms of availability and affordability. This chapter demonstrated a strategy of feedstock blending and densification to address the supply chain challenges. Blending takes advantage of low-cost feedstock to avoid the prohibitive costs incurred through reliance on a single feedstock resource, while densification produces feedstocks with increased bulk density and desirable feed handling properties, as well as reduced transportation cost. We also review recent research on the blending and densification dealing with various types of feedstocks with a focus on the impacts of these preprocessing steps on biochemical conversion, that is, various thermochemical pretreatment chemistries and enzymatic hydrolysis, into fermentable sugars for biofuel production.

Keywords: blending, densification, conversion performance, advanced feedstock supply system, preprocessing

1. Introduction

Global demands for energy, finite petroleum reserves, and growing concerns over climate change have prompted considerable interest in lignocellulosic biomass as a sustainable alternative to fossil-derived sources for the production of transportation fuels. The Renewable

Fuel Standard (RFS2) [1] mandates the use of 36 billion gallons of renewable fuels by 2022 under the U.S. Energy Independence and Security Act (EISA) of 2007 [2]. Biomass availability and quantity pose significant barriers to the realization of large-scale production of lignocellulose-derived biofuels. The U.S. Department of Energy's (DOE) 2016 Billion Ton Report has projected the potential for more than one billion tons of biomass in the form of agricultural, forestry, waste, and algal materials capable of displacing approximately 30% of U.S. petroleum consumption without adverse environmental effects or negative impacts to production of food and agricultural products [3].

The conversion of biomass into affordable bio-based fuels and chemicals aims to displace all of the products currently made from a barrel of oil. Research and development efforts focused on the production of bio-derived hydrocarbon fuels and products seek to mobilize the bioeconomy in order to diversify energy resources that enable energy production. However, development of biomass as a sustainable energy resource for fuels and chemicals will require advances aimed at solving logistical challenges to ensure a cost-effective and consistent feedstock supply to the biorefinery [4–7]. Efficient utilization of the available resources for biofuels production requires considerations of supply chain cost, feedstock quality and conversion performance that dictates overall process economics. Logistical operations like harvest, collection, preprocessing, storage, and transportation have a significant impact on biomass availability and feedstock cost and quality [6, 8]. The large-scale deployment of lignocellulosic biomass for energy production has been severely limited by the high cost associated with the feedstock supply chain and technology barriers in conversion to fuel [8–10].

Initial development of the biofuels industry has centered around high-productive, single-resource areas that rely on sufficient quantity to enable selection and sourcing of suitable materials for conversion processes. However, as the bioeconomy grows and production moves away from highly productive, resource-rich areas, the impact of the spatial and temporal variability inherent to biomass feedstocks [6] cannot be managed solely by passive means in order to meet requirements for quality and quantity. The expansion of the industry will necessitate the adoption of "advanced" concepts within the supply system in order to meet cost, quality, and quantity requirements.

In addition, the "conventional system" currently employed by the cellulosic biofuel industry relies on a vertically integrated feedstock supply system where a single biomass feedstock is procured through contracts with local farmers, harvested and stored locally, and delivered in a low-density, baled format to the conversion facility [7]. This system has been demonstrated to work in high-yield regions, such as the U.S. Corn Belt; however, recent analyses have shown that conventional systems fail to meet feedstock cost targets outside of highly productive regions [11]. Realization of large-scale production of lignocellulosic biofuels will require modification to the current system in order to enable a consistent, cost-effective, and continuous supply of biomass to the biorefinery [10]. In comparison, the advanced feedstock supply system (AFSS) employs a wide range of preprocessing techniques, such as feedstock blending and densification in distributed biomass depots, and shows great promise for enabling improvements in handling and quality, consistency and uniformity, facilitating access to resources, and stability in storage [4, 7, 12, 13].

Biorefineries that rely upon a single feedstock to meet tonnage requirements are vulnerable to significant risks, in terms of both availability and affordability. Diversification of biomass supply has the potential to reduce risk [7], in some cases by as much as 80% [14], while enabling the lowest delivery cost [15]. Achieving a continuous, year-round supply of a single biomass resource is unlikely given the seasonal availability of most agricultural crops. Furthermore, climate change poses an inevitable risk to biomass supply systems for a developing bioenergy industry. Langholtz et al. [16] highlight the risk of extreme weather events to the bioenergy supply chain that are certain to cause reductions in feedstock production and increased price for agricultural commodities and biofuels. Other work has shown that drought has a significant, negative impact on biomass quality, in addition to biomass production yields [17]. Recent studies suggest a blended feedstock strategy to enable supply chain resilience may provide a solution to reliance on a single biomass resource [10, 18].

Low-density biomass feedstocks also pose a significant challenge to supply chain operations that translate to difficulties in storage, transportation, handling, and feeding [19], which hinder the large-scale use of biomass for biofuel production. Large volumes of low-density feedstocks require more resources for transportation and shipping. The size of the transportation resources needed to reach the 2050 target of 50% reduction of greenhouse gas via biofuels, biopower, and bioproducts exceeds the resources used to move the 2010 world grain and oil seed resources by 6- to 10-fold [20]. Densification processes, such as pelletization would increase the bulk and energy density of raw biomass, improve stability during storage and handling, create flowable feedstocks that are compatible with existing handling systems, and improve transport efficiency and cost [19].

The use of blended and densified feedstocks in conversion pathways instead of conventionally ground biomass from a single source addresses several challenges in the current biomass supply chain, including availability, transportation, storage, cost, quality, and supply variability [7, 19, 21–23]. This chapter provides a glimpse into the potential for preprocessing options, for example, blending and densification, to provide benefits to both biomass cost and conversion.

2. Addressing feedstock supply challenges

2.1. Feedstock blending strategy

A promising strategy to reduce supply risk is to blend different biomass feedstocks. Blending has been used by many industries (e.g., coal and animal feed) to affect the quality of the feedstock [24]. In the coal industry, different grades of coal are blended in order to meet emission targets and minimize ash production during power generation [25–27]. In the animal feed industry, a variety of feedstocks are blended to meet the desired nutritional requirements for a specific target animal [28]. Similarly, the concept of blending can be extended to the biofuel and bioproducts industry.

Formulating a designed feedstock through blending and other preprocessing methods allows low-cost and typically low-quality biomass to be blended with biomass of higher cost and typically higher quality to achieve the specifications required at the in-feed of a conversion facility (note that different conversion processes may require different specifications, and the cost required to meet those specifications will vary). The use of low-cost biomass allows the supply chain to implement additional preprocessing technologies that actively control feedstock quality, while also bringing more biomass into the system. This analysis and design approach is referred to as the "least-cost formulation" strategy [29]. In addition, recent work has shown that blended biomass feedstocks demonstrate improved flowability behavior [30], suggesting the potential for blending to extend benefits from the supply chain to feeding systems at the refinery.

The farmgate price is used to describe the economic availability of biomass resources and includes the cost of production and harvest [3]. **Figures 1** and **2** represent the cost of corn stover as a function of availability by state or region; these figures illustrate the increase in farmgate price with increasing demand. By blending feedstocks, the biorefinery can take advantage of the lower end of each supply curve to reduce cost. For example, **Figure 3** shows supply curves for switchgrass and corn stover from a 12-county region in northwestern Kansas, approximately 90 miles by 120 miles in size. In this region, only 700,000 tons per year of switchgrass (red curve) are available at $50/ton which could not support a biorefinery (capacity of ≥800,000 tons per year). There is sufficient corn stover to supply 1.6 million tons but at a farmgate price of $58/dry ton (blue curve). Thus, the strategy of combining the two feedstocks (green curve) shows that 1.6 million tons could be supplied to a biorefinery for a lower farmgate price of $48/dry ton.

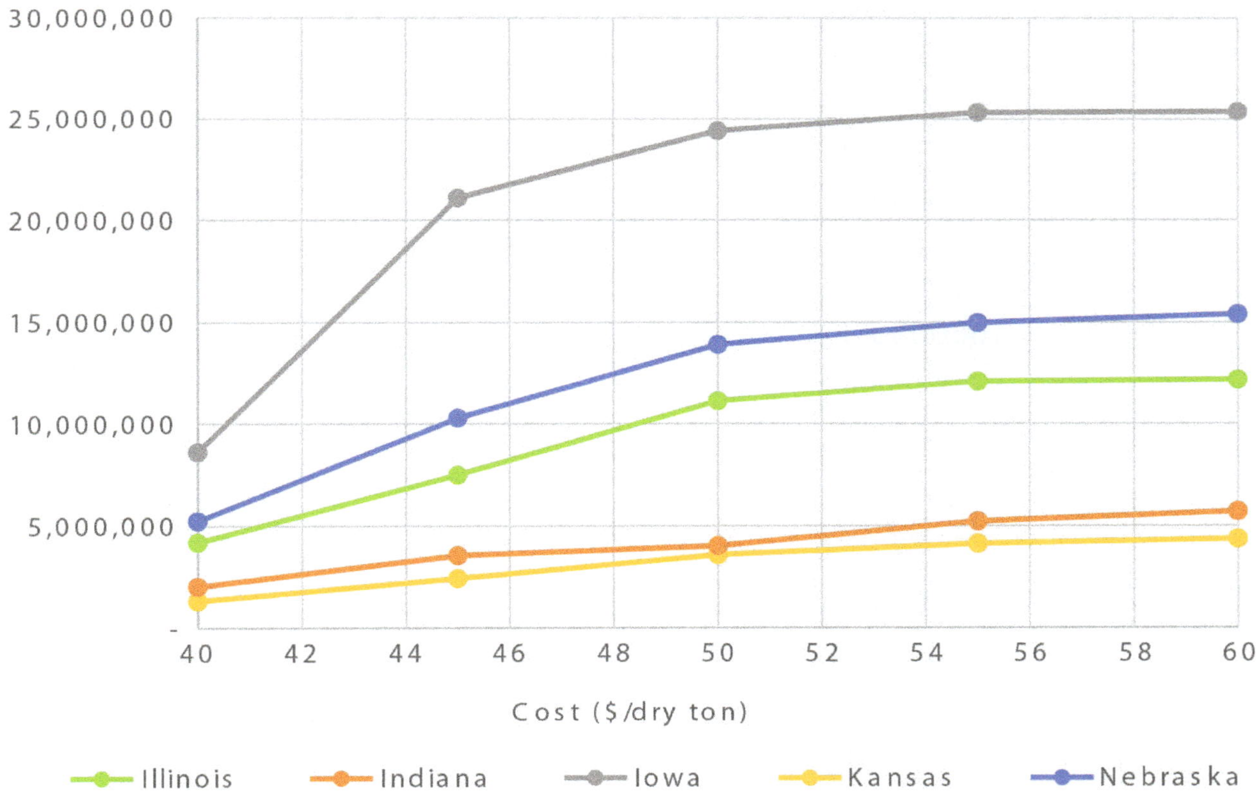

Figure 1. Corn stover availability by state as a function of farmgate price [33]. Availability and cost data assume base case corn stover yields for 2015. The data for each state is the sum of available corn stover for each county at a given farm gate price.

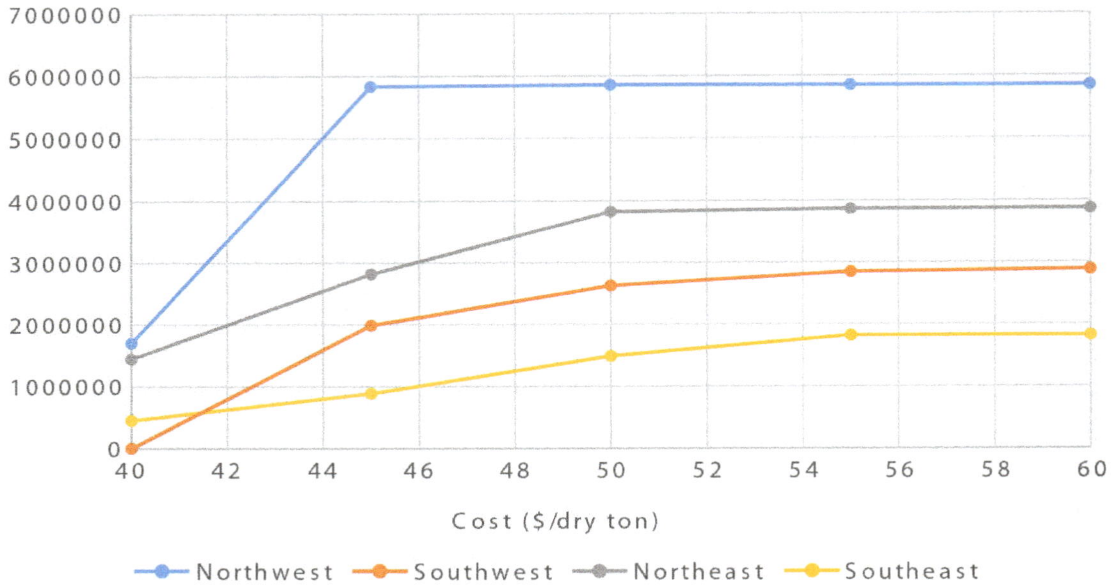

Figure 2. Feedstock supply curves for various locations in Iowa [33]. Availability and cost assume base case corn stover yields for 2015 for 61 counties in Iowa.

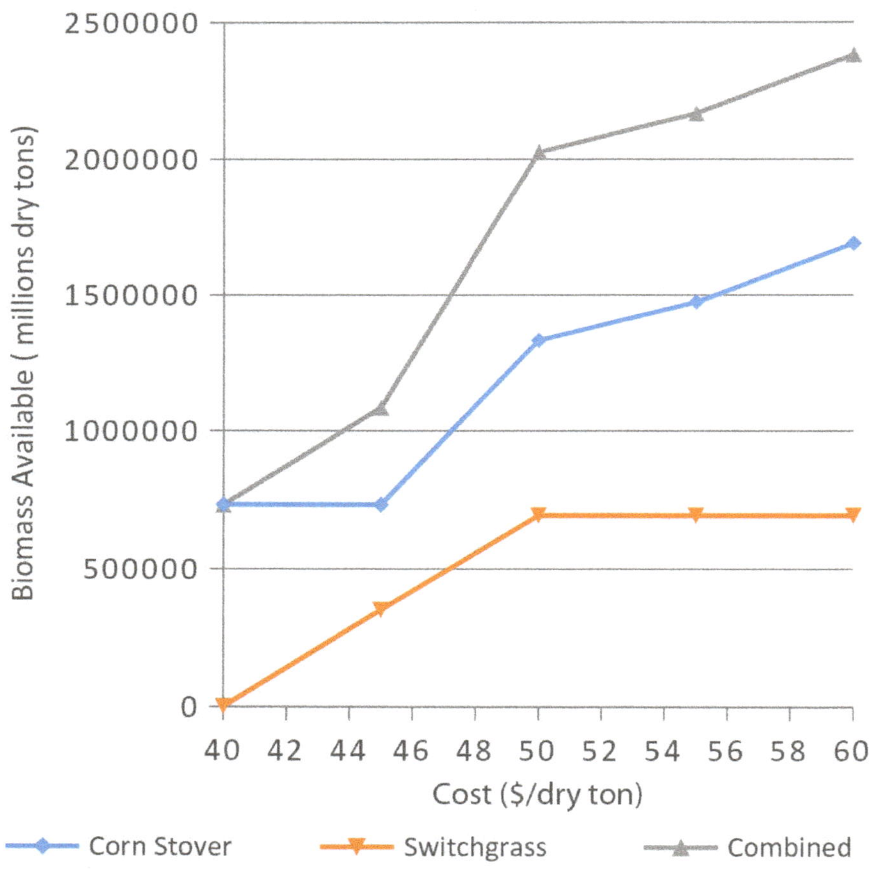

Figure 3. Biomass availability in northwestern Kansas as a function of farmgate price; prices and quantities shown are for 2015 and assume base case yields [34]. Supply curves illustrate biomass availability as a function of cost for a 12-county region in northwestern Kansas, with an approximate area of 90 miles × 120 miles. Curves are shown for corn stover (blue), switchgrass (red), and combined for a blended supply of corn stover and switchgrass (green).

Feedstock blending allows a biorefinery to utilize less of a single and expensive biomass type by collecting a variety of biomass (e.g., corn stover, switchgrass, sorghum, yard waste) and effectively moving down the cost versus supply curve and paying a lower average price for each feedstock. This does not change the supply versus cost curves for each resource; instead, it describes a system where purchasers are using a combination of least-cost resources and blending them to meet feedstock specifications for a subsequent biomass conversion process [29]. Costs may be further reduced by contraction of the draw radius for material collection, which reduces transportation cost. Feedstock formulation enabled through blending and other preprocessing strategies allows low-cost, low-quality biomass to be blended with higher cost and higher quality to achieve the in-feed specifications at the conversion facility. Blending feedstocks of differing quality results in a feedstock that has properties representative of the proportions of the materials that were blended together. Final price and quality are basically a weighted average of the price and quality of the components. It is important to realize a balance must be maintained and cost benefits may be not be linearly related to quality impacts. For many feedstock blends, there is likely a threshold quality level that cannot be surpassed to realize equal economic benefit. Biomass quality is a key consideration when analyzing biomass cost and availability. In combination with densification, wider sourcing areas can be tapped (including resources that are considered stranded using conventional supply systems).

Combining different biomass resources into the supply system also creates cost benefits by reducing overall grower payments [12]. The blended feedstock strategy relies on the availability of multiple feedstock resources that can be blended in an economical supply radius [31], which, in turn, decreases grower payment by reducing the required amount of any single biomass resource. In this manner, blending has the potential to expand the regionally available, biomass resource pool to include feedstocks of marginal quality at lower cost. In addition, a blended strategy offers the potential for feedstock quality upgrades and reduced variability [6, 21]. Blending high-quality feedstocks with low-cost, low-quality feedstocks is a strategy that can be used to meet quality specifications [21] at the biorefinery, in addition to achieving volume and cost targets in the supply chain [32]. An analysis by Maung and colleagues [18] has shown that a multi-crop cellulosic feedstock strategy lowers transportation costs compared to reliance on a single resource. Additionally, sourcing multiple feedstocks for cellulosic biofuel production mitigates supply risks associated with policies that govern crop residue removal. Further, Maung et al. suggest that a multi-feedstock strategy enhances understanding of the links between environmental policy, economies of density, economics of geography, transportation, risk and diversification in the biomass feedstock supply chain.

2.2. Densification

Reducing transportation costs while producing feedstocks with desirable (and consistent) physical properties such as increased bulk density and enhanced handling and processing characteristics requires densification of low-bulk density biomass. Commodity production for renewable fuels and chemicals requires large-scale biomass resources managed through AFSS and distributed biomass depots. These depots can provide feedstock stability, size reduction, and managed moisture [20]. Distributed biomass depots can reduce transportation and shipping costs and improve feedstock stability and consistency by employing strategies such as size reduction, moisture management, blending, and densification. This allows greater access to stranded feedstocks and can reduce grower payment through feedstock blending [4, 11, 35]. Reducing transportation costs while producing feedstocks with desirable (and consistent) physical properties such as increased bulk density and enhanced handling and processing characteristics requires densification of low-bulk density biomass.

Pelleted biomass is produced from raw, ground material that is conditioned with heat and/or moisture, compacted, and extruded through a die [2, 3]. The economics and physical properties of densified biomass formats produced from agricultural residues have been explored in several studies [36–39]. Pelleting of biomass can increase unit density of raw biomass resources by as much as 10-fold [19], resulting in a flowable and durable product that is compatible with existing biomass supply system infrastructure. It has been shown that activation of the natural binders in biomass, such as lignin, through combined moisture and temperature effects during the process of densification is key to the development of particle-particle bonding that is required for durability [9]. The extent of lignification contributes significantly to biomass recalcitrance [4], and lignin alteration during the process of densification may impact biomass reactivity to pretreatment and enzymatic hydrolysis [12].

Currently, there are many types of densification systems available: pellet mills, piston and roller presses, tabletizers, and extruders [19]. Pelleting of biomass can increase unit density of raw biomass resources by as much as 10-fold [1], producing a uniform, durable product with free-flowing characteristics that may be more compatible with biorefinery operations. The economics and physical properties of densified biomass formats produced from agricultural residues have been explored in several studies [2, 9–11]. Industrial pelleting has developed into a well-established process using wood and wood chips. Global pellet demand has reached 23-million metric tons [40].

3. Impact of blending and densification on pretreatment processes in biochemical conversion pathways

Thermochemical pretreatment processes are used in biochemical conversion pathways to facilitate enzymatic access to cellulose and enable conversion of complex carbohydrate polymers into fermentable sugars. These promising processes include ammonia fiber expansion (AFEX), dilute acid, alkaline and ionic liquid (IL) pretreatment technologies. Specifically, AFEX is a physicochemical pretreatment process performed under high pressure (200–400 psi) and moderate temperature (80–150°C) with concentrated ammonia for a brief residence time (5–30 min) before pressure release [41]. AFEX pretreatment facilitates enzymatic access to cellulose by breaking down the cellulose crystalline structure and depolymerizing the lignin. Dilute-acid (DA) pretreatment relies on the combined effect of dilute sulfuric acid (0.25–2 wt.%), temperature (140–200°C), and time (seconds to minutes) to solubilize hemicellulose and improve enzymatic accessibility to cellulose [42, 43]. Alkaline pretreatment technologies focus on lignin solubilization and deacetylation under relative mild conditions (60–180°C) with NaOH or ammonium hydroxide (i.e., soaking in aqueous ammonia, namely SAA) as catalyst [44–46]. Recently, ILs are receiving significant attention as a class of novel environmental benign "green solvents" to dissolve and disrupt the biomass cell wall, reduce cellulose crystallinity and lignin content, and increase the porosity and surface area for enhanced enzymatic digestibility [26, 47–52]. In addition, this pretreatment technique shows great capability of fractionating wide range of feedstocks [50, 52, 53].

Although significant efforts have been focused on pretreatment of single lignocellulosic biomass in loose and ground format, recently researchers started to look into the application of biomass blending and densification for biochemical conversion into fermentable sugars. The details of biomass blending and quality improvement, characteristics of various densification formats of diverse feedstock types, and their impacts on conversion performance are discussed below.

3.1. Impact of biomass blending

Feedstock blending is one approach offering promising solution to overcome current challenges on biomass supply such as significant compositional variations [21, 22]. Therefore, it is imperative to develop conversion technologies that can process blended biomass feedstocks with minimal negative impact in terms of overall performance of the relevant biochemical pathway unit operations: pretreatment, fermentable sugar production, fermentation, and fuel titers.

Ionic liquid (IL) pretreatment has shown uniqueness in efficiently handling wide range of feedstocks; thus, this technology was investigated on the feasibility to process mixed feedstocks. It was firstly demonstrated in a US patent that the two or more feedstocks, including softwood, hardwood, grass, agricultural residues, and byproducts, can be combined for IL pretreatment with equivalent sugar conversion in comparison with single feedstocks [54]. Shi et al. evaluated the efficiency of feedstock blending along with the densification coupled with IL pretreatment to address the issues of feedstock diversity and compositional variations [55]. The IL 1-ethyl-3-methylimmidizolium acetate can process mixtures of pine, eucalyptus, switchgrass, and corn stover (in 1:1:1:1 ratios) and result in fast saccharification by reaching 90% digestibility within 24 h, which is comparable to any single feedstock type among the four starting biomass materials [52, 55]. A continuation study was further performed to investigate the IL pretreatment of the same mixture of four biomass in both flour and pellet formats, in comparison with dilute acid (DA) and soaking in aqueous ammonia (SAA) pretreatment methods, for simultaneous saccharification and fermentation into advanced biofuel isopentenol [26]. Their results show significant variations on the chemical composition, crystallinity, and enzymatic digestibility of the pretreated feedstock across the three different pretreatment technologies studied. IL pretreatment liberated the highest sugar titers from mixed biomass either in flour or pellets and is capable of handling mixed feedstocks with equal efficiency, and thus outperformed DA and SAA pretreatment methods which are more effective in pretreating herbaceous biomass feedstock and less effective in woody biomass for the mixed feedstock utilization. The high sugar production from IL process in turn led to the highest isopentenol titers in fermentation as compared to DA and SAA pretreatments.

While these three studies focused on the blends of various feedstock types that are agriculturally derived, researchers also looked into the utilization of the municipal solid waste as blending agent with lignocellulose to provide lower cost of biorefinery feedstock inputs [51]. The MSW/ corn stover blends (ratio varying from 1:1 to 1:9 on the dry weight basis) went through two types of IL pretreatment for sugar conversion, one is pretreatment by IL 1-ethyl-3-methylimidazolium acetate followed by enzymatic hydrolysis, and another is enzyme-free acidolysis in IL 1-ethyl-3-methylimidazolium chloride with addition of mineral acid. Both processes show promising sugar conversion with glucose yield over 80% and xylose yield over 75%, suggesting the great potential to use MSW for biofuel production while maintaining performance and lowering cost.

Since the data from these four studies of biomass blends were obtained at low solid loading and milliliter level of operations, which cannot be directly transferred to industrially relevant scales, Li et al. performed the process scale-up and integration of IL1-ethyl-3-methylimidazolium acetate pretreatment on herbaceous (switchgrass) and woody (eucalyptus) blends (1:1 ratio) by 30-fold at 10% solid loading [50]. In comparison with single feedstocks, this biomass blend recovered similar yields of glucan, xylan, and lignin as switchgrass and eucalyptus at 6-L scale operation. The pretreated mixed feedstock was further enzymetically hydrolyzed at 2-L scale with 96% sugar yield [50, 56]. Additionally, the same group also investigated the scale-up of IL acidolysis using 1-ethyl-3-methylimidazolium chloride and mineral acid on MSW/ corn stover blends and obtained sugar conversion yields that are comparable to small-scale studies [51, 53, 57]. These results indicate that mixed feedstocks, either agriculturally derived or MSW blended, are viable and valuable resources to consider when assessing biomass availability and affordability demands of the biorefineries. These initial scale-up evaluations demonstrate that IL-based pretreatment is feedstock agnostic, and there is no fundamental issues in terms of performance associated with the larger operations. This early-stage, 6-L scale-up process development integrates the unit operations of pretreatment, homogenization, continuous washing/separation, and product recovery for simplified feedstock handling, reduced water consumption and mitigation of IL inhibition, all of which can be further connected with downstream microbial fermentation for advanced biofuel production.

A few studies have examined the impact of blended or mixed biomass feedstocks on sugar yields from biochemical conversion using other pretreatment technologies. Karki et al. [58] reported on the enzymatic hydrolysis of mixtures of switchgrass and tall wheatgrass following dilute-acid and aqueous ammonia pretreatments. Switchgrass and tall wheatgrass were similar in composition before and after dilute-acid pretreatment, although tall wheatgrass had significantly higher glucose yields from enzymatic hydrolysis. Mixtures of the two species produced glucose yields that were higher than switchgrass and lower than tall wheatgrass following dilute-acid pretreatment and enzymatic hydrolysis. This study also demonstrated hydrolysis yields for mixtures could be predicted based on results of the individual components.

Brodeur-Campbell et al. [59] reported on the effects of biomass mixtures on sugar recovery from combined dilute-acid pretreatment and enzymatic hydrolysis. Aspen, a hardwood species that is suitable for efficient biochemical processing, was chosen as a model species in this study. Balsam, representing a high-lignin, softwood species, and switchgrass, a herbaceous energy crop with high ash, were chosen for comparative studies using 1:1 mixtures of aspen:balsam and aspen:switchgrass. No synergistic or antagonistic effects were identified in this study for three different pretreatment severities and three enzyme loadings examined. Again, total sugar recoveries for mixtures could be predicted by linear interpolation (±4%) from sugar yields of the pure biomass species. Similarly, Wolfrum et al. examined the effect of blending combined with densification on sugar yields from blends of corn stover, switchgrass, and *Miscanthus* biomass feedstocks [60, 61]; results showed the pelleting had a slightly positive, although not significant effect on total sugar yield. As in the previous studies, sugar yields could be predicted with reasonable accuracy from knowledge of the pure biomass feedstocks.

These studies demonstrate the efficient conversion of blended feedstocks to fermentable sugars and highlight the great potential for blending to expand the available resources for biofuel production. Biomass blending strategy certainly provides equivalent conversion performance as compared with single feedstock, in addition to its economic benefits toward the future development.

3.2. Impact of biomass densification

Lignocellulosic biomass with low bulk and energy density requires relatively high energy to transport, store, and distribute the feedstock from the field to the biorefinery gate for conversion, and the loose ground materials often pose problems of material feeding and handling in the reactors. Biomass densification typically involves exposing the biomass to elevated pressures and temperatures to remove excess water and compress the biomass. This process acts as a mild thermochemical pretreatment and can also impact the composition and structure of the biomass [55]. Several densification forms have been demonstrated recently, and this section reviews and compares the impact of densification on various thermochemical pretreatment.

3.2.1. Pellets

Recently, a growing body of literature has assessed the impact of pelletizing herbaceous and woody materials on the bioconversion process when combined various pretreatment technologies. Pelleted biomass is produced from raw, ground material that is conditioned with heat and/or moisture, compacted, and extruded through a die [2, 3]. It has been shown that activation of the natural binders in biomass, such as lignin, through combined moisture and temperature effects during the process of densification is key to the development of particle-particle bonding that is required for durability [9]. The extent of lignification contributes significantly to biomass recalcitrance [4], and lignin alteration during the process of densification may impact biomass reactivity to pretreatment and enzymatic hydrolysis [12].

Published reports evaluating the impact of pelletization on the bioconversion of corn stover, sorghum, wheat straw, big bluestem grasses, softwood, and switchgrass have shown positive trends using lower severity alkaline pretreatment. Similar or slightly higher sugar release and ethanol yield were observed in the pelleted format when compared to the nonpelleted format after pretreatment and enzymatic hydrolysis. Guragain et al. [62] evaluated the effect of alkaline pretreatment on sugar release and ethanol production in pelleted and nonpelleted wheat straw, corn stover, big bluestem, and sorghum stalk; mass recovery after alkali pretreatment increased by 14, 11, 2, and 5%, respectively, compared to nonpelleted biomass. Volumetric sugar production increased for all feedstocks except sorghum, although final sugar yields were not significantly different between the pelleted and non-pelleted biomass. Nahar and Pryor [63] reported that combining pelleting and pretreatment with SAA treatment reduced cellulase loading to achieve 90% glucose yield at 10 FPU per g glucan in switchgrass. Pelleting the switchgrass did not affect the feedstock composition compared to the non-pelleted switchgrass. Hoover et al. [64] evaluated the effect of physical properties resulting from pelleting AFEX-pretreated corn stover. Comparing grind size, die speed, and preheating on pellet properties on the sugar release after enzymatic hydrolysis showed the following: Die speed had no effect on sugar yield, while a larger grind size (4 mm vs. 6 mm) had a similar or lower effect on sugar yields after enzymatic hydrolysis. Overall, pelleting AFEX-treated biomass increased or had no effect on sugar yields at low or high ammonia loadings. Bals et al. [65] tested the susceptibility of AFEX-treated, corn stover pellets to enzymatic hydrolysis at high solids loading (18–36%). Pelletization slightly increased the initial rate of hydrolysis relative to raw biomass, enabled mixing, and resulted in higher glucose yields at 18% solids loading relative to unpelletized biomass (68% vs. 61%). Similarly, Rijal et al. [66] demonstrated that DA-treated switchgrass did not impact glucose yield in the finer ground, and pelleted materials compared to the native material. However, glucose yields from aqueous ammonia pretreatment, followed by enzymatic hydrolysis, were higher for both powder and pelleted materials compared to the native material. Glucose yield for the DA- and SAA-treated and pelleted switchgrass was 98 and 79%, respectively.

Theerarattananoon et al. [67] evaluated the impact of pelleting conditions on sugar release and chemical composition of corn stover, wheat straw, sorghum stalk, and big bluestem grass. Dilute-acid pretreatment and subsequent enzymatic hydrolysis increased glucan content in the pretreated solids compared to the nonpelleted companion feedstock for corn stover, wheat straw, and big bluestem prairie grass. Glucan content in the pretreated pelleted sorghum stalks was slightly less than nonpelleted sorghum stalks. Enzymatic hydrolysis results suggested that pelleting increased cellulose yield for all feedstocks. While wheat straw had the highest cellulose yield (94.1%), Ray et al. [68] evaluated the impact of densification on the bioconversion of corn stover, ground and pelleted format. The low solids dilute-acid pretreatment resulted in higher theoretical ethanol yields from the pelleted versus the non-pelleted format of 84 and 69%, respectively. Pelleted and ground corn stover was pretreated at higher solids loading at multiple pretreatment severities and showed slightly increased reactivity across three of the five severities tested.

Similar to other pretreatment technologies, conversion of biomass feedstocks with low energy and bulk density using ILs is not an economic process. To address this issue, Shi et al. investigated and compared the IL pretreatment of switchgrass, lodgepole pine, corn stover, and eucalyptus in both flour and densified pellet formats with 1-ethyl-3-methylimmidizolium acetate at 160°C and 10% solid loading for 3 h [55]. There was no significant difference between the physio-chemical properties, that is, composition and cellulose crystallinity, of the pretreated flour and pellets. The subsequent enzymatic digestibility results show that sugar yields from both formats reach 90% conversion within 24 h, suggesting densifying a wide range of feedstocks may be a competitive solution with no significant adverse impacts, provided that they are coupled with the appropriate conversion technology. Although significant improvements in terms of IL cost and recycling need to be resolved before this technology is commercially viable, biomass densification certainly provides the economic benefits toward the future development.

3.2.2. ComPAKo briquettes

Additional studies have been performed quantifying ethanol yields from densified AFEX-pretreated corn stover, switchgrass, and prairie cordgrass. Rijal et al. [69] studied the effect of initial particle size (2, 4, 8 mm) and densification on ethanol production. They employed a novel densification method, ComPAKo that uses a gear, mesh system to produce compacted biomass briquettes (1 inch × 0.5 inch × 4 inch). The ComPAKo equipment operates at lower temperatures (30–60°C) and pressures, lowering energy costs. Also, the capital investment for ComPAKo is less than half that of a pellet mill. The bulk density of the briquettes ranged between 380 and 460 kg/m^3 with moisture content of 11–15%. The AFEX-pretreated biomass was used for both simultaneous saccharification fermentation (SSF) at 4% glucan loading and separate hydrolysis with fermentation (SHF) at 1% glucan loading with an enzyme loading of 15 FPU and 64 CBU/g of glucan for hydrolysis. Results demonstrated that 2-mm densified corn stover briquettes yielded approximately 5% higher ethanol than 8-mm densified material. They also showed that grinding the densified 8-mm briquettes to 2 mm prior to SSF studies did not result in significant ethanol yield differences, but the 2-mm densified corn stover showed 4% higher yield than post-grinding the 8-mm briquettes prior to hydrolysis. The ethanol yields from the SSF did not significantly differ for the AFEX-treated corn stover or switchgrass when compared to the densified AFEX-treated material, but they noted a negative impact for the prairie cordgrass densified material by 16%. This was attributed to the observation that densified AFEX-treated prairie cordgrass was stronger and did not break apart during mixing and hydrolysis. Upon grinding of the AFEX-treated densified prairie cordgrass, the ethanol yields were 35% less than with the nondensified material, indicating that prairie cordgrass densification is not beneficial. The negative impacts of densification on AFEX-treated prairie cordgrass may be attributed to amount or structure of lignin in this feedstock. Sugar yields during SHF were not impacted for the corn stover or switchgrass densified material, but they were significantly diminished for prairie cordgrass. However, when comparing SHF ethanol yields, switchgrass densified material gave significantly lower yields, while yields from densified corn stover were only slightly higher, but the densified prairie cordgrass produced higher yields than either corn stover or switchgrass. The results support AFEX as an effective pretreatment technology for ComPAKo densification processes, thus reducing the need for additional particle size reduction for effective hydrolysis. These technologies however will produce different sugar and ethanol yields dependent on feedstock choice and subsequent hydrolysis and fermentation. AFEX-treated densified corn stover yielded the better quality briquettes of the three biomass types tested in the AFEX-treated, ComPAKo processes.

Biersbach et al. [70] also studied ethanol yields from briquettes of AFEX-treated corn stover, switchgrass, and prairie cordgrass and assessed the impact enzyme loading has during SSF and SHF, and they tested storage of these densified materials. They used the ComPAKo method to produce briquettes of 1–2 cm with a bulk density range between 380 and 460 kg/m^3 and moisture content of 11–15%. Ethanol yield was improved for all AFEX-pretreated biomass tested regardless of enzyme dose or fermentation regimen (SSF or SHF). They found that ComPAKo densified AFEX-treated biomass did not consistently have an impact on ethanol yields in most of the conditions they tested, but in three of the corn stover tested conditions, densification increased ethanol yields up to 13%. For experiments using switchgrass and prairie cordgrass densified material, the densification caused 7 and 22% reduction in ethanol yields, respectively. They concluded that the ethanol yield differences of the various feedstocks could be attributed to the glucan content and pretreatment efficiency. They also found that the higher enzyme dose (15 FPU Spezyme CP, 64 CBU Novozyme 188) during enzymatic hydrolysis generally increased ethanol yields in the range of 18–317% for SHF and 28–62.5% for SSF, dependent on the feedstock. When the AFEX-treated densified briquettes were stored for 6 months, there was an increase in ethanol yield of 12–17%, with the exception of the prairie cordgrass which gave a 55% reduction in ethanol yield when SHF was performed, but not with SSF.

3.2.3. *Extrusion pelleting*

Extrusion pelleting is another densification technology that Sundaram and Muthukumarappan [71] used to evaluate AFEX-pretreated corn stover, switchgrass, and prairie cordgrass. They tested the effects of various parameters during laboratory-scale single-screw extrusion pelleting and the impact of those parameters on pellet bulk density, hardness, and sugar recovery from enzymatic hydrolysis. The parameters tested included moisture content (5, 10, and 15%), hammer mill particle size (2, 4, and 8 mm), and extrusion barrel temperature (75, 100, and 125°C). In general, the bulk density of the AFEX-treated biomass particles decreased as the particle size increased, and the bulk density increased with increasing the moisture content. Similar to other studies, the AFEX-pretreated material increased the pellet bulk density for each feedstock (650.6 kg/m^3 for corn stover, 680.1 kg/m^3 for prairie cord grass, and 627.7 kg/m^3 for switchgrass) compared to untreated material (453.0, 463.2, and 433.9 kg/m^3, respectively). The moisture content significantly impacted pellet bulk density with higher moisture content causing an increase. However, particle size of AFEX-pretreated material had no impact on pellet bulk density, but it inversely affected the untreated pellets; likewise extrusion temperature did not significantly impact AFEX-pretreated pellet bulk density but did negatively impact the untreated material. Pellet hardness was also determined for AFEX-pretreated pellets of corn stover, switchgrass, and prairie cordgrass with maximum hardness values of 2342.8, 2424.3, and 1298.6 N for each feedstock, respectively. The hardness of the AFEX-treated pellets was not significantly different at different barrel temperatures, indicating that good quality pellets can be achieved at 75°C, thus reducing costs. Moisture content correlated with pellet hardness for treated and untreated materials which is typical for extrusion pelleting and in combination with moisture content, particle size impacted pellet hardness, with 2 and 4 mm particles yielding maximum hardness. The percent glucose released form AFEX-pretreated pellets ranged from 88.9 to 94.9% for corn stover, 90.1 to 94.9% for prairie cord grass, and 87.0 to 92.9% for switchgrass. These glucose yields were 1.6, 2.1 and 2.3 fold higher than those from untreated pellets, respectively and xylose yields increased 1.6, 1.4, and 2.0 fold for AFEX-treated pellets compared to untreated pellets, respectively. Neither glucose yields nor xylose yields were significantly impacted by the extrusion temperatures or the particle sizes tested during extrusion pelleting, again indicating a low temperature of 75°C can be used to achieve quality pellets for conversion. Finally, the results show the extrusion pelleting process can be performed at low temperatures and larger particle size without significantly impacting sugar yields, thus reducing pellet processing costs.

These key findings suggest that densification of biomass does not negatively affect its composition and downstream conversion and may actually increase bioconversion or perhaps reduce the requirements for a given conversion level. However, many of these evaluations involving herbaceous feedstocks were conducted under low-solids, non-mixed, and batch conditions, which make extrapolations to more process-relevant conditions difficult.

4. Summary

This book chapter evaluates the potential of preprocessing options, that is, blending and densification, for uniform, consistent, quality-controlled, and cost-effective feedstock development, and reviews their impacts on feedstock supply chain logistic and downstream conversion performance. The use of blended and densified feedstocks in conversion pathways instead of conventionally ground biomass from a single source addresses several challenges in the current biomass supply chain, including availability, transportation, storage, cost, quality, and supply variability. Review and summary of recent research further demonstrate that a biomass blending strategy provides an efficient way to meet quality and conversion performance specifications in comparison with the conversion of single feedstock. Densified formats can perform equivalent to non-densified formats in terms of sugar and ethanol biochemical conversion performance. Both blending and densification provide great promise to enable more cost-effective downstream processing.

Acknowledgements

This work is supported by the U.S. Department of Energy, Office of Energy Efficiency and Renewable Energy, Bioenergy Technologies Office, under DOE Idaho Operations Office Contract DE-AC07-05ID14517. The authors would like to extend thanks to AAE for thoughtful review of this chapter.

Author details

Allison E. Ray[1*], Chenlin Li[1], Vicki S. Thompson[1], Dayna L. Daubaras[1], Nicholas J. Nagle[2] and Damon S. Hartley[1]

*Address all correspondence to: allison.ray@inl.gov

1 Idaho National Laboratory, Idaho Falls, ID, USA

2 National Renewable Energy Laboratory, Golden, CO, USA

References

[1] USEPA, March 17, 2016 [cited 2016 July 14]; Available from: https://www.epa.gov/renewable-fuel-standard-program.

[2] EISA, *Energy Independence and Security Act (EISA) of 2007*. 2007 January 4, 2007 [cited 2012 September 27, 2012]; Available from: http://www.gpo.gov/fdsys/pkg/BILLS-110hr6enr/pdf/BILLS-110hr6enr.pdf.

[3] USDOE, *2016 Billion-ton report: advancing domestic resources for a thriving bioeconomy*, in M.H. Langholtz, B.J. Stokes, and L.M. Eaton, Editors, Economic Availability of Feedstocks, vol 1. 2016, Oak Ridge National Laboratory: Oak Ridge, TN, USA.

[4] Hess, J.R., et al., *Commodity-scale production of an infrastructure-compatible bulk solid from herbaceous lignocellulosic biomass*, in Uniform-Format Bioenergy Feedstock Supply System Design Report. 2009, Idaho National Laboratory: Idaho Falls, ID.

[5] Hess, J.R., C.T. Wright, and K.L. Kenney, *Cellulosic biomass feedstocks and logistics for ethanol production*. Biofuels, Bioproducts and Biorefining, 2007. **1**(3): pp. 181–190.

[6] Kenney, K.L., et al., *Understanding biomass feedstock variability*. Biofuels, 2013. **4**(1): pp. 111–127.

[7] Lamers, P., et al., *Strategic supply system design—a holistic evaluation of operational and production cost for a biorefinery supply chain*. Biofuels, Bioproducts and Biorefining, 2015. **9**(6): pp. 648–660.

[8] Rentizelas, A.A., I.P. Tatsiopoulos, and A. Tolis, *An optimization model for multi-biomass tri-generation energy supply*. Biomass and Bioenergy, 2009. **33**(2): pp. 223–233.

[9] Banerjee, S., et al., *Commercializing lignocellulosic bioethanol: technology bottlenecks and possible remedies*. Biofuels, Bioproducts and Biorefining, 2010. **4**(1): pp. 77–93.

[10] Oke, M.A., M.S.M. Annuar, and K. Simarani, *Mixed feedstock approach to lignocellulosic ethanol production—prospects and limitations*. Bioenergy Research, 2016: pp. 1–15.

[11] Argo, A.M., et al., *Investigation of biochemical biorefinery sizing and environmental sustainability impacts for conventional bale system and advanced uniform biomass logistics designs*. Biofuels, Bioproducts and Biorefining, 2013. **7**(3): pp. 282–302.

[12] Jacobson, J., et al., *Techno-economic analysis of a biomass depot*. 2014, Idaho National Laboratory: Idaho Falls, ID.

[13] Searcy, E. and J.R. Hess, Uniform-Format Feedstock Supply System: A Commodity-Scale Design to Produce an Infrastructure-Compatible Biocrude from Lignocellulosic Biomass. 2010, Idaho National Laboratory: Idaho Falls, ID.

[14] Hansen, J., J. Jacobson, and M. Roni. Quantifying Supply Risk at a Cellulosic Biorefinery. in 33rd International Conference of the System Dynamics Society. 2015. Cambridge, MA, USA.

[15] Sultana, A. and A. Kumar, *Optimal configuration and combination of multiple lignocellulosic biomass feedstocks delivery to a biorefinery*. Bioresource Technology, 2011. **102**(21): pp. 9947–9956.

[16] Langholtz, M., et al., *Climate risk management for the U.S. cellulosic biofuels supply chain*. Climate Risk Management, 2014. **3**: pp. 96–115.

[17] Emerson, R., et al., *Drought effects on composition and yield for corn stover, mixed grasses, and Miscanthus as bioenergy feedstocks*. Biofuels, 2014. **5**(3): pp. 275–291.

[18] Maung, T.A., et al., *The logistics of supplying single vs. multi-crop cellulosic feedstocks to a biorefinery in southeast North Dakota*. Applied Energy, 2013. **109**: pp. 229–238.

[19] Tumuluru, J.S., et al., *A review of biomass densification systems to develop uniform feedstock commodities for bioenergy application*. Biofuels, Bioproducts and Biorefining, 2011. **5**(6): pp. 683–707.

[20] Richard, T.L., *Challenges in scaling up biofuels infrastructure*. Science, 2010. **329**(5993): pp. 793–796.

[21] Williams, C.L., et al., *Sources of biomass feedstock variability and the potential impact on biofuels production*. Bioenergy Research, 2016. **9**(1): pp. 1–14.

[22] Li, C., et al., *Impact of feedstock quality and variation on biochemical and thermochemical conversion*. Renewable and Sustainable Energy Reviews, 2016. **65**: pp. 525-536..

[23] Tumuluru, J.S., et al., *Formulation, pretreatment, and densification options to improve biomass specifications for co-firing high percentages with coal*. Industrial Biotechnology, 2012. **8**(3): pp. 113–132.

[24] Hill, L.D., Grain Grades and Standards: Historical Issues Shaping the Future. 1990, Chicago, IL, USA: University of Illinois Press.

[25] Sami, M., K. Annamalai, and M. Wooldridge, *Co-firing of coal and biomass fuel blends*. Progress in Energy and Combustion Science, 2001. **27**(2): pp. 171–214.

[26] Shi, J., et al., *Impact of pretreatment technologies on saccharification and isopentenol fermentation of mixed lignocellulosic feedstocks*. BioEnergy Research, 2015. **8**(3): pp. 1004–1013.

[27] Boavida, D., et al., *A study on coal blending for reducing NOx and N_2O levels during fluidized bed combustion*. Clean Air, 2012. **5**: pp. 175–191.

[28] Reddy, D.V. and N. Krishna, *Precision animal nutrition: a tool for economic and eco-friendly animal production in ruminants*. Livestock Research for Rural Development, 2009. **21**(3): p. 36.

[29] USDOE, Bioenergy Technologies Office Multi-Year Program Plan. 2016: Washington, D.C.

[30] Crawford, N.C., et al., *Evaluating the pelletization of "pure" and blended lignocellulosic biomass feedstocks*. Fuel Processing Technology, 2015. **140**: pp. 46–56.

[31] Kenney, K.L., et al., Feedstock Supply System Design and Economics for Conversion of Lignocellulosic Biomass to Hydrocarbon Fuels. Conversion Pathway: Biological Conversion of Sugars to Hydrocarbons: The 2017 Design Case. 2013, Idaho National Laboratory: Idaho Falls, ID.

[32] Thompson, V.S., et al., *Assessment of municipal solid waste as a blend feedstock to lower biomass feedstock costs*, in 36th Symposium on Biotechnology for Fuels and Chemicals. 2014: Clearwater Beach, FL.

[33] USDOE, *Bioenergy knowledge discovery framework*. 2016 [cited 2015 November 4, 2015]; Available from: www.bioenergykdf.net.

[34] USDOE, *Bioenergy knowledge discovery framework*. 2015 [cited 2015 November 4, 2015]; Available from: www.bioenergykdf.net.

[35] Lamers, P., et al., *Techno-economic analysis of decentralized biomass processing depots*. Bioresource Technology, 2015. **194**: pp. 205–213.

[36] Kaliyan, N. and R.V. *Morey, Natural binders and solid bridge type binding mechanisms in briquettes and pellets made from corn stover and switchgrass*. Bioresource Technology, 2010. **101**(3): pp. 1082–1090.

[37] Larsson, S.H., et al., *High quality biofuel pellet production from pre-compacted low density raw materials*. Bioresource Technology, 2008. **99**(15): pp. 7176–7182.

[38] Mani, S., L.G. Tabil, and S. Sokhansanj, *Effects of compressive force, particle size and moisture content on mechanical properties of biomass pellets from grasses*. Biomass and Bioenergy, 2006. **30**(7): pp. 648–654.

[39] Mani, S., L.G. Tabil, and S. Sokhansanj, *Specific energy requirement for compacting corn stover.* Bioresource Technology, 2006. **97**(12): pp. 1420–1426.

[40] Greene, J., *Q1 2015 Recap: Global Wood Pellet Demand Creates US Opportunities*. 2015 April 14, 2015 [cited 2016 September 11, 2016]; Available from: http://blog.forest2market.com/wood-pellet-demand-creates-opportunity.

[41] Bals, B., et al., *Ammonia fiber expansion (AFEX) treatment of eleven different forages: Improvements to fiber digestibility in vitro*. Animal Feed Science and Technology, 2010. **155**(2–4): pp. 147–155.

[42] Himmel, M.E., et al., *Biomass recalcitrance: engineering plants and enzymes for biofuels production*. Science, 2007. **315**(5813): pp. 804–807.

[43] Weiss, N.D., J.D. Farmer, and D.J. Schell, *Impact of corn stover composition on hemicellulose conversion during dilute acid pretreatment and enzymatic cellulose digestibility of the pretreated solids*. Bioresource Technology, 2010. **101**(2): pp. 674–678.

[44] Chen, Y., et al., *Understanding of alkaline pretreatment parameters for corn stover enzymatic saccharification*. Biotechnology for Biofuels, 2013. **6**(1): pp. 1–10.

[45] Chen, X., et al., *A highly efficient dilute alkali deacetylation and mechanical (disc) refining process for the conversion of renewable biomass to lower cost sugars*. Biotechnology for Biofuels, 2014. **7**(1): pp. 1–12.

[46] Tao, L., et al., *Process and technoeconomic analysis of leading pretreatment technologies for lignocellulosic ethanol production using switchgrass*. Bioresource Technology, 2011. **102**(24): pp. 11105–11114.

[47] Li, C.L., et al., *Comparison of dilute acid and ionic liquid pretreatment of switchgrass: Biomass recalcitrance, delignification and enzymatic saccharification*. Bioresource Technology, 2010. **101**(13): pp. 4900–4906.

[48] Passos, H., M.G. Freire, and J.A.P. Coutinho, *Ionic liquid solutions as extractive solvents for value-added compounds from biomass*. Green Chemistry, 2014. **16**(12): pp. 4786–4815.

[49] Brandt, A., et al., *Deconstruction of lignocellulosic biomass with ionic liquids*. Green Chemistry, 2013. **15**(3): pp. 550–583.

[50] Li, C., et al., *Scale-up of ionic liquid-based fractionation of single and mixed feedstocks*. BioEnergy Research, 2015: pp. 1–10.

[51] Sun, N., et al., *Blending municipal solid waste with corn stover for sugar production using ionic liquid process*. Bioresource Technology, 2015. **186**: pp. 200–206.

[52] Li, C.L., et al., *Comparing the recalcitrance of eucalyptus, pine, and switchgrass using ionic liquid and dilute acid pretreatments*. Bioenergy Research, 2013. **6**(1): pp. 14–23.

[53] Li, C., et al., *Scale-up and process integration of sugar production by acidolysis of municipal solid waste/corn stover blends in ionic liquids*. Biotechnol Biofuels 2017. In Press.

[54] Singh, S., et al., *Mixed feedstocks processing using an ionic liquid*. 2013, US20130183739 A1.

[55] Shi, J., et al., *Impact of mixed feedstocks and feedstock densification on ionic liquid pretreatment efficiency*. Biofuels, 2013. **4**(1): pp. 63–72.

[56] Tanjore, D., et al., *Resolving process scale-up issues of ionic liquid pretreatment and saccharification of biomass to monomeric sugars*, in 35th Symposium on Biotechnology for Fuels and Chemicals. 2013: Portland, Oregon.

[57] Li, C., et al., *Scale-up and process integration of sugar production by acidolysis of single and mixed feedstocks in ionic liquids*, in 37th Symposium on Biotechnology for Fuels and Chemicals. 2015: San Diego, CA.

[58] Karki, B., N. Nahar, and S.W. Pryor, *Enzymatic hydrolysis of switchgrass and tall wheatgrass mixtures using dilute sulfuric acid and aqueous ammonia pretreatments*. Biological Engineering Transactions, 2011. **3**(3): pp. 163–171.

[59] Brodeur-Campbell, M., J. Klinger, and D. Shonnard, *Feedstock mixture effects on sugar monomer recovery following dilute acid pretreatment and enzymatic hydrolysis*. Bioresource Technology, 2012. **116**: pp. 320–326.

[60] Wolfrum, E., et al., *The effect of feedstock densification on structural sugar release and yield in and biofuel feedstock and feedstock blends*, in Symposium on Biotechnology for Fuels and Chemicals. 2015: San Diego, CA.

[61] Wolfrum, E.J., et al., *The effect of biomass densification on structural sugar release and yield in biofuel feedstock and feedstock blends*. BioEnergy Research, 2017. Accepted.

[62] Guragain, Y.N., et al., *Evaluation of pelleting as a pre-processing step for effective biomass deconstruction and fermentation*. Biochemical Engineering Journal, 2013. **77**: pp. 198–207.

[63] Nahar, N. and S.W. Pryor, *Reduced pretreatment severity and enzyme loading enabled through switchgrass pelleting*. Biomass and Bioenergy, 2014. **67**: pp. 46–52.

[64] Hoover, A.N., et al., *Effect of pelleting process variables on physical properties and sugar yields of ammonia fiber expansion pretreated corn stover.* Bioresource Technology, 2014. **164**: pp. 128–135.

[65] Bals, B.D., et al., *Enzymatic hydrolysis of pelletized AFEX™-treated corn stover at high solid loadings.* Biotechnology and Bioengineering, 2014. **111**(2): pp. 264–271.

[66] Rijal, B., et al., *Combined effect of pelleting and pretreatment on enzymatic hydrolysis of switchgrass.* Bioresource Technology, 2012. **116**: pp. 34–41.

[67] Theerarattananoon, K., et al., *Effects of the pelleting conditions on chemical composition and sugar yield of corn stover, big bluestem, wheat straw, and sorghum stalk pellets.* Bioprocess and Biosystems Engineering, 2012. **35**(4): pp. 615–623.

[68] Ray, A.E., et al., *Effect of pelleting on the recalcitrance and bioconversion of dilute-acid pretreated corn stover under low- and high-solids conditions.* Biofuels, 2013. **4**(3): pp. 271–284.

[69] Rijal, B., et al., *Effect of initial particle size and densification on afex-pretreated biomass for ethanol production.* Applied Biochemistry and Biotechnology, 2014. **174**(2): pp. 845–854.

[70] Biersbach, G., et al., *Effects of enzyme loading, densification, and storage on AFEX-pretreated biomass for ethanol production.* Applied Biochemistry and Biotechnology, 2015. **177**(7): pp. 1530–1540.

[71] Sundaram, V. and K. Muthukumarappan, *Impact of AFEX™ pretreatment and extrusion pelleting on pellet physical properties and sugar recovery from corn stover, prairie cord grass, and switchgrass.* Applied Biochemistry and Biotechnology, 2016. **179**(2): p. 202–219.

10

Reaction Behaviors of Bagasse Modified with Phthalic Anhydride in 1-Allyl-3-Methylimidazolium Chloride with Catalyst 4-Dimethylaminopyridine

Hui-Hui Wang, Xue-Qin Zhang, Yi Wei and Chuan-Fu Liu

Abstract

The modification of lignocellulose with cyclic anhydrides could confer stronger hydrophilic properties to lignocellulose, which could be used in many industrial fields. To elucidate the modification mechanism of lignocellulose, bagasse was phthalated comparatively with its three main components in 1-allyl-3-methylimidazolium chloride (AmimCl) using 4-dimethylaminopyridine as catalyst and phthalic anhydride as acylation reagent in the present study. From FT-IR and 2D HSQC analyses, the skeleton of bagasse and the fractions were not significantly changed during phthalation in AmimCl. 2D HSQC results suggested that the reactive hydroxyls in bagasse were partially phthalated, and the reactivity of the hydroxyls in anhydroglucose units followed the order C-6 > C-2 > C-3. Similarly, the reactivity order of hydroxyls in anhydroxylose units was C-2 > C-3. For lignin, the predominant diesterification occurred during the homogeneous modification, and both aliphatic and aromatic hydroxyls were phthalated. The reactivity order of phenolic hydroxyls was S-OH > G-OH > H-OH, which was distinct from that without catalyst. In addition, it was found that the thermal stability of phthalated bagasse was affected by the disruption of cellulose crystallinity and the degradation of components. The thermal stability of the phthalated bagasse decreased upon chemical modification and regeneration.

Keywords: bagasse, phthalic anhydride, esterification, reaction behavior, 2D HSQC NMR, ^{31}P NMR, AmimCl

1. Introduction

Nowadays, the growing environmental concerns resulting from petroleum-based plastic materials and excessive use of fossil fuel have driven a strong global interest in renewable bio-based polymers and composites derived from natural resources [1, 2]. Biomass resource is one of the most abundant, inexpensive, and currently underutilized products from agricultural and biorefinery industries [3]. Biomass resources, especially agricultural residues with great renewability, biocompatibility, and biodegradability, are directing the development of next generation of chemicals, materials, energy, and processes [4]. Therefore, it could be particularly advisable to shift society's dependence away from fossil resources to biomass resources.

Bagasse, sugar industry residues, is mainly composed of lignocellulose, which represents one of the most abundant renewable resources on earth. Chemical modification represents an efficient process to introduce new functionalities into lignocellulose. Because of the poor dissolubility of lignocellulose in common molecular solvent, the heterogeneous modification has been studied for decades. However, heterogeneous modification is always accomplished at low efficiency and nonuniformity, and accompanied with the occurrence of side reactions [5, 6]. Even worse, the solvent or reaction medium for heterogeneous modification could not be recovered after modification [7], which caused environmental pollution and wastage of resources. It is a never-ending endeavor to explore homogeneous reaction systems to compensate the deficiency of heterogeneous systems.

Fortunately, it has been found that several dual solvent systems and ionic liquids (ILs) could dissolve lignocellulose [8]. Compared with other solvent systems, ILs exhibit negligible vapor pressure, theoretically making infinite recycling possible. In addition, the high electrochemical thermal stability, nonflammability, and designability made ILs widely used in diverse fields, such as solvent for synthesis and catalysis, separation, analytical chemistry, and biomass refinery [9–11]. The bridge between lignocellulose and ILs were built by Swaltloshi et al., who reported the nonderivatizating dissolution of cellulose in BmimCl [12]. In 2007, it was reported that woods with different hardness could be readily dissolved in various imidazium-based ILs under mild conditions [13, 14], which opened a new window of opportunities in homogeneous functionalization of lignocellulose. Since then, the chemical modification of lignocellulose in ILs has been expanding fast. Due to the existence of abundant hydroxyls in lignocellulose, the modification of lignocellulose with carboxylic anhydride, acid chloride, and isocyanides could be achieved using ILs as reaction media [15].

The chemical modification of lignocellulose with linear carboxylic acid anhydrides or acid chloride can produce the corresponding carboxylic acid or HCl as a by-product. However, the modification of lignocellulose with dicarboxylic acid anhydride is an effective process and can attach carboxyl groups onto lignocellulose. The attachment of carboxyl group can greatly improve the compatibility between modified lignocellulose and other polymer materials by producing new chemical linkages and the effect of hydrogen bonds [16]. Besides, the introduction of carbonyl group could also increase the hydrophilicity [17], achieve ion exchange [18], and remove heavy mental ion [19]. Our previous works showed that the homogeneous chemical modification of the isolated cellulose from bagasse with succinic anhydride and phthalic anhydride without catalyst could be achieved, and the optimal reaction conditions (reaction temperature 90°C and reaction time 90 min) for preparing modified cellulose with maximum degree of substitution (DS) were also screened [20, 21]. In 2010, Li et al. reported chemical modification of cellulose in BmimCl with catalyst DMAP could remarkably improve the DS of modified products [22]. However, the structural parameters of lignocellulosic derivatives were complex and difficult to control due to the complicated mixture of cell wall components. Therefore, it is very essential to elucidate the modification mechanism of lignocellulose to produce tunable materials directly from lignocellulose, although some simple investigations have been published. Especially, the application of catalyst DMAP could obviously increase the degree of substitution of lignocellulose modified with cyclic anhydride, which is beneficial to produce polymer composites based on esterified lignocellulose. As far as the authors are aware, there have been no reports of the detailed structural changes of lignocellulose during the homogeneous phthalation with catalyst DMAP. We therefore investigated the reaction behaviors of lignocellulose during the homogenous phthalation with catalyst DMAP.

In the present study, bagasse was modified with phthalic anhydride with catalyst DMAP in AmimCl. Three main components cellulose, hemicelluloses, and lignin were isolated from bagasse, and were phthalated under the same conditions to elucidate the modification mechanism. The physicochemical properties of phthalated samples were characterized by FT-IR, CP/MAS ^{13}C NMR, ^{1}H NMR, liquid-state ^{13}C NMR, and 2D HSQC NMR. The thermal stability of the unmodified, regenerated, and modified materials was also investigated using TG.

2. Material and methods

2.1. Materials

Bagasse was obtained from a local factory (Jiangmen, China). It was dried in sunlight and then cut into small pieces. The cut bagasse was ground and screened to prepare 40–60 mesh size particles (450–900 μm). The dried ground samples were dewaxed with toluene-ethanol (2:1 v/v) and then dried in a cabinet oven with air circulation at 50°C for 24 h [15]. Ionic liquid AmimCl was purchased from the Shanghai Cheng Jie Chemical Co., Ltd. and used as received. All other chemicals used were of analytical grade and purchased from Guangzhou Chemical Reagent Factory (Guangzhou, China).

2.2. Isolation of cellulose, hemicelluloses and lignin from bagasse

Cellulose and hemicelluloses were isolated according to the previous literatures [21, 23]. Briefly, the extractive-free bagasse was delignified at 75°C for 2 h with sodium chlorite at pH 3.8–4.0, followed by the extraction with 10% NaOH with the ratio of solid to liquid at 1:20 g/mL for 10 h at room temperature (four times). The solid residues were filtered out, washed thoroughly with distilled water, then washed with ethanol and dried in an oven with air circulation at 50°C for 24 h to obtain cellulose. The filtrate from NaOH extraction (the first time) was neutralized, concentrated, and transferred into three volumes of 95% ethanol with agitation. The precipitates were filtered out, washed with 70% ethanol, and freeze-dried to obtain hemicelluloses.

Lignin was isolated from the extractive-free bagasse according to the previous literatures [24, 25]. The obtained crude lignin with high content of carbohydrates was purified as follows: dried crude lignin (1 g) was dissolved in 2 mL of acetone/water (9:1, v/v) in a beaker. The obtained lignin solution was added dropwise into 200 mL of distilled water with stirring, and the resulted suspension was further centrifuged. The obtained solid residues were washed with distilled water (thrice, total 90 mL), and then freeze-dried. The dried residues were dissolved into 2 mL of 1,2-dichloroethane/ethanol (2:1, v/v). The obtained solution was added dropwise into 200 mL of anhydrous ether in a beaker with stirring at room temperature, and the resulted suspension was further agitated for 30 min. The obtained suspension was centrifuged, and the solid residues was washed with anhydrous ether (thrice, total 90 mL) and freeze-dried.

2.3. Homogeneous phthalation of bagasse and the isolated fractions (cellulose, hemicelluloses and lignin)

The homogeneous phthalation of bagasse and its isolated fractions was performed according to the previous literature [26]. The extractive-free ground bagasse (12 g) was first finely ball-milled for 4 h in a planetary mill (Grinder BM4, China, Beijing) at 400 rpm using two 500 mL ZrO2 jars (500 g, 1 cm diameter). Cellulose isolated from bagasse was also ball-milled for 4 h under the same conditions, while hemicelluloses and lignin were directly used in the present study.

Typically, the prepared materials (0.2 g) was added to 10 g of AmimCl at room temperature under nitrogen atmosphere for 5 min and heated at 90°C under stirring for 4 h to obtain a clear solution. Phthalic anhydride (the weight ratio of phthalic anhydride to material, 4:1) and DMAP (the weight ratio of DMAP to phthalic anhydride, 5%) were added to the solution. The flask was continuously purged with nitrogen atmosphere for 5 min. The reaction was performed at 90°C for 90 min with agitation. Subsequently, the resulted solution was precipitated with ethanol (99 wt%, 200 mL) under agitation. The suspension was further stirred for 24 h. The solid residues were filtered out, washed with ethanol (three times, total 600 mL) to remove unreacted phthalic anhydride, catalyst DMAP, and AmimCl, and free-dried for further characterization.

2.4. Determination of WPG and DS

WPG [15] was calculated according to Eq. (1):

$$WPG = \frac{M_1 - M_0}{M_0} \times 100 \qquad (1)$$

where M_0 and M_1 are the vacuum-dried weights of the bagasse or fractions before and after chemical modification, respectively.

The phthalation degree of phthalated samples was evaluated by DS, according to Eq. (2):

$$DS = \frac{n_{OH'}}{n_{OH}} \times 100 \qquad (2)$$

where n_{OH}' is the substituted hydroxyls contents, and n_{OH} is the hydroxyl content of unmodified sample.

The substituted hydroxyl content of the phthalated samples was determined based on the equivalent volume of known molarity NaOH and HCl by back-titration method [21]. A known weight of the sample was dissolved in 10 mL 0.1 mol/L NaOH. The excess of NaOH was back-titrated with 0.025 mol/L HCl by using automatic potentiometric titrator, and the final pH of end titration point was set as 7.0. The titration was repeated three times, and the average value of HCl volume was used as the calculations of the substituted hydroxyl content according to Eq. (3):

$$n_{OH'} = \frac{c_1V_1 - c_2V_2}{2} \times \frac{1000}{m} \quad (3)$$

where n_{OH}' is the substituted hydroxyls contents, m (g) is the dry weight of sample analyzed, c_1 (mol/L) is the molarity of NaOH, V_1 (mL) is the volume of NaOH, c_2 (mol/L) is the molarity of HCl, and V_2 (mL) is the equivalent volume of known molarity HCl.

Based on the assumption that the cellulose and hemicelluloses are composed of AGU and AXU, respectively, the theoretical hydroxyl content of unmodified cellulose and hemicelluloses were calculated from their macromolecular structure according to Eqs. (4) and (5), respectively. The total hydroxyl content of unmodified lignin was 6.31 mmol/g by ^{31}P NMR analysis. The hydroxyl content of unmodified bagasse was calculated from Eq. (6), according to the contents of cellulose, hemicelluloses, and lignin in the extractive-free bagasse:

$$n_C = \frac{1000}{162} \times 3 \quad (4)$$

where n_C (mmol/g) is the theoretical hydroxyl content of unmodified cellulose, 162 g/mol is the molar mass of AGU, and 3 is the number of hydroxyl groups on each AGU, and

$$n_H = \frac{1000}{132} \times 2 \quad (5)$$

where n_H (mmol/g) is the theoretical hydroxyl content of unmodified hemicelluloses, 132 g/mol is the molar mass of AXU, and 2 is the number of hydroxyl groups on each AXU, and

$$n_B = n_C \times 44.85 + n_H \times 33.13 + n_L \times 19.24 \quad (6)$$

where n_B and n_L (mmol/g) are the hydroxyl contents of unmodified bagasse and lignin, respectively, and 44.85%, 33.13%, and 19.14% are the contents of cellulose, hemicelluloses, and lignin, respectively, in the extractive-free bagasse according to the standard NREL methods.

2.5. Characterization

FT-IR spectra were collected on an FT-IR spectrophotometer (Nicolet 510) using a KBr disk containing approximately 1% finely ground samples. Thirty-two scans were taken for each sample with a resolution of 2 cm^{-1} in transmittance mode in the range of 4000–400 cm^{-1}.

^{1}H NMR, ^{13}C NMR, ^{31}P NMR, and 2D HSQC spectra were recorded on Bruker AV-III HD 600 spectrometer (Germany). The ^{1}H NMR spectra were recorded at 298.0 K from the dried samples (40 mg) dissolved in 0.5 mL DMSO-d_6. The detailed collecting and processing parameters for ^{1}H NMR analysis were as follows: number of scans, 16; receiver gain, 31; acquisition time, 2.7263 s; relaxation delay, 1.0 s; pulse width, 11.0 s; spectra frequency, 600.17 Hz; and spectra width, 12019.2 Hz.

For CP/MAS ^{13}C NMR analysis, dried samples (200 mg) were tested at 303.0 K. The detailed collecting and processing parameters were as follows: number of scans, 2000; receiver gain, 567; acquisition time, 0.0344 s; relaxation delay, 1.5 s; pulse width, 4.55 s; spectra frequency, 100.61Hz; and spectra width, 29761.9 Hz.

For liquid-state ^{13}C NMR analysis, dried samples (40 mg) was dissolved in 0.5 mL DMSO-d_6 at 298.0 K, the detailed collecting and processing parameters were as follows: number of scans, 15,000; receiver gain, 1290; acquisition time, 1.2583 s; relaxation delay, 1.5 s; pulse width, 12.00 s; spectra frequency, 100.61 Hz; and spectra width, 26041.7 Hz.

For 2D HSQC NMR analysis, the spectra were acquired using 40 mg samples in the 0.5 mL DMSO-d_6 at 299.9 K. The detailed collecting and processing parameters were as follows: number of scans, 16; receiver gain, 187; acquisition time, 0.1420 s; relaxation delay, 2.0 s; pulse width, 11.0 s; spectra frequency, 600.17/150.91 Hz; and spectra width, 7211.51/24875.6 Hz.

The hydroxyls contents of lignin samples were determined by ^{31}P NMR spectroscopy, according to the reported method [25]. The detailed collecting and processing parameters were as follows: number of scans, 300; receiver gain, 187; acquisition time, 0.3408 s; relaxation delay, 2.0 s; pulse width, 12.0 s; spectra frequency, 242.95 Hz; and spectra width, 96,153.8 Hz.

TG of samples was performed on a simultaneous thermal analyzer (SDT Q600, TA Instrument). The sample weighted between 8 and 10 mg. The scans were first run from room temperature to 100°C, were equilibrated at 100°C for 5 min, and then were cooled to 40°C at the cooling rate of 20°C/min. The scans were again heated from 40°C to 700°C at the heating rate of 10°C/min under nitrogen flow.

3. Results and discussion

3.1. Homogeneous phthalation of bagasse and its fractions

The reaction behaviors of bagasse during homogeneous phthalation in AmimCl were investigated comparatively with the isolated fractions under the same conditions. The contents of cellulose, hemicelluloses, and lignin in the extractive-free bagasse were determined as 44.85%, 33.13%, and 19.14%, respectively, according to the standard NREL methods [27].

As shown in **Table 1**, the DS of phthalated cellulose, hemicelluloses, and lignin were 19.88%, 13.67%, and 27.10%, respectively, indicating the reactivity of lignin was the highest among the three main components during bagasse homogeneous modification in AmimCl. The DS of phthalated bagasse (10.80%) was lower than phthalated fractions, which was probably due to the complicated linkages between bagasse fractions. Besides, the WPG of phthalated cellulose, hemicelluloses, and lignin were 17.04%, 5.98%, and -49.70%, respectively, suggesting that the

	Phthalation conditions				n_{OH}' (mmol/g)	DS (%)	WPG (%)
Sample no	Molar ratio[a]	DMAP[b]	Time (min)	Temperature (°C)			
B_s	4:1	5%	90	60	1.57	10.80	-10.51%
C_s	4:1	5%	90	60	3.68	19.88	17.04%
H_s	4:1	5%	90	60	2.07	13.67	5.98%
L_s	4:1	5%	90	60	1.71	27.10	-49.70%

[a]Represents the weight ratio (w/w) of phthalic anhydride/materials.
[b]Represents the dosage of catalyst dimethylaminopyridine (DMAP) by weight ratio (w/w).
B_s, C_s, H_s, and L_s represent the phthalated bagasse, cellulose, hemicelluloses, and lignin, respectively.

Table 1. The substituted hydroxyl content, weight percent gain, and degree of substitution of samples.

significant degradation of lignin occurred during homogeneous phthalation. The WPG of phthalated bagasse was -10.51%, probably resulting from the degradation of lignin during bagasse homogeneous modification.

3.2. ^{31}P NMR analysis

To investigate the detailed information of lignin hydroxyls during homogeneous modification, ^{31}P NMR spectroscopy was applied in the present study to quantify the hydroxyl contents of lignin samples [28, 29]. The ^{31}P NMR spectra of unmodified lignin (L_o, spectrum a) and phthalated lignin (L_s, spectrum b) are shown in **Figure 1**. According to the reported literatures [30, 31], the assignments of signals in ^{31}P NMR spectra were as followed: international standard (IS), 145.50–144.80 ppm; aliphatic OH, 149.50–145.50 ppm; phenolic S-OH, 143.50–142.50 ppm; phenolic G-OH, 140.00–138.50 ppm; phenolic H-OH 138.00–137.50 ppm; COOH, 135.50–134.50 ppm.

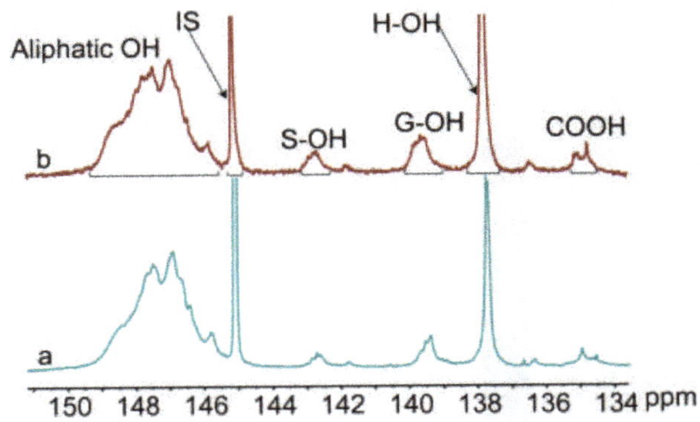

Figure 1. ^{31}P NMR spectra of unmodified lignin (L_o, spectrum a) and phthalated lignin (L_s, spectrum b).

The hydroxyl contents of unmodified and phthalated lignin are present in **Table 2**, and the decreased percentages of lignin hydroxyls are shown in **Figure 2**. After phthalation, the aliphatic hydroxyl content decreased from 4.71 mmol/g in unmodified lignin to 3.33 mmol/g in phthalated lignin, and the total phenolic hydroxyl content was also reduced from 1.60 (L_o) to 1.27 mmol/g (L_s). These results indicated that both aromatic and phenolic hydroxyls were phthalated during lignin modification in AmimCl. Besides, the total hydroxyl content of lignin decreased from 6.31 (L_o) to 4.60 mmol/g (L_S), while the carboxyl content slightly increased from 0.08 (L_o) to 0.09 (L_S) mmol/g. The result suggested that the predominant diester form [32] during lignin homogeneous modification in AmimCl. For lignin phenolic hydroxyls, the content of phenolic S-OH decreased from 0.16 to 0.09 mmol/g after phthalation; the content of phenolic G-OH decreased from 0.40 (L_o) to 0.25 (L_S) mmol/g; the decrease in the content of phenolic H-OH from 1.04 (L_r) to 0.93 (L_S) mmol/g was also observed. These results indicated phenolic S-, G-, and H-OH were phthalated during lignin modification in AmimCl.

Sample	L_o	L_s
Aliphatic OH (mmol/g)	4.71	3.33
Phenolic OH (mmol/g)	1.60	1.27
S-OH (mmol/g)	0.16	0.09
G-OH (mmol/g)	0.40	0.25
H-OH (mmol/g)	1.04	0.93
COOH (mmol/g)	0.08	0.09
S/G ratio (mmol/g)	0.40	0.38

Table 2. The hydroxyl contents of lignin samples.

As shown in **Figure 2(a)**, the decreased percentage of aromatic hydroxyls was higher than that of phenolic hydroxyls, indicating the higher reactivity of aliphatic hydroxyls during lignin homogeneous phthalation. For phenolic hydroxyls, the decreased percentage of phenolic S-OH, G-OH, and H-OH were present (**Figure 2b**). Therefore, the phthalation reactivity of phenolic hydroxyls followed the order of S-OH > G-OH > H-OH.

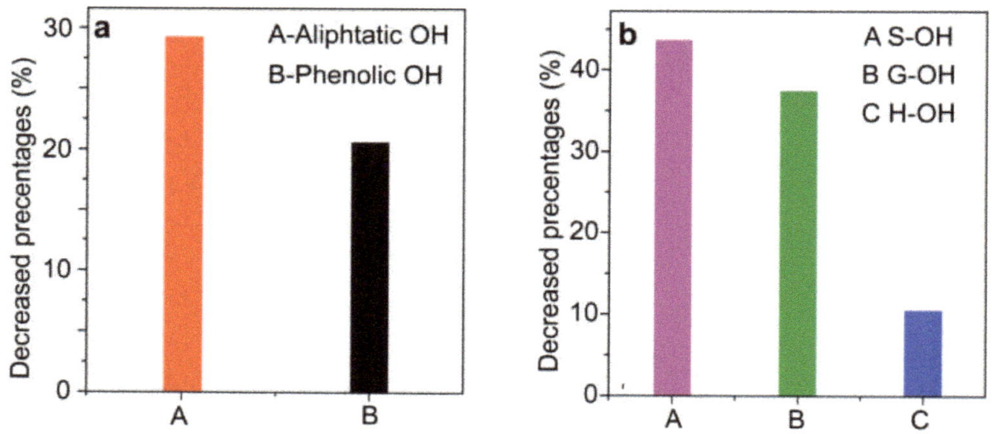

Figure 2. The decreased percentages of lignin hydroxyls.

3.3. FT-IR

The FT-IR spectra of unmodified cellulose (C_o, spectrum a), regenerated cellulose (C_r, spectrum b), and phthalated cellulose (C_s, spectrum c) are shown in **Figure 3**. The absorbance at 3433,

2895, 1646, 1376, 1165, and 1057 cm^{-1} were associated with unmodified cellulose [20]. These peaks present in regenerated and phthalated cellulose indicated that the skeleton of cellulose was stable during cellulose dissolution and modification in AmimCl. Compared with the spectrum of unmodified cellulose, the spectrum of the phthalated cellulose (spectrum c) provides evidence of phthalation by showing the presence of three new bands at 1718, 1580, and 749 cm^{-1} [21], corresponding to the stretching of carbonyl group, antisymmetric stretching of carboxylic anions, and the out-of-plane C-H bending of ortho disubstituted benzene, respectively. In addition, the intensity of peak at 1057 cm^{-1} for C-O-C stretching in spectrum c slightly decreased indicating the degradation of cellulose macromolecules during dissolution and modification. This degradation was also reported in the previous literature [21].

Figure 4 illustrates the FT-IR spectra of unmodified hemicelluloses (spectrum a), regenerated hemicelluloses (spectrum b), and phthalated hemicelluloses (spectrum c). Similarly, the

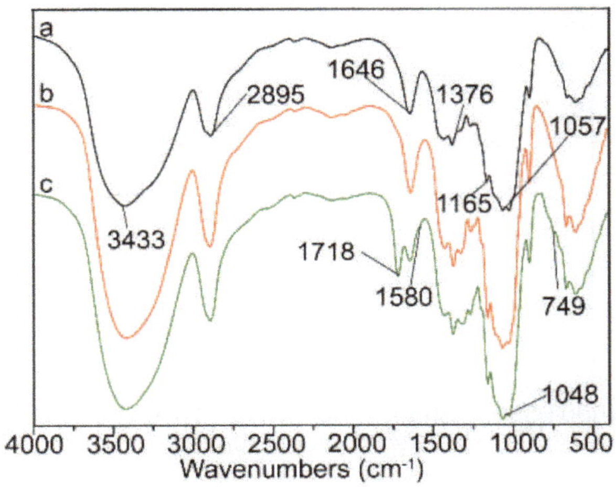

Figure 3. FT-IR spectra of unmodified cellulose (C_o, spectrum a), regenerated cellulose (C_r, spectrum b), and phthalated cellulose (C_s, spectrum c).

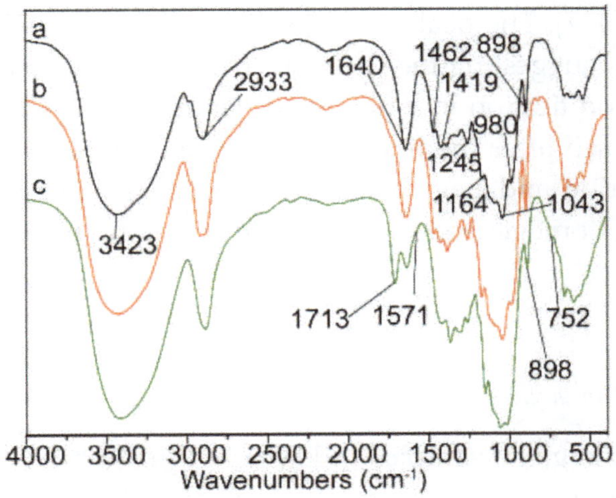

Figure 4. FT-IR spectra of unmodified hemicelluloses (H_o, spectrum a), regenerated hemicelluloses (H_r, spectrum b), and phthalated hemicelluloses (H_s, spectrum c).

skeleton of hemicelluloses remained unchanged after dissolution and modification in AmimCl. Compared with unmodified hemicelluloses, the new bands at 1713, 1571, and 752 cm^{-1} in phthalated hemicelluloses indicated the phthalation of hemicelluloses [33].

The FT-IR spectra of unmodified lignin (spectrum a), regenerated lignin (spectrum b), and phthalated lignin (spectrum c) are present in **Figure 5**. The bands at 3418, 2930, 1692, 1504, 1595, 1267, 1118, 1031, 824, and 747 cm^{-1} are related to unmodified lignin [34, 35]. Similar to cellulose and hemicelluloses, the skeleton of lignin was also stable after dissolution and modification in AmimCl. Compared with unmodified lignin, the intensity of bands at 1702 and 748 cm^{-1} in phthalated lignin obviously increased, suggesting the occurrence of esterification between lignin and phthalic anhydride [36].

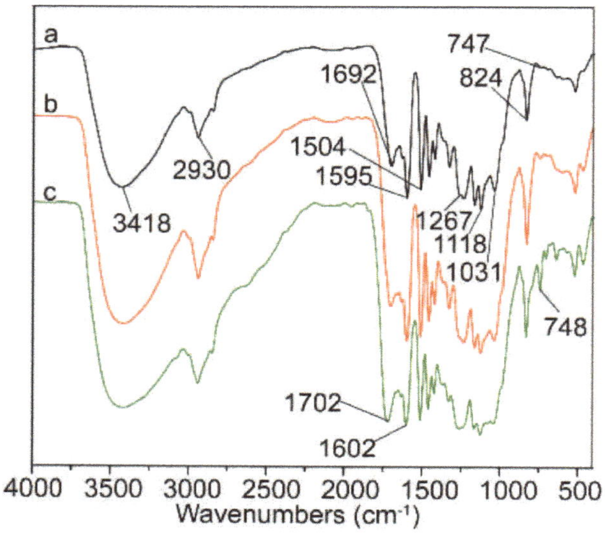

Figure 5. FT-IR spectra of unmodified lignin (L_o, spectrum a), regenerated lignin (L_r, spectrum b), and phthalated lignin (L_s, spectrum c).

FT-IR spectra of unmodified bagasse (B_o, spectrum a), regenerated bagasse (B_r, spectrum b), and phthalated bagasse (B_s, spectrum c) are shown in **Figure 6**. The bands were assigned according to the previous literature [37]. The peaks at 1605, 1515, 1375, and 897 cm^{-1} remained predominant in modified bagasse, suggesting the skeleton of bagasse fractions were unchanged during bagasse homogeneous modification in ionic liquid AmimCl [15]. The intensities of the peak at 1725 and 744 cm^{-1} increased after modification, providing the direct evidence of esterification between phthalic anhydride and bagasse. In conclusion, the phthalation of bagasse and its fractions occurred during homogeneous modification according to FT-IR analyses.

3.4. NMR analysis

3.4.1. Solid-state ^{13}C NMR of bagasse

The poor dissolubility of bagasse samples in DMSO-d_6 made high-resolution liquid-state 2D HSQC NMR spectroscopy impossible, and CP/MAS ^{13}C NMR spectroscopy was applied for bagasse samples analysis, as shown in **Figure 7**. The signals in CP/MAS ^{13}C NMR spectra were assigned based on the reported literatures [15, 21]. The signals at 191.2, 170.5, and 130.5 ppm

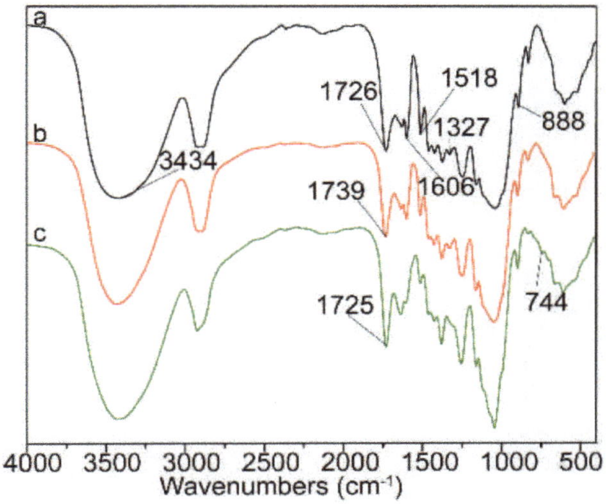

Figure 6. FT-IR spectra of unmodified bagasse (B_o, spectrum a), regenerated bagasse (B_r, spectrum b), and phthalated bagasse (B_s, spectrum c).

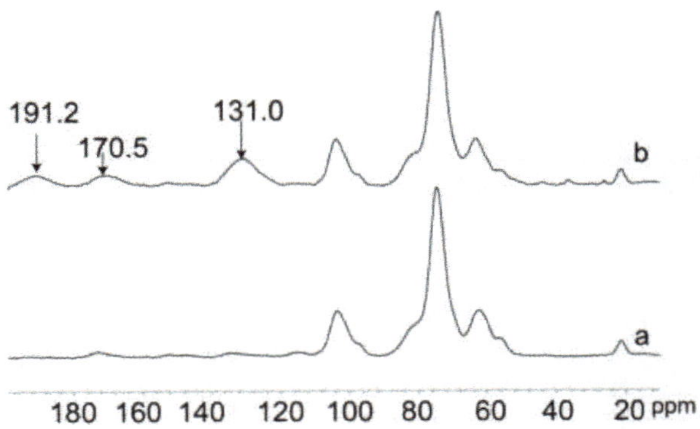

Figure 7. Solid-state ^{13}C NMR spectra of unmodified bagasse (B_o, spectrum a) and phthalated bagasse (B_s, spectrum b).

correspond to benzene ring, carbonyl group, and carboxylic groups, respectively, in phthalated bagasse. Compared with unmodified bagasse, the intensities of the three signals in phthalated bagasse remarkably increased, indicating the attachment of phthaloyl group onto bagasse.

The signals between 60 and 105 ppm relate to the carbons of carbohydrates. The chemical shifts and intensities of signals in this region remained basically unchanged, indicating the carbon skeleton of bagasse polysaccharides (cellulose and hemicelluloses) was stable during homogeneous modification, which was consistent with the results from FT-IR. Due to the low resolution of CP/MAS ^{13}C NMR spectroscopy, the detailed information of bagasse fractions during homogeneous phthalation need to be further revealed with 2D HSQC NMR spectroscopy.

3.4.2. Detailed reaction behaviors of cellulose during homogeneous phthalation

The HSQC spectra of unmodified cellulose (spectrum a) and phthalated cellulose (spectrum b) are present in **Figure 8**. The primary correlations of cellulose were well assigned, as reported in the previous literatures [20, 21]. The cross-peaks of cellulose were clearly observed in **Figure 8(a)** and **(b)** at 73.49/3.06 [C-C_2 (C_2/H_2)], 75.33/3.36 [C-C_3 (C_3/H_3)], 80.85/3.33 [C-C_4 (C_4/H_4)], 77.14/

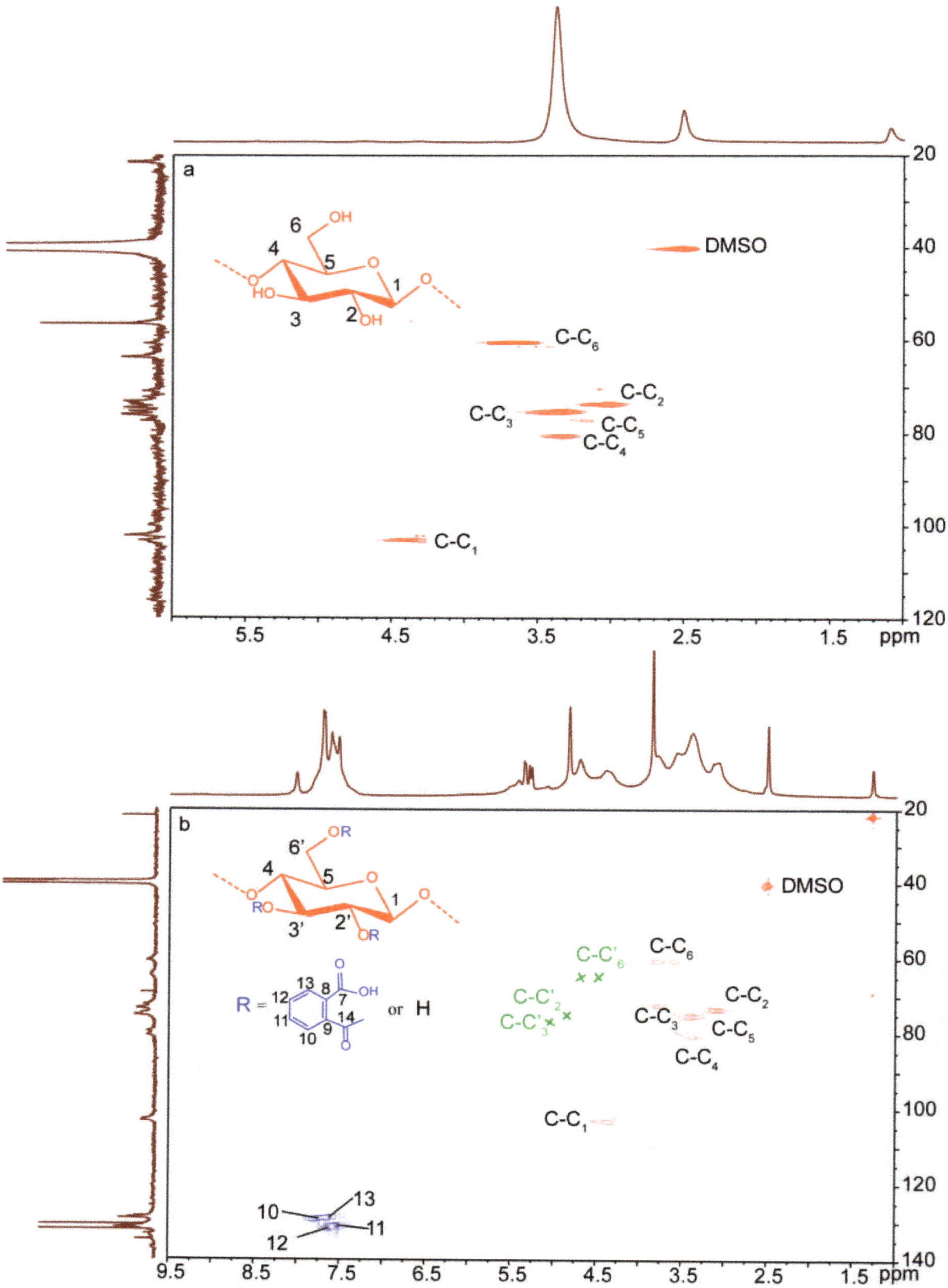

Figure 8. 2D HSQC spectra of unmodified cellulose (C_o, spectrum a) and phthalated cellulose (C_s, spectrum b).

3.18 [C-C_5 (C_5/H_5)], and 103.44/4.33 [C-C_1 (C_1/H_1)] ppm; the two C-C_6 (C_6/H_6) peaks were also distinctively located at 60.77/3.79 and 60.77/3.58 ppm, respectively. These results confirmed that the skeleton of cellulose was stable during homogeneous phthalation in AmimCl, consistent with the FT-IR analysis.

The cross-peaks at 131.49/7.62, 131.11/7.52, 129.43/7.78, and 128.94/7.61 ppm were associated with C_{10}/H_{10}, C_{11}/H_{11}, C_{12}/H_{12}, and C_{13}/H_{13} in the aryl group of phthalated cellulose, respectively. The presence of these peaks in phthalated cellulose confirmed the attachment of phthaloyl group onto cellulose. More importantly, two peaks from substituted C-6 in AGU C-C_6' (C_6/H_6) appeared at 64.39/4.39 and 64.50/4.67 ppm, and peaks from substituted C-2 [C-C_2' (C_2/H_2)] and C-3 [C-C_3' (C_3/H_3)] in AGU were located at 74.32/4.60 and 75.30/4.83 ppm, respectively. These results suggested the successful phthalation of hydroxyls at C-6, C-2, and C-3 positions in AGU. The DS of hydroxyls on different positions could be evaluated upon the integral area of the characteristic substituted correlations. By integration, the DS of C_6-OH, C_2-OH, and C_3-OH in AGU were 12.05, 2.54, and 1.84%, respectively. Therefore, the DS of hydroxyls in AGU followed the order C-6 > C-2 > C-3. The phthalated cellulose was previously prepared under the similar experimental conditions without catalyst in our laboratory, and the DS of hydroxyls at C-6, C-2, and C-3 positions were 6.30, 2.01, and 0% in AGU, respectively. By contrast, it could be concluded that the addition of catalyst DMAP could improve the uniformity and DS of phthalated cellulose.

3.4.3. Detailed reaction behaviors of hemicelluloses during homogeneous phthalation

The 2D HSQC spectra of unmodified hemicelluloses (spectrum a) and phthalated hemicelluloses (spectrum b) are present in **Figure 9**. According to the previous literatures [38, 39], the correlations of xylan were at 102.12/4.25 [X-C_1 (C_1/H_1)], 73.39/3.03 [X-C_2 (C_2/H_2)], 74.60/3.23 [X-C_3 (C_3/H_3)], and 75.84/3.4 [X-C_4 (C_4/H_4)] ppm, respectively, and the cross-peaks from X-C_5 (C_5/H_5) appeared at 63.59/3.16 and 63.59/3.86 ppm in both unmodified and phthalated hemicelluloses. Besides, the cross-peaks of arabinose were also well resolved at 107.76/5.24 [A-C_1 (C_1/H_1)], 86.62/3.89[A-C_4 (C_4/H_4)], 81.32/3.76[A-C_2 (C_2/H_2)], 78.32/3.57 [A-C_3 (C_3/H_3)], and 62.13/3.40 [A-C_5 (C_5/H_5)] ppm, respectively. These results indicated that the hemicelluloses isolated from bagasse were mainly composed of arabinose and xylan, consistent with the previous literatures [25, 27]. Compared with unmodified hemicelluloses, the cross-peaks at 128.81/7.66, 129.34/7.76, 130.99/7.51, and 131.31/7.63 ppm were for C_{12}/H_{12}, C_9/H_9, C_{10}/H_{10}, and C_{11}/H_{11}, respectively, in the phthaloyl group of phthalated hemicelluloses. These results further confirmed the occurrence of the esterification between hemicelluloses and phthalic anhydride during homogeneous modification in AmimCl. The cross-peaks at 74.95/4.70 and 76.58/5.02 ppm relate to the phthalated hydroxyls at C-2 and C-3 positions in AXU. The DS of hydroxyls at C-2 and C-3 positions in AXU were 3.56 and 1.54% by integral evaluation, indicating the higher reactivity of C_2-OH in AXU during hemicelluloses homogeneous phthalation.

3.4.4. Detailed reaction behaviors of lignin during homogeneous phthalation

The 2D HSQC NMR spectra of unmodified lignin (spectrum a) and phthalated lignin (spectrum b) are illustrated in **Figure 10**, and the structure of identified lignin subunits with

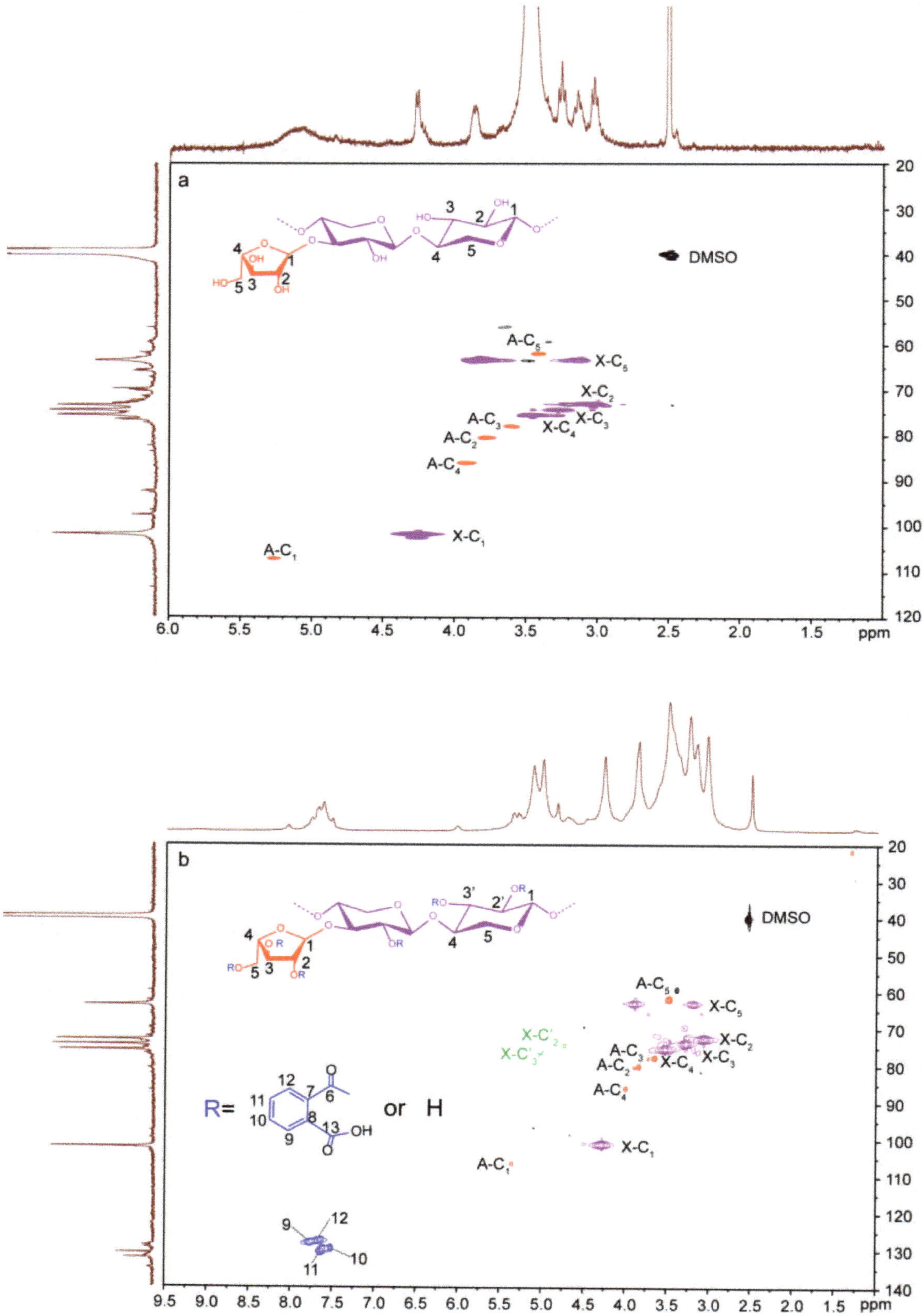

Figure 9. 2D HSQC spectra of unmodified hemicelluloses (H_o, spectrum a) and phthalated hemicelluloses (H_s, spectrum b).

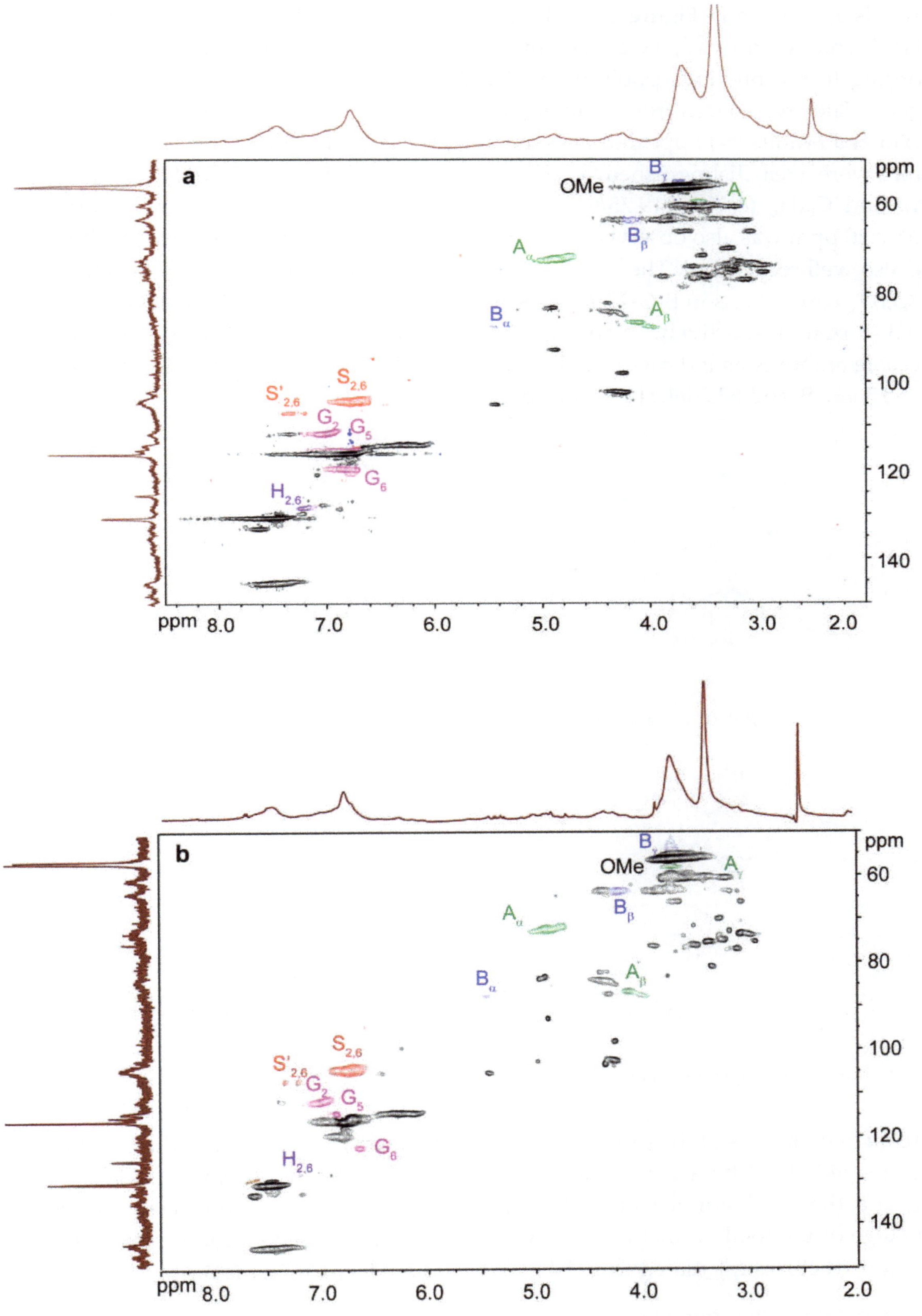

Figure 10. 2D HSQC spectra of unmodified lignin (Lo, spectrum a) and phthalated lignin (Ls, spectrum b).

hydroxyls are shown in **Figure 11**, including aryl ether (β-O-4′, A), phenylcoumaran (β-5′, B), guaiacyl unit (G-unit, C), syringyl unit (S-unit, S), and *p*-hydrophenyl unit (H-unit, H). According to the previous publications [24, 25], the correlations of these signals were well assigned. The cross-peaks from S-units $C_{2,6}/H_{2,6}$ was at 104.68/6.72 ppm, and the signals for C_α-oxidized S-units (S') appeared at δ_C/δ_H 107.15/7.34 ppm. The cross-peaks originated from G-units were well distinguished: C_2/H_2 (δ_C/δ_H 111.69/7.00 ppm), C_5/H_5 (δ_C/δ_H 115.33/6.71 ppm), and C_6/H_6 (δ_C/δ_H 119.75/6.80 ppm). The $C_{2,6}/H_{2,6}$ correlations in H-units at δ_C/δ_H 118.66/7.20 ppm was also observed. For lignin side-chains, lignin substructures with hydroxyls were also well recognized. The C_α/H_α, C_β/H_β (linked to G/H units), C_β/H_β (linked to S units), and C_γ/H_γ correlations in β-*O*-4′ linkages were at δ_C/δ_H 72.43/4.84, 84.16/4.35, 86.69/4.09, and 60.35/3.48 ppm, respectively. In addition, phenylcoumaran (β-5′, substructure B) appeared in noticeable amounts as indicated by the C_α/H_α, C_β/H_β, and C_γ/H_γ correlations at δ_C/δ_H 87.67/5.43, 55.36/3.70, and 63.50/4.21 ppm, respectively.

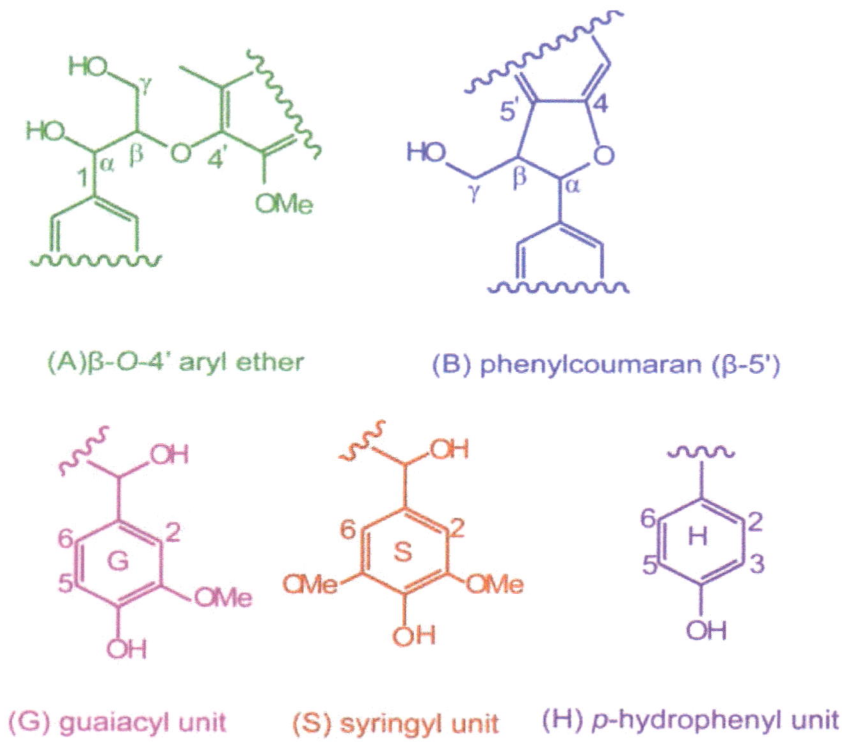

Figure 11. The structure of lignin subunits.

The relative quantities of the primary substructures were calculated according to the previous literatures [40, 41]. The aromatic lignin S-, G-, and H-units were expressed as a fraction of 100%, and the relative molar quantities of aryl ether, phenylcoumaran were expressed as a percentage of the total aromatic S-, G-, and H-units. The detailed information of quantitative lignin substructures calculated from HSQC spectra are listed in **Table 3**.

After phthalation, the content of aryl ether increased from 57.80/100Ar (L_o) to 58.33/100Ar (L_s), and the content of phenylcoumaran also increased from 2.76/100Ar (L_o) to 3.13/100Ar (L_s). These increases were probably due to the decreased intensity of $C_{2,6}/H_{2,6}$ correlations from

Sample	L_o	L_s
Aliphatic percentages of (S+G+H)		
Aryl ether (A)	57.80/100Ar	58.33/100Ar
Phenylcoumaran (B)	2.76/100Ar	3.13/100Ar
Aromatic molar percentages (S+G+H)		
Syringyl (S)	49.54%	57.81%
Guaiacyl (G)	45.78%	37.50%
p-hydroxyphenyl (H)	4.59%	4.69%
S/G	1.08	1.54

Table 3. Quantitative information of unmodified lignin (L_o) and phthalated lignin (L_s) determined with 2D HSQC NMR.

aromatic units resulting from the condensation of lignin [41] during the homogeneous phthalation. The relative percentages of G-units decreased from 45.78% to 37.50%, while the increases in the relative percentage of S- and H-units were observed. This result suggested that the G-units were easily degraded during lignin homogeneous modification [42, 43]. Besides, the increase in the S/G ratio from 1.08 (L_o) to 1.54 (L_s) further confirmed the degradation of G-units during lignin homogeneous modification.

3.5. Thermal analysis

In the present study, the effect of dissolution, regeneration, and modification in AmimCl on the thermal stability of samples was investigated by comparing unmodified, regenerated, and phthalated samples. **Figure 12** illustrates the TG curves of cellulose (A), hemicelluloses (B), lignin (C), and bagasse (D), respectively.

The onset and midpoint degradation temperature of regenerated cellulose were higher than those of unmodified cellulose, indicating the improved thermal stability after dissolution and regeneration in IL. This was probably because cellulose fractions with low molecular weight could not be regenerated from IL [15]. As shown in **Table 4**, the onset and midpoint degradation temperature of phthalated cellulose were lower than those of regenerated cellulose, suggesting the thermal stability of phthalated cellulose decreased after homogeneous modification. The decreased thermal stability of phthalated cellulose was probably due to partial hydrolysis, degradation of cellulose, and decrystallization of crystalline cellulose macromolecules [21].

Similar to cellulose, the thermal stability of regenerated hemicelluloses were slightly higher than that of unmodified hemicelluloses, probably resulting from the loss of hemicellulosic fractions with low molecular weights during dissolution and regeneration in AmimCl. Compared with regenerated hemicelluloses, the onset and midpoint degradation temperature of phthalated hemicelluloses decreased to 212 and 265°C, respectively, indicating the decreased thermal stability of phthalated hemicelluloses after homogeneous modification. This decrease was also possibly resulted from the degradation of hemicelluloses during homogeneous modification [44].

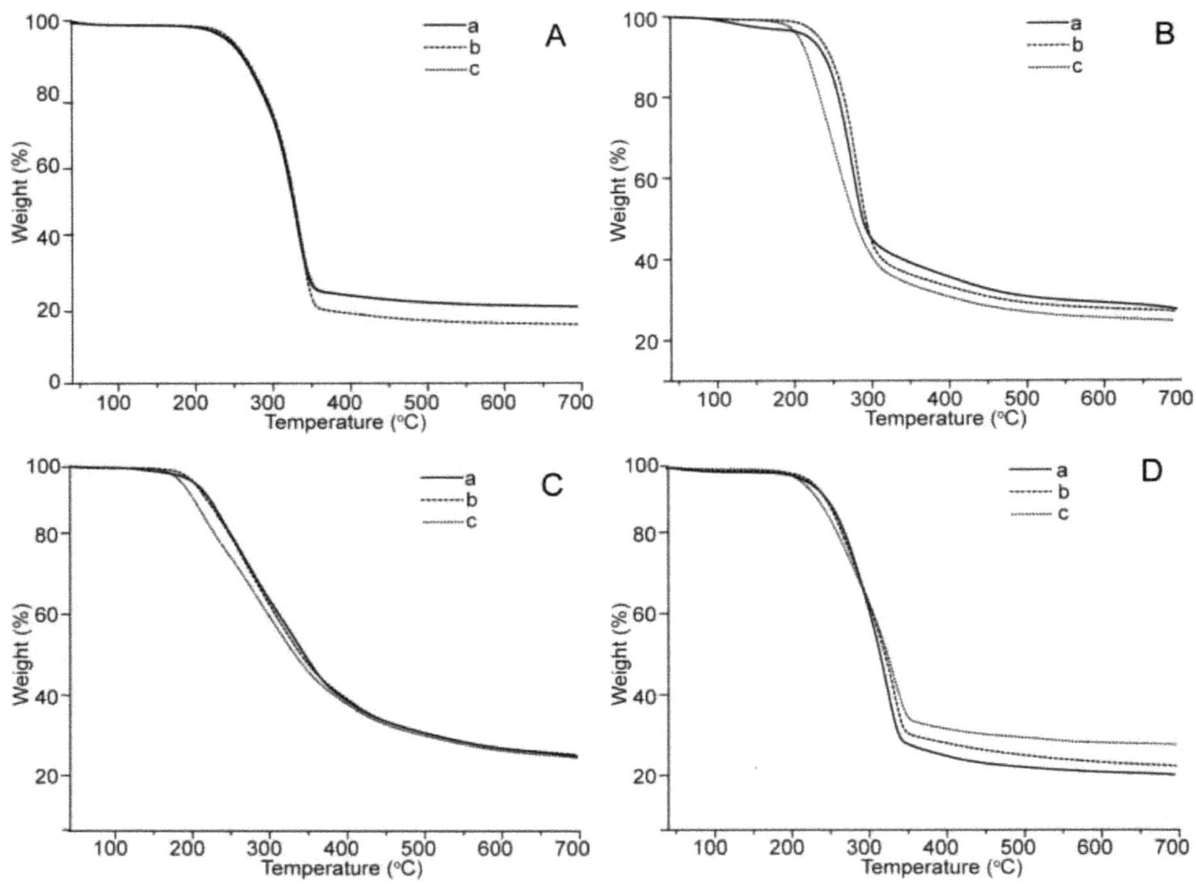

Figure 12. TG curves of cellulose (A), hemicelluloses (B), lignin (C), and bagasse (D), respectively. The curves a, b, and c represent unmodified, regenerated, and phthalated samples, respectively.

Sample	Onset T (°C)	Midpoint T (°C)
C_o	279	315
C_r	292	320
C_s	282	319
H_o	240	281
H_r	243	282
H_s	212	265
L_o	200	358
L_r	196	345
L_s	174	345
B_o	256	305
B_r	248	302
B_s	244	304

Table 4. The onset and midpoint degradation temperature of samples.

The onset and midpoint degradation temperature of regenerated lignin were both lower than those of unmodified lignin, as listed in **Table 4**, indicating the decreased thermal stability of lignin after dissolution and regenerating. Compared with regenerated lignin, the thermal stability of phthalated lignin after homogeneous modification was also decreased. This decrease was probably resulted from the degradation of lignin during dissolution, regeneration, and modification in AmimCl.

For bagasse, the thermal stability of bagasse samples decreased after dissolution and regeneration in ionic liquid AmimCl. This was probably resulted from the degradation of lignin during dissolution and regeneration. Compared with regenerated bagasse, the onset degradation temperature of phthalated bagasse decreased, while the midpoint degradation temperature of phthalated bagasse increased. These results suggested the thermal stability of lignocellulose was affected by the loss of fractions with low molecular weights and the degradation of phthalated samples. The former led to the increase in the thermal stability, while the later resulted in the decreased thermal stability of phthalated lignocellulose.

4. Conclusions

The phthalation of bagasse and its fractions occurred during homogeneous modification in AmimCl, and the reactivity of the three main components followed the order: lignin > cellulose > hemicelluloses. The FT-IR and NMR analyses suggested the skeleton of bagasse remained stable during dissolution, regeneration, and modification; the reactivity of hydroxyls in AGU followed the order: C-6 > C-2 > C-3; and the reactivity order in AXU was as follows: C-2 > C-3. For lignin, both aromatic and phenolic hydroxyls were phthalated; the reactivity of aromatic hydroxyls was higher than phenolic hydroxyls; and the reactivity order of lignin phenolic hydroxyls was as follows: S-OH > G-OH > H-OH. The thermal stability of modified lignocellulose was affected by the disruption of cellulose crystallinity and the degradation reaction of the modified components.

Acknowledgments

This work was financially supported by the National Natural Science Foundation of China (31170550, 31170555), the Fundamental Research Funds for the Central Universities (2014ZG0046), and the National Program for Support of Top-notch Young Professionals.

Nomenclature

AmimCl	1-allyl-3-methylimidazolium chloride
DMAP	4-dimethylaminopyridine
ILs	ionic liquids

DS	degree of substitution
FT-IR	Fourier transform infrared resonance
CP/MAS	solid-state cross polarization/magnetic angle spinning
NMR	nuclear magnetic resonance
2D HSQC	two dimensional heteronuclear single quantum correlation
AGU	anhydroglucose units
AXU	anhydroxylose units
S	syringyl
G	guaiacyl
H	*p*-hydroxyphenyl
DMSO-d_6	perdeutero-dimethylsulfoxide
BmimCl	1-butyl-3-methylimicazium chloride
TG	thermogravimetric analysis
WPG	weight percentage gain

Author details

Hui-Hui Wang, Xue-Qin Zhang, Yi Wei and Chuan-Fu Liu*

*Address all correspondence to: chfliu@scut.edu.cn

State Key Laboratory of Pulp and Paper Engineering, South China University of Technology, Guangzhou, PR China

References

[1] Nagarajan V, Mohanty AK, Misra M. Sustainable green composites: Value addition to agricultural residues and perennial grasses. ACS Sustainable Chemistry & Engineering. 2013;**1**:325–333. DOI: 10.1021/sc300084z

[2] Ragauskas AJ, Williams CK, Davison BH, Britovsek G, Cairney J, Eckert CA, Frederick WJ, Hallett JP, Leak DJ, Liotta CL. The path forward for biofuels and biomaterials. Science. 2006;**311**:484–489. DOI: 10.1126/science.1114736

[3] Liam B, Philip O. Biofuels from microalgae—A review of technologies for production, processing, and extractions of biofuels and co-products. Renewable and Sustainable Energy Reviews. 2010;**14**:557–577. DOI: 10.1016/j.rser.2009.10.009

[4] Minjares-Fuentes R, Femenia A, Garau MC, Candelas-Cadillo MG, Simal S, Rosselló C. Ultrasound-assisted extraction of hemicelluloses from grape pomace using response surface methodology. Carbohydrate Polymers. 2016;**138**:180–191. DOI: 10.1016/j.carbpol.2015.11.045

[5] Massimo B, Giovanna F, Mariastella S, Antonino L. Surface chemical modification of natural cellulose fibers. Journal of Applied Polymer Science. 2002;**83**:38–45. DOI: 10.1002/app.2229

[6] Freire CSR, Silvestre AJD, Neto CP, Belgacem MN, Gandini A. Controlled heterogeneous modification of cellulose fibers with fatty acids: Effect of reaction conditions on the extent of esterification and fiber properties. Journal of Applied Polymer Science. 2006;**100**:1093–1102. DOI: 10.1002/app.23454

[7] Viera RGP, Rodrigues Filho G, de Assunção RMN, Meireles CdS, Vieira JG, de Oliveira GS. Synthesis and characterization of methylcellulose from sugar cane bagasse cellulose. Carbohydrate Polymers. 2007;**67**:182–189. DOI: 10.1016/j.carbpol.2006.05.007

[8] Fasching M, Schröder P, Wollboldt RP, Weber HK, Sixta H. A new and facile method for isolation of lignin from wood based on complete wood dissolution. Holzforschung. 2008;**62**:15–23. DOI: 10.1515/HF.2008.003

[9] Sheldon RA. Green solvents for sustainable organic synthesis: State of the art. Green Chemistry. 2005;**7**:267–278. DOI: 10.1039/b418069k

[10] Olivier-Bourbigou H, Magna L, Morvan D. Ionic liquids and catalysis: Recent progress from knowledge to applications. Applied Catalysis A: General. 2010;**373**:1–56. DOI: 10.1016/j.apcata.2009.10.008

[11] Zakrzewska ME, Bogel-Łukasik E, Bogel-Łukasik R. Ionic liquid-mediated formation of 5-hydroxymethylfurfural? A promising biomass-derived building block. Chemical Reviews. 2010;**111**:397–417. DOI: 10.1021/cr100171a

[12] Swatloski RP, Spear SK, Holbrey JD, Rogers RD. Dissolution of cellose with ionic liquids. Journal of the American Chemical Society. 2002;**124**:4974–4975. DOI: 10.1021/ja025790m

[13] Kilpeläinen I, Xie HB, King AS, Granstrom M, Heikkinen S, Argyropoulos DS. Dissolution of wood in ionic liquids. Journal of Agricultural and Food Chemistry. 2007;**55**:9142–9148. DOI: 10.1021/jf071692e

[14] Fort DA, Remsing RC, Swatloski RP, Moyna P, Moyna G, Rogers RD. Can ionic liquids dissolve wood? Processing and analysis of lignocellulosic materials with 1-n-butyl-3-methylimidazolium chloride. Green Chemistry. 2007;**9**:63–69. DOI: 10.1039/b607614a

[15] Chen MJ, Chen CY, Liu CF, Sun RC. Homogeneous modification of sugarcane bagasse with maleic anhydride in 1-butyl-3-methylimidazolium chloride without any catalysts. Industrial Crops and Products. 2013;**46**:380–385. DOI: 10.1016/j.indcrop.2013.02.023

[16] Ibrahim MM, Dufresne A, El-Zawawy WK, Agblevor FA. Banana fibers and microfibrils as lignocellulosic reinforcements in polymer composites. Carbohydrate Polymers. 2010;**81**:811–819. DOI: 10.1016/j.carbpol.2010.03.057

[17] Xie HB, King A, Kilpelainen I, Granstrom M, Argyropoulos DS. Thorough chemical modification of wood-based lignocellulosic materials in ionic liquids. Biomacromolecules. 2007;**8:**3740–3748. DOI: 10.1021/bm700679s

[18] Nada AA, Hassan ML. Ion exchange properties of carboxylated bagasse. Journal of Applied Polymer Science. 2006;**102:**1399–1404. DOI: 10.1002/app.24255

[19] Anirudhan TS, Radhakrishnan PG. Chromium (III) removal from water and wastewater using a carboxylate-functionalized cation exchanger prepared from a lignocellulosic residue. Journal of Colloid and Interface Science. 2007;**316:**268–276. DOI: 10.1016/j.jcis.2007.08.051

[20] Liu CF, Sun RC, Zhang AP, Ren JL, Wang XA, Qin MH, Chao ZN, Luo W. Homogeneous modification of sugarcane bagasse cellulose with succinic anhydride using a ionic liquid as reaction medium. Carbohydrate Research. 2007;**342:**919–926. DOI: 10.1016/j.carres.2007.02.006

[21] Liu CF, Sun RC, Zhang AP, Qin MH, Ren JL, Wang XA. Preparation and characterization of phthalated cellulose derivatives in room-temperature ionic liquid without catalysts. Journal of Agricultural and Food Chemistry. 2007;**55:**2399–2406. DOI: 10.1021/jf062876g

[22] Li WY, Lan W, Chen D, Liu CF, Sun RC. DMAP-catalyzed phthalylation of cellulose with phthalic anhydride in [bmim] Cl. BioResources. 2011;**6:**2375–2385.

[23] Peng F, Ren JL, Xu F, Bian J, Peng P, Sun RC. Fractional study of alkali-soluble hemicelluloses obtained by graded ethanol precipitation from sugar cane bagasse. Journal of Agricultural and Food Chemistry. 2009;**58:**1768–1776. DOI: 10.1021/jf9033255

[24] Moghaddam L, Zhang Z, Wellard RM, Bartley JP, O'Hara IM, Doherty WOS. Characterisation of lignins isolated from sugarcane bagasse pretreated with acidified ethylene glycol and ionic liquids. Biomass and Bioenergy. 2014;**70:**498–512. DOI: 10.1016/j.biombioe.2014.07.030

[25] Zhang AP, Liu CF, Sun RC, Xie J. Extraction, purification, and characterization of lignin fractions from sugarcane bagasse. BioResources. 2013;**8:**1604–1614.

[26] Chen MJ, Shi QS. Transforming sugarcane bagasse into bioplastics via homogeneous modification with phthalic anhydride in ionic liquid. ACS Sustainable Chemistry & Engineering. 2015;**3:**2510–2515. DOI: 10.1021/acssuschemeng.5b00685

[27] Lan W, Liu CF, Sun RC. Fractionation of bagasse into cellulose, hemicelluloses, and lignin with ionic liquid treatment followed by alkaline extraction. Journal of Agricultural and Food Chemistry. 2011;**59:**8691–8701. DOI: 10.1021/jf201508g

[28] Argyropoulos DS. Quantitative phosphorus-31 NMR analysis of lignins, a new tool for the lignin chemist. Journal of Wood Chemistry and Technology. 1994;**14:**45–63.

[29] Crestini C, Argyropoulos DS. Structural analysis of wheat straw lignin by quantitative 31P and 2D NMR spectroscopy. The occurrence of ester bonds and α-O-4 substructures. Journal of Agricultural and Food Chemistry. 1997;**45:**1212–1219. DOI: 10.1021/jf960568k

[30] Pu YQ, Cao SL, Ragauskas AJ. Application of quantitative 31P NMR in biomass lignin and biofuel precursors characterization. Energy & Environmental Science. 2011;**4:**3154–3166. DOI: 10.1039/c1ee01201k

[31] Jasiukaityte E, Kunaver M, Crestini C. Lignin behaviour during wood liquefaction—Characterization by quantitative ^{31}P, ^{13}C NMR and size-exclusion chromatography. Catalysis Today. 2010;**156:**23–30. DOI: 10.1016/j.cattod.2010.02.001

[32] Vaidya AA, Gaugler M, Smith DA. Green route to modification of wood waste, cellulose and hemicellulose using reactive extrusion. Carbohydrate Polymers. 2016;**136:**1238–1250. DOI: 10.1016/j.carbpol.2015.10.033

[33] Peng XW, Ren JL, Sun RC. Homogeneous esterification of xylan-rich hemicelluloses with maleic anhydride in ionic liquid. Biomacromolecules. 2010;**11:**3519–3524. DOI: 10.1021/bm1010118

[34] Sun YC, Xu JK, Xu F, Sun RC. Efficient separation and physico-chemical characterization of lignin from eucalyptus using ionic liquid–organic solvent and alkaline ethanol solvent. Industrial Crops and Products. 2013;**47:**277–285. DOI: 10.1016/j.indcrop.2013.03.025

[35] Wang Y, Song H, Hou JP, Jia CM, Yao S. Systematic isolation and utilization of lignocellulosic components from sugarcane bagasse. Separation Science and Technology. 2013;**48:**2217–2224. DOI: 10.1080/01496395.2013.791855

[36] Chen MJ, Zhang XQ, Liu CF, Sun RC, Lu FC. Approach to renewable lignocellulosic biomass film directly from bagasse. ACS Sustainable Chemistry & Engineering. 2014;**2:**1164–1168. DOI: 10.1021/sc400555v

[37] Liu CF, Sun RC, Qin MH, Zhang AP, Ren JL, Ye J, Luo W, Cao ZN. Succinoylation of sugarcane bagasse under ultrasound irradiation. Bioresource Technology. 2008;**99:**1465–1473. DOI: 10.1016/j.biortech.2007.01.062

[38] Lin Y, King JY, Karlen SD, Ralph J. Using 2D NMR spectroscopy to assess effects of UV radiation on cell wall chemistry during litter decomposition. Biogeochemistry. 2015;**125:**427–436. DOI: 10.1007/s10533-015-0132-1

[39] Guan Y, Zhang B, Qi XM, Peng F, Yao CL, Sun RC. Fractionation of bamboo hemicelluloses by graded saturated ammonium sulphate. Carbohydrate Polymers. 2015;**129:**201–207. DOI: 10.1016/j.carbpol.2015.04.042

[40] Sette M, Wechselberger R, Crestini C. Elucidation of lignin structure by quantitative 2D NMR. Chemistry–A European Journal. 2011;**17:**9529–9535. DOI: 10.1002/chem.201003045

[41] Wen JL, Yuan TQ, Sun SL, Xu F, Sun RC. Understanding the chemical transformations of lignin during ionic liquid pretreatment. Green Chemistry. 2014;**16:**181–190. DOI: 10.1039/c3gc41752b

[42] Kim JY, Shin EJ, Eom IY, Won K, Kim YH, Choi D, Choi IG, Choi JW. Structural features of lignin macromolecules extracted with ionic liquid from poplar wood. Bioresource Technology. 2011;**102:**9020–9025. DOI: 10.1016/j.biortech.2011.07.081

[43] Varanasi P, Singh P, Arora R, Adams PD, Auer M, Simmons BA, Singh S. Understanding changes in lignin of *Panicum virgatum* and *Eucalyptus globulus* as a function of ionic liquid pretreatment. Bioresource Technology. 2012;**126:**156–161. DOI: 10.1016/j.biortech.2012.08.070

[44] Wang HT, Yuan TQ, Meng LJ, She D, Geng ZC, Sun RC. Structural and thermal characterization of lauroylated hemicelluloses synthesized in an ionic liquid. Polymer Degradation and Stability. 2012;**97:**2323–2330. DOI: 10.1016/j.polymdegradstab.2012.07.033

Permissions

All chapters in this book were first published in BVEVE, by InTech Open; hereby published with permission under the Creative Commons Attribution License or equivalent. Every chapter published in this book has been scrutinized by our experts. Their significance has been extensively debated. The topics covered herein carry significant findings which will fuel the growth of the discipline. They may even be implemented as practical applications or may be referred to as a beginning point for another development.

The contributors of this book come from diverse backgrounds, making this book a truly international effort. This book will bring forth new frontiers with its revolutionizing research information and detailed analysis of the nascent developments around the world.

We would like to thank all the contributing authors for lending their expertise to make the book truly unique. They have played a crucial role in the development of this book. Without their invaluable contributions this book wouldn't have been possible. They have made vital efforts to compile up to date information on the varied aspects of this subject to make this book a valuable addition to the collection of many professionals and students.

This book was conceptualized with the vision of imparting up-to-date information and advanced data in this field. To ensure the same, a matchless editorial board was set up. Every individual on the board went through rigorous rounds of assessment to prove their worth. After which they invested a large part of their time researching and compiling the most relevant data for our readers.

The editorial board has been involved in producing this book since its inception. They have spent rigorous hours researching and exploring the diverse topics which have resulted in the successful publishing of this book. They have passed on their knowledge of decades through this book. To expedite this challenging task, the publisher supported the team at every step. A small team of assistant editors was also appointed to further simplify the editing procedure and attain best results for the readers.

Apart from the editorial board, the designing team has also invested a significant amount of their time in understanding the subject and creating the most relevant covers. They scrutinized every image to scout for the most suitable representation of the subject and create an appropriate cover for the book.

The publishing team has been an ardent support to the editorial, designing and production team. Their endless efforts to recruit the best for this project, has resulted in the accomplishment of this book. They are a veteran in the field of academics and their pool of knowledge is as vast as their experience in printing. Their expertise and guidance has proved useful at every step. Their uncompromising quality standards have made this book an exceptional effort. Their encouragement from time to time has been an inspiration for everyone.

The publisher and the editorial board hope that this book will prove to be a valuable piece of knowledge for researchers, students, practitioners and scholars across the globe.

List of Contributors

Nuno Couto and Valter Silva
INEGI-FEUP, Faculty of Engineering, University of Porto, Porto, Portugal

Frank Huess Hedlund
COWI, Kongens Lyngby, Denmark
Technical University of Denmark (DTU/Compute), Kongens Lyngby, Denmark

Øssur Jarleivson Hilduberg
Danish Maritime Accident Investigation Board (DMAIB), Valby, Denmark

Luis C. Fernández Linares, Kevin Á. González Falfán and Citlally Ramírez-López
Professional Interdisciplinary Biotechnology Unit of the National Polytechnic Institute, Mexico City, Mexico

Inyinbor Adejumoke Abosede and Oluyori Abimbola Peter
Department of Physical Sciences, Landmark University, Omu Aran, Nigeria

Adelani-Akande Tabitha Adunola
Department of Biological Sciences, Landmark University, Omu Aran, Nigeria

Sibel Irmak
Biological Systems Engineering, Industrial Agricultural Products Center, University of Nebraska-Lincoln, Nebraska, USA

WeiSheng Zeng
Academy of Forest Inventory and Planning, State Forestry Administration, Beijing, China

Carlos Escudero-Oñate
Norwegian Institute for Water Research (NIVA), Gaustadalléen, Oslo, Norway

Núria Fiol and Isabel Villaescusa
Chemical Engineering Department, EPS, University of Girona, Girona, Spain

Jordi Poch
Applied Mathematics Department, EPS, University of Girona, Girona, Spain

C. Luke Williams, Rachel M. Emerson, Jaya Shankar Tumuluru, Allison E. Ray, Chenlin Li, Vicki S. Thompson, Dayna L. Daubaras and Damon S. Hartley
Idaho National Laboratory, Idaho Falls, ID, USA

Nicholas J. Nagle
National Renewable Energy Laboratory, Golden, CO, USA

Hui-Hui Wang, Xue-Qin Zhang, Yi Wei and Chuan-Fu Liu
State Key Laboratory of Pulp and Paper Engineering, South China University of Technology, Guangzhou, PR China

Index

R

S

T

V

W

CPSIA information can be obtained
at www.ICGtesting.com
Printed in the USA
LVHW060854030122
707669LV00002B/143

9 781639 870769